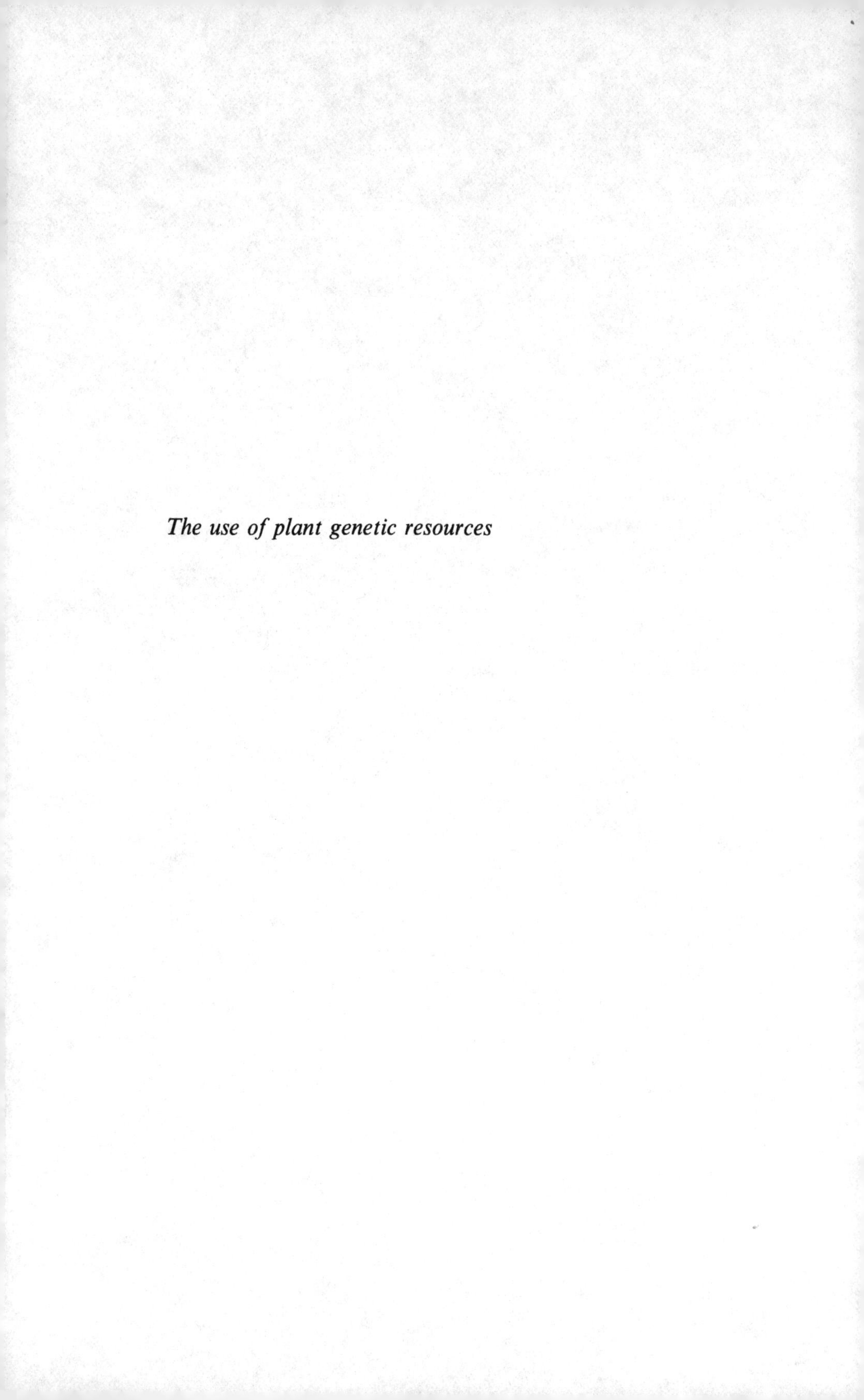

The use of plant genetic resources

The use of plant genetic resources

Edited by
A. H. D. BROWN
CSIRO Division of Plant Industry, Canberra, Australia
O. H. FRANKEL
CSIRO Division of Plant Industry, Canberra, Australia
D. R. MARSHALL
Waite Agricultural Research Institute, University of Adelaide, Australia
J. T. WILLIAMS
International Board for Plant Genetic Resources, FAO, Rome, Italy

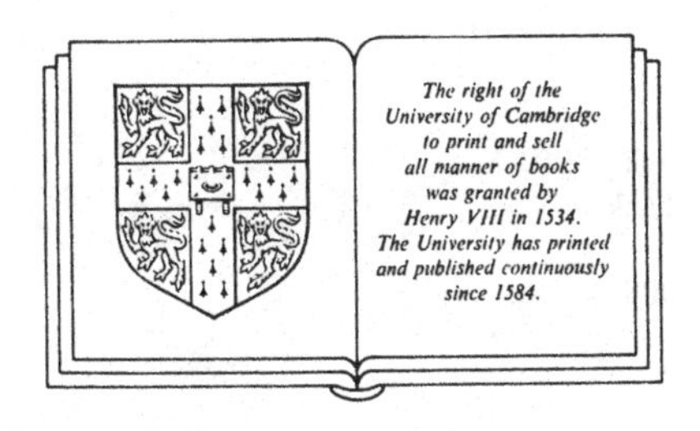
The right of the
University of Cambridge
to print and sell
all manner of books
was granted by
Henry VIII in 1534.
The University has printed
and published continuously
since 1584.

CAMBRIDGE UNIVERSITY PRESS
Cambridge
New York New Rochelle
Melbourne Sydney

CAMBRIDGE UNIVERSITY PRESS
Cambridge, New York, Melbourne, Madrid, Cape Town, Singapore, São Paulo

Cambridge University Press
The Edinburgh Building, Cambridge CB2 8RU, UK

Published in the United States of America by Cambridge University Press, New York

www.cambridge.org
Information on this title: www.cambridge.org/9780521345842

© International Board for Plant Genetic Resources 1989

This publication is in copyright. Subject to statutory exception
and to the provisions of relevant collective licensing agreements,
no reproduction of any part may take place without the written
permission of Cambridge University Press.

First published 1989

A catalogue record for this publication is available from the British Library

Library of Congress Cataloguing in Publication data
The use of plant genetic resources/edited by A. H. D. Brown . . [et al.].
p. cm.
1. Germplasm resources, Plant. 2. Germplasm resources, Plant-
Utilization. I. Brown, A. H. D.
SB123.3.U84 1988
631.5′23—dc19 88-12292

ISBN 978-0-521-34584-2 hardback
ISBN 978-0-521-36886-5 paperback

Transferred to digital printing 2008

Contents

Contributors

R. Bernatsky, Department of Plant Breeding and Biometry, 252 Emerson Hall, Cornell University, Ithaca, New York, 14853, USA.

A. H. D. Brown, CSIRO Division of Plant Industry, GPO Box 1600, Canberra, ACT, 2601 Australia.

J. J. Burdon, CSIRO Division of Plant Industry, GPO Box 1600, Canberra, ACT, 2601 Australia.

T. T. Chang, International Rice Research Institute, PO Box 933, Manila, Philippines.

C. G. D. Chapman, International Board of Plant Genetic Resources, FAO/LNOR, 101 22nd Street, Washington DC 20437, USA.

A. B. Damania, ICARDA, PO Box 5466, Aleppo, Syria.

D. N. Duvick, Pioneer Hi-Bred International Inc., 700 Capital Square, 400 Locust Street, Des Moines, Iowa, 50309, USA.

G. Fischbeck, Technische Universität München, Lehrstühl für Pflanzenbau, 8050 Freising, Weihenstephan, West Germany.

O. H. Frankel, CSIRO Division of Plant Industry, GPO Box 1600, Canberra, ACT, 2601 Australia.

R. W. Gibbons, ICRISAT, Patancheru PO, Andhra Pradesh, 502 324, India.

K. S. Gill, Punjab Agricultural University, Ludhiana-141404, Punjab, India.

S. Hamon, Institut Français de Recherche Scientifique pour le Developpement en Cooperation (ORSTOM). Boîte Postale V-51, Abidjan, Côte d'lvoire.

J. G. Th. Hermsen, Institute of Plant Breeding, Agricultural University, Wageningen, The Netherlands.

A. M. Jarosz, CSIRO Division of Plant Industry, GPO Box 1600, Canberra, ACT, 2601 Australia.

G. Ladizinsky, Faculty of Agriculture, Hebrew University, Rehovot, Israel.

D. R. Marshall, University of Sydney Plant Breeding Institute, Sydney, NSW, Australia.

M. H. Mengesha, ICRISAT, Patancheru PO, Andhra Pradesh, 502 324, India.

J. P. Moss, ICRISAT, Patancheru PO, Andhra Pradesh, 502 324, India.

R. G. Palmer, USDA, Department of Genetics, Iowa State University, Ames, Iowa, 50011, USA.

W. J. Peacock, CSIRO Division of Plant Industry, GPO Box 1600, Canberra, ACT, 2601 Australia.

P. M. Perret, European Cooperative Programme for Genetic Resources, IBPGR Headquarters, FAO, 00100 Rome, Italy.

K. E. Prasada Rao, ICRISAT, Patancheru, Andhra Pradesh, 502 324, India.

V. Ramanatha Rao, ICRISAT, Patancheru, Andhra Pradesh, 502 324, India.

V. G. Reddy, ICRISAT, Patancheru PO, Andhra Pradesh, 502 324, India.

J. S. C. Smith, Pioneer Hi-Bred International Inc., 7301 NW 62nd Avenue, Johnston, Iowa, 50131, USA.

J. P. Srivastava, ICARDA, PO Box 5466, Aleppo, Syria.

S. D. Tanksley, Department of Plant Breeding and Biometry, 252 Emerson Hall, Cornell. University, Ithaca, New York, 14853, USA.

D. H. van Sloten, International Board for Plant Genetic Resources, IBPGR Headquarters, FAO, 00100 Rome, Italy.

J. T. Williams, International Board for Plant Genetic Resources. IBPGR Headquarters, FAO, 00100 Rome, Italy.

P. H. Williams, Department of Plant Pathology, University of Wisconsin, Madison, Wisconsin, 53706, USA.

L. A. Withers, Department of Agriculture and Horticulture, University of Nottingham, School of Agriculture, Sutton Bonington, Loughborough, LE12 5RD, UK. Currently: International Board for Plant Genetic Resources, IBPGR Headquarters, FAO, 00100 Rome, Italy.

Preface

There can be little doubt that plant breeders have now a greater range of genetic diversity available to them than ever before. Moreover, it is available to breeders anywhere in the world, subject to some technical constraints. This is due to the co-operation of national and international institutions in an international network promoted and co-ordinated by the International Board for Plant Genetic Resources (IBPGR). Each of the important crops or groups of crops, including all the major and many minor ones, is represented in one or more facilities which act as 'base collections' charged with responsibility for long-term conservation of the genetic resources of one or more crops. Associated 'active collections' provide the link with the users of collections.

Needless to say, the usefulness of collections to plant breeders and other users, including evolutionists, plant pathologists, taxonomists and other experimental biologists, depends in the first instance on the extent to which they are geographically and ecologically representative and on the presence of genes of particular interest to plant breeders. Collections have been enriched by greatly increased collecting activities in recent years, many of which were stimulated or organised by IBPGR. Indeed, collections of many more crops are a great deal more comprehensive than ever before.

Then why are they not used by breeders to a greater extent than they appear to be? Various reasons have been suggested. Breeders tend to use breeding materials with which they are familiar and which are reasonably adapted to their environment, as against alien materials requiring a lengthy programme of pre-adaptation. Further, users require information on collections to be presented in a manner that will allow them to identify accessions of potential use in their projects. This process involves the description, or 'characterisation', of the material, and its 'evaluation' for characteristics of particular concern to plant breeders. Both characterisation and evaluation have been defined by IBPGR and descriptor lists for many crops have been published. The work involved in these operations is considerable, in many instances exceeding the capacity of national collections which are close – hence most relevant to users.

This book explores the factors that are likely to limit or to facilitate the utilisation of plant germplasm. It grew out of a workshop convened by the IBPGR Programme Committee at St Mathieu de Treviers, near Montpellier, France, 9–12 September 1986.

The book has six parts: in the first part three users representing public and private plant breeders and experimental biologists define the role of collections and suggest ways to enhance their usefulness. The second part presents three case histories of collections and discusses limitations to effective use and how they could be remedied. Large collections are contrasted in the third part with the recently proposed representative core collections. A chapter on smaller collections in Europe shows how the association of national collections stimulates collaboration between breeders and genebank managers. The fourth part describes the evaluation system in three widely differing collections. There follows an assessment of the state of management in germplasm collections, and a discussion of the principles of characterisation and evaluation and of the roles of curators, specialists in relevant fields, and plant breeders. With the higher priority now accorded to wild crop relatives, the fifth part examines how collections are to be broadened and made more representative by the inclusion of crop-related species. The final section outlines recently developed techniques which are beginning to open up new approaches in all areas of genetic resources work.

W. J. Peacock
Chairman
International Board for Plant Genetic Resources

Acknowledgements

The editors gratefully acknowledge:

1. The initiative of the International Board for Plant Genetic Resources in convening and funding the workshop *Genetic Resources and the Plant Breeder*, Montpellier, 9–12 September, 1986, which gave rise to this book;
2. The contribution of Dr R. J. Gorringe, who prepared the index.

Acknowledgements

The editors gratefully acknowledge:
The initiative of the International Board for Plant Genetic Resources in conceiving and launching the [illegible] Genetic Resources and the Plant Breeder [illegible] 12 September 1986, which gave rise to this book.
The contribution of Dr [illegible] in [illegible]

Part I

Role of genetic resource collections in research and breeding

1

Germplasm collections and the public plant breeder

K.S. GILL

The role of germplasm in the improvement of cultivated plants has been well recognised (Frankel & Hawkes, 1975; Hawkes, 1981; Holden & Williams, 1984). However, the use of germplasm collections, particularly in the developing countries is still limited despite this wide recognition (Gill, 1984; 1985). The lack of effective utilisation of germplasm by public plant breeders in developing countries raises several interrelated questions:

1. To what extent have public plant breeders used existing collections for crop improvement in developing countries?
2. What do plant breeders, especially those in developing countries, want in collections?
3. How would breeders use such collections?
4. Why have breeders not made greater use of existing collections?

These questions are considered in this paper. As a result, a series of recommendations for enhancing the effectiveness and use of germplasm collections by public breeders emerge.

Germplasm utilisation

Smith and Duvick (this volume) have analysed the role of private plant breeders in the evaluation and utilisation of germplasm. However, private plant breeders have not played a major role in breeding improved varieties in India and other developing countries. Rather, varietal improvement in these countries is generally carried out by public breeders working in agricultural universities and other government-funded research institutions.

In India, various breeders have developed limited germplasm collections of their respective crops and related species for utilisation in breeding

Table 1.1. *Utilisation of germplasm in the development of improved varieties by the PAU (1962–86)*

Crop	Number of varieties	Number of parental accessions				
		Indian landraces	Indian elite lines	Exotic landraces	Exotic elite lines	Others
Field crops						
Bread wheat	20	—	20	—	32	—
Durum wheat	2	—	—	—	6	—
Triticale	2	—	1	—	5	—
Barley	3	—	3	—	1	1 (Induced mutant)
Rice	11	—	7	—	10	1 (Induced mutant)
Maize	10	1	31	—	10	—
Pearl Millet	7	—	52	—	13	—
Grain legumes (Pulses)	27	12	17	3	1	
Oilseeds	19	6	8	3	3	3 (Induced mutants)
Cotton	10	2	12	—	4	
Sugarcane	7	—	7	—	—	
Forages	16	6	14	2	1	1 (Interspecific cross)
Vegetable crops	42	20	15	—	15	2 (One induced mutant and one interspecific cross)
Horticultural crops	55	24	15	—	18	—
Total	231	71	202	8	119	8

programmes through personal exploration of indigenous germplasm and correspondence with fellow national and international breeders and institutes. Previously known as the Plant Introduction Division of the Indian Agricultural Research Institute, New Delhi, but since 1976 the National Bureau of Plant Genetic Resources (NBPGR), has co-ordinated the acquisition, exchange, evaluation and maintenance of crop germplasm. The NBPGR has extensive germplasm collections numbering thousands and a very effective introduction programme.

At the Punjab Agricultural University (PAU), Ludhiana, plant breeders maintain 30,244 accessions of various field (28,193), vegetable (1,350) and horticultural crops (701). These accessions are evaluated and utilised in breeding programmes by a team of scientists comprising breeders, pathologists, agronomists, entomologists and so on for each crop. At the PAU, during the past 25 years, 134 varieties of field crops, 42 of vegetables and 55 of horticultural crops have been developed through the direct and indirect use of germplasm. The origin of the parental accessions used in the development of these varieties is given in Table 1.1.

The varieties of field and vegetable crops developed in the initial stages were mostly selected from indigenous landraces. Subsequently, varieties were developed using indigenous and exotic collections in hybridisation and selection. In only four cases have related species of cultivated crop plants been used in wide crosses to develop new varieties. On the other hand, varieties of the horticultural plants have been mostly developed from local collections or exotic introductions.

Breeders working in other State agricultural universities and State and Central Institutes have used germplasm to develop improved varieties the same way as the PAU. In all the major field and vegetable crops, the

Table 1.2. *Wheat accessions evaluated in India in 1985–86*

Trial/Nursery	Number of accessions
Yield trials	1,057
National genetic stock nursery	312
Alkalinity/Salinity tolerance nursery	225
National yield nursery	125
Quality component nursery	75
Plant pathological screening nurseries (rusts, Karnal bunt, powdery mildew, hill bunt, foot rot, loose smut, flag smut and foliar diseases)	1,224
Drought resistance screening nursery	23
Heat tolerance nursery	16
	3,057

superior lines developed by various breeders are tested across the country in the All India Co-ordinated trials or nurseries over a number of years. As an example, as many as 3,057 wheat entries from different institutions in India were evaluated in 1985–86 at different locations (Table 1.2). These nurseries and trials are excellent means by which the available germplasm is evaluated and exchanged.

Since the NBPGR does not evaluate and characterise germplasm exhaustively, PAU has also introduced diverse germplasm of wheat and related species (6,563 accessions) from various national and international organisations under a PL 480 project entitled 'Collection, maintenance and evaluation of germplasm of wheat and related species and utilisation of useful variability'. Useful variability for resistance to diseases and quality traits has been identified (Gill & Aujla, 1986). The identified donors have been successfully used to derive elite germplasm including synthesis of amphiploids (*Triticum durum* × *T. monococcum*) which are resistant to Karnal bunt and possess high protein. Some genes conferring resistance (*Lr1, Lr3, Lr10* and *Lr14b*) to leaf rust have been identified. Genes *kr1* and *kr2* for high crossability have been transferred from cv. Chinese Spring to a high-yielding and well-adapted bread wheat variety WL 711. The high level of tolerance of *T. monococcum* to herbicide Isoproturon has been transferred to *T. durum* through triploid hybrid bridge. Recently, private breeders in India have also undertaken limited breeding work to develop high yielding hybrids of crops such as maize, sorghum, pearl millet, cotton, forages and some vegetable crops. Some private organisations (e.g. Tata Energy Research Institute, Hindustan Lever Ltd) have shown interest during the past two or three years in funding tissue culture and genetic engineering research to develop improved germplasm of crop species.

Germplasm collections wanted by the public plant breeders

The main interest of both the public or private breeder in any country is in germplasm which can be utilised for attaining specific breeding objectives. The plant breeder, especially in the developing countries, requires the following characteristics in the cultivated germplasm of the crops of interest.

Classified working collection

The breeder is usually interested in using only a small fraction of the total germplasm at any one time to meet immediate and often pressing objectives. Therefore, optimum working germplasm collections should consist of different types of superior genetic variability for yield, yield components, plant height, maturity, resistance to disease, pests and other stress conditions, and quality.

Genetic variability in specific traits

On the basis of the author's experience, a few important characteristics can be listed for which useful variability is limited in germplasm collections of cultivated species of various crops:

Crop	Resistance and tolerances for which variability is limited or lacking
Bread wheat	Karnal bunt (*Neovossia indica*), downy mildew (*Sclerophthora macrospora*), yellow dwarf (barley yellow dwarf virus), alkalinity
Rice	Sheath blight (*Corticum sasakii*)
Peal millet	Ergot (*Claviceps microcephala*)
Maize	Drought
Chickpea	Gram blight (*Ascochyta rabei*)
Pigeon pea	Sterility mosaic (Sterility mosaic virus)
Moong	Yellow mosaic (Yellow mosaic virus)
Cotton	Pink boll worm (*Heliothus* spp.)
Brassica spp.	Alternaria blight (*Alternaria brassicae*), aphids, frost
Okra	Yellow vein (Yellow vein mosaic virus)
Muskmelon	Downy mildew (*Pseudoperonospora cubensis*)
Tomato	Late blight (*Phytophthora infestans*), leaf curl (Leaf curl virus)
Potato	Late blight (*P. infestans*)
Onion	Purple blotch (*Alternaria pori*), downy mildew (*Peronospora destructor*)
Mango	Malformation (*Fusarium moniliformae*)

Some of the genes such as those for male sterility, dwarfness and disease resistance, have been exploited by the breeders in developing varieties which are cultivated over large areas. Use of such germplasm with a narrow genetic base can be hazardous. Variability is also limited in those crops for which there are no specific international centres and co-ordination as is the case for horticultural and forage crops. The breeder would thus want diverse sources of useful variability for these specific traits in the germplasm collection to meet his requirements.

Variability in desirable agronomic background

For effective use of germplasm, the breeder would want the required variability to be available in an agronomically desirable background. In some cases, the stocks possessing useful variability are in agronomically poor backgrounds. These include some of the *Yr* and *Lr* genes for resistance to stripe rust and leaf rust in wheat, *hiproly* gene for high lysine and high protein in barley, and *ph* mutant gene for induced homoeologous recombination in wheat. Similarly the breeder would want the genetic variability from the wild species transferred into the background of cultivated species. Ideally, some centrally sponsored institutes should continuously accomplish this according to the requirement of breeders.

Genetic basis of a desirable trait

The available genetic collections may have a number of duplications for the same trait. The breeder, therefore, wants to have information on the genetic architecture of the parental material with respect to the desired character so that effective use of the same can be planned.

How should the public plant breeder use germplasm collections?

Some of the important points related to the effective use of germplasm by the plant breeder are given below:

Identification of useful donors

The available germplasm needs to be screened for identification of useful donors using artificial epiphytotics/hot spots/sick plots/replicated trials by a specialised team/centres in line with the specific requirements of breeders. On the basis of this identification, the germplasm can be characterised into appropriate groups and conveniently used in hybridisation.

The plant breeder should have a dynamic working collection and add new genetic stocks and 'elite' strains from time to time possessing desirable genetic variability.

Pyramiding of genes

The breeder should attempt pyramiding of genes conditioning low levels of different components of resistance to diseases, pests and tolerance to salt and other environmental stresses where a high level of resistance does not exist in any one accession. Such 'modified' genetic stocks can be released as varieties or used in further breeding.

Breeding for a few specific traits at a time

Single genotypes combining all the desirable traits are often rare in collections. It would be desirable to transfer one or a few characters such as disease and pest resistance (controlled by major genes) at a time in a desirable agronomic background, before recombining them. However, this may be difficult for yield components and other polygenically controlled traits.

Use of diversified sources

The plant breeder should seek additional sources of variability where possible. For example, additional sources must be sought for cytoplasmic male sterility for hybrid varieties and for dwarfness in order to breed semi-dwarf varieties. Then they should be used to widen the genetic base to overcome vulnerability to diseases and pests.

Exploitation of inter-specific group diversity

In many cases, plant breeders have confined their efforts to using genetic variation within defined intraspecific groups. As an example, breeding of winter wheat has been mainly done by using only winter germplasm and breeding of spring wheat by using only spring germplasm. Similar examples occur in other crops, such as rice, groundnut, chickpea, rape seed and mustard. The breeder should seek to exploit the greater genetic diversity available through hybridisation of parents selected from diverse groups within a species.

Wide hybridisation

To use the genetic variation from related species and genera, the use of new technologies such as embryo rescue, *in vitro* pollination, somatic cell hybridisation and recombinant DNA techniques need to be encouraged. Wide hybridisation in crop improvement has to be carried out in collaboration with scientists engaged in cytogenetics and tissue culture research to obtain and use species hybrids in breeding programmes.

Basic information on genomic relationship, crossability barriers and the potential to recover desirable recombinants from cross combinations involving diverse germplasm must be collected. On the basis of this information the breeder should select parents from the total collection.

Inter-disciplinary approach

The effective use of germplasm requires an inter-disciplinary approach in which the plant breeder and other scientists work together. This collaboration is particularly necessary for efficient characterisation of germplasm. It has been generally observed that in disease and other stress screening nurseries, only the concerned scientist is responsible. The involvement of a plant breeder in such screening nurseries would help him in selecting suitable donors. Although the concept of an inter-disciplinary approach exists in various institutes, there is still a need to further strengthen this in practice.

Limited utilization of the existing germplasm collections

The germplasm collections in many developing countries may be extensive but they are not put to effective use by public breeders. In some cases, the existing collections may be inadequate and may not represent the useful genetic variability required by the plant breeder. Some of the pertinent reasons for limited use of the existing germplasm collections are listed below.

Lack of proper evaluation of germplasm

One of the main reasons for lack of proper evaluation of germplasm for stress conditions is inadequate facilities for artificial epiphytotics and for simulation of drought, heat, frost and salinity stresses. Similarly, screening of germplasm for quality traits is frequently hampered because of lack of facilities.

Limited genetic characterisation of the germplasm

Information on the genetic make-up of the useful variability in the existing germplasm is lacking. Only limited information is available on the number of genes controlling particular traits, allelic relationships, linkages, etc. Moreover, in many cases, the available germplasm has not been properly classified for components of yield, disease and quality which is essential for its utilisation by breeders.

Difficulty to meet the specific flowering requirements of the cultivated germplasm

Cultivated germplasm of a species from different latitudes and altitudes has specific temperature and photoperiod requirements. Examples include winter wheat, *japonica* rice and sugarcane which normally do not flower under conditions in northern India. Most plant breeders lack facilities for the maintenance of such germplasm.

Lack of useful mutants in desirable agronomic backgrounds

Many desirable genetic traits occur in agronomically poor backgrounds. Breeders are thus reluctant to use such stocks in their hybridisation programmes due to the difficulty in recovering the desirable genotypes.

Non-ultilisation of cultivated parents of the cultivated polyploids

The cultivated parents of polyploid species, such as wheat, cotton and *Brassica*, are known to possess useful variability for various characteristics. However, the parent species have not been used extensively by plant breeders due to chromosomal imbalance in interspecific crosses and limited training of the plant breeders in cytogenetics. Repeated backcrosses are required. As a consequence, breeders hesitate to use such germplasm to isolate derivatives from the hybrid progenies.

Difficulty in maintenance of germplasm of cross-pollinated crops

It is very laborious and expensive to maintain the germplasm of cross-pollinated crops either through controlled selfing or sib-pollination. Maintenance of small seed samples of cross-pollinated crops can lead to genetic drift.

Lack of facilities for data storage and retrieval

Although the International Board for Plant Genetic Resources (IBPGR) has introduced the use of standard descriptors for a variety of crops, these are not being used by many scientists for the classification and evaluation of germplasm. It is also often difficult to isolate the desired accession from a large collection for want of adequate data storage and retrieval facilities.

Insufficient trained staff and funds

In developing countries, breeders are responsible both for maintenance and utilisation of germplasm. The maintenance of large germplasm collections itself requires funds and trained staff which are often limited. Moreover, due to lack of long-term storage facilities, the required year-to-year maintenance is very expensive and time-consuming.

Difficulty in producing wide crosses

Wild germplasm is known to possess useful variability for various characteristics. However, such germplasm is being used to a limited extent by breeders because of various problems such as incompatibility barriers, undesirable linkages, lack of chromosome pairing and recombination.

Restricted gene flow in apomictic, polyembryonic and asexually propagated species

In certain crop species including sugarcane, potato and fruit crops, apomixis, polyembryony and predominant asexual reproduction are barriers to gene flow and these account for limited utilisation of their germplasm. In horticultural crops, long generation time can also be responsible for limited evaluation and utilisation.

Delay in quarantine clearance

Although the NBPGR has made extensive arrangements for speedy exchange of germplasm, delay in the clearance of germplasm as well as rejection of certain stocks during quarantine, still restricts the utilisation of germplasm resources.

Recommendations for enhancing the effectiveness and utilisation of germplasm collections

Rapid evaluation and characterisation of germplasm collections

In general, evaluation lags behind collection of germplasm. Even preliminary evaluation and characterisation have not kept pace with collection. Highest priority should be given to characterisation and evaluation of germplasm collections for its effective utilisation. For this, it is essential to establish regional nurseries in hot spot areas. The testing of germplasm for disease and other stress conditions in National Screening Nurseries is being done in a few field crops such as wheat, rice and maize. There is an urgent need to strengthen similar facilities for characterisation of germplasm of other crops such as pulses, oilseeds, cotton and vegetables.

Development and use of efficient screening techniques

Efficient laboratory and field screening techniques are necessary, not only for the characterisation of large germplasm collections but also for its utilisation. There is, therefore, an urgent need to develop such facilities. A few years ago, difficulties in breeding Karnal bunt resistant wheat varieties were experienced due to lack of screening techniques. Now, we have developed a technique for artificial inoculation of Karnal bunt both under laboratory and field conditions which can be successfully used for screening of germplasm for resistance to this serious disease (Gill & Aujla, 1986). Similar techniques need to be developed for other situations concerning diseases, pests and stress conditions.

Gene cataloguing

Very limited information is available on the genetic make-up of the superior lines from germplasm collections. For example, we have isolated a number of wild wheat and *Aegilops* lines resistant to virulent races of stripe rust and powdery mildew under artificial epiphytotic conditions but we do not know if these possess the same or different genes for resistance to these diseases. Therefore, it is essential to carry out genetic studies for determining the genetic constitution of the desirable germplasm after screening. This information should be made available to plant breeders by a team of scientists entrusted with the duty.

Transfer of useful genetic variability into agronomically desirable backgrounds

As mentioned above, most of the genetic stocks possessing useful variability and other mutant genes are agronomically undesirable. Concerted efforts are required to transfer useful genes into agronomically desirable varieties and such improved germplasm should be given to plant breeders for utilisation. Such work is often lacking even in the developed countries.

Strengthening facilities for maintenance and utilisation of wild species

For many traits, useful variability in the cultivated germplasm is limited or lacking. Under such situations, there is a need to utilise genetic variability from wild species. However, most of the breeders have limited facilities for controlled temperature and light conditions, embryo rescue, etc. There is therefore an urgent need to strengthen facilities at various national centres for the utilisation of wild species for introgression of alien genes into cultivated species.

Establishing germplasm resources centres

A separate unit for handling the germplasm collections of various breeding departments should be established in each agricultural university in developing countries. It should work in close collaboration with the breeders, geneticists and biologists and provide up-to-date information on the available germplasm. This system of maintenance would not only relieve the breeders so that they can concentrate on breeding work but would also save the germplasm from being lost as accessions of no immediate value usually do not receive due attention from the breeders and are discarded. Such units could co-ordinate the germplasm work at the university level and with the national centre such as NBPGR in India.

Establishment of regional centres for germplasm storage under natural conditions

It is very laborious and expensive to maintain and grow germplasm year after year. There is, therefore, an urgent need to establish

Fig. 1.1. A proposed scheme for effective utilisation of germplasm by the public plant breeder in developing countries.

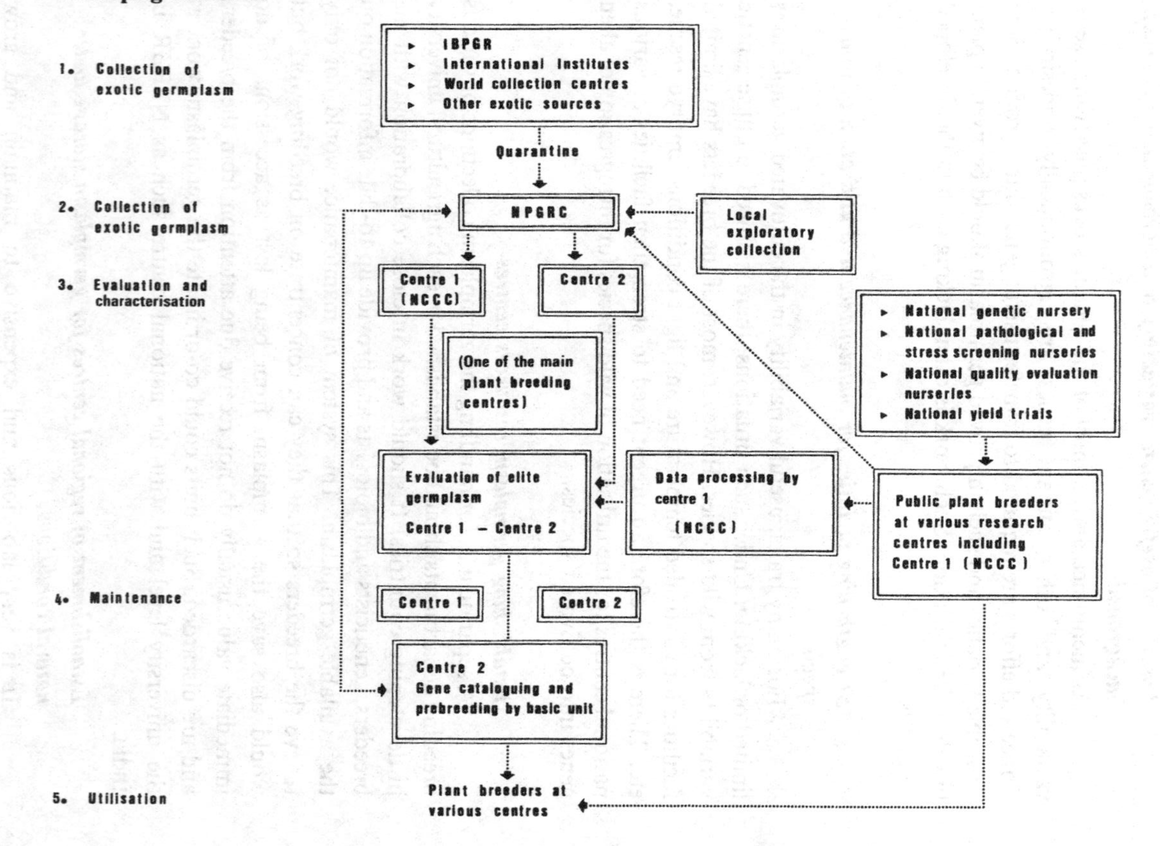

centres for storage of germplasm under natural conditions requiring minimum additional facilities. We have made some efforts to store germplasm of wheat and allied species at Keylong in Himachal Pradesh, India, where the relative humidity and temperatures are suitable for short- and medium-term storage of germplasm. Such centres need to be identified in appropriate locations for storage of germplasm.

Cryopreservation of germplasm of vegetatively propagated species

The germplasm of vegetatively propagated genetic stocks and crops with recalcitrant seeds need special attention for maintenance and utilisation (see Withers, this volume). Such collections must be maintained in duplicate and *in vitro* conservation adopted in collaboration with tissue culture laboratories.

Strengthening facilities for computerised gene banks for data recording, processing and retrieval

Plant breeders would always be interested in accessions possessing multiple disease resistance which are superior in yield components or other specific traits which cannot be easily assessed. To achieve this, record systems should include information on the conserved stocks, systematic description of various morphological and agronomic traits; results of evaluation and maintenance and distribution records. A uniform recording system would be needed for collection and retrieving information originating at different centres.

Training in genetic resources

Courses on genetic resources at undergraduate and postgraduate level, like those at the University of Birmingham (UK) could be useful to train young scientists. Short-term refresher courses for training technicians on the collection and utilisation of germplasm should be developed by the responsible national centres within countries.

A proposed scheme for effective utilisation of germplasm collections by the public plant breeder

A scheme is proposed for effective utilisation of germplasm collections by plant breeders in developing countries (Fig. 1.1). The germplasm collections received from various sources by the NPGRC should be made available to two centres: First, a National Crop Co-ordinating Centre (NCCC) and, second, one of the selected plant breeding centres (Centre 2). Two centres are suggested to minimise losses due to unforeseen hazards. Centre 1 should arrange the testing of these materials at various hot spots, sick plots, stress nurseries, etc and in replicated trials. Elite germplasm can be identified in collaboration with the respective crop scientists including Centre 1 in the region/country. The NCCC on the basis of multi-location data should identify 'elite' germplasm, which should be made available to the interested breeders for utilisation. The elite germplasm should be maintained in duplicate at the two centres. The public breeders need then only maintain working collections.

Utilisation of germplasm collections

As shown in Fig. 1.1, the elite germplasm identified on the basis of evaluation of Centre 1 (NCCC) and Centre 2 must be made available to the public breeders working at various research stations for utilisation in different breeding programmes. The germplasm collection, characterised into different components of yield, disease, quality and stress conditions, could be used by the breeders in crop improvement programmes. Promising materials developed by the use of such collection by breeders should also form a part of the existing germplasm collection and should be made available to the NCCC as well as the National Plant Genetic Resources Centre.

A basic unit should be established at Centre 2 for gene cataloguing and pre-breeding work.

References

Frankel, O. H. & J. G. Hawkes eds. (1975). *Crop Genetic Resources for Today and Tomorrow*. Cambridge University Press, Cambridge.

Gill, K. S. (1984). Research imperatives beyond the green revolution in the third world. In *Human Fertility, Health and Food – Impact on Molecular Biology and Technology*, pp. 195–231, D. Puett (ed.), United Nations Fund for Population Activities. New York.

Gill, K. S. (1985). Exchange and utilization of genetic resources for crop improvement. In *Regional Conference on Plant Quarantine Support for Agricultural Development*. ASPEAN Serdang, Malaysia (in press).

Gill, K. S. & S. S. Aujla. (1986). Research on breeding for Karnal bunt resistance in wheat. In *Fifth Biennial Smut and Karnal bunt Workers Workshop*. CIMMYT, Mexico.

Hawkes, J. G. (1981). Germplasm collection, preservation and use. In *Plant Breeding II*, pp. 57–83 (K. J. Frey ed.) Iowa State University Press, Ames.

Holden, J. H. W. & J. T. Williams eds. (1984). *Crop Genetic Resources: Conservation and Evaluation*. George Allen & Unwin, London.

2

Germplasm collections and the private plant breeder

J.S.C. SMITH AND D. N. DUVICK

Introduction

A breeding programme that is genetically broadly based should be able to provide ideal results: steady gains under selection and the ability to respond readily to changed environments, diseases and economic trends (Simmonds, 1962). In contrast, a narrowly based programme would provide slow response to selection and increase the likelihood of crises triggered by outbreaks of disease and insect pests. A lack of genetic variability across breeding programmes could exacerbate these deficiencies nationally or internationally, conceivably threatening the usefulness of available varieties and, of longer term significance, the usefulness of breeding stocks.

Exotic germplasm in plant breeding

Historical and recent usage

Foreign genes from varieties that are exotic to a particular region, including wild, weedy and alien species, can provide increased genetic diversity to the currently used germplasm base. Indeed, the evolutionary origins of all the major crop plants proclaim the absolute dependence of the most productive agricultural regions upon species that once were foreign (Hawkes, 1983).

It is difficult to obtain figures that portray accurately the degree of usage of exotic germplasm in current breeding programmes or in commercially released varieties. There are few currently available measures of germplasm identity that can be applied to a range of cultivated species. Therefore, such estimates must be based upon pedigree information. Unfortunately, such estimates can be rather subjective and are, most likely, unavailable from private breeding programmes.

In a survey of private maize breeders, Goodman (1985) found that from

0.4 per cent to 1.9 per cent of the United States maize acreage contained eight per cent to 37 per cent temperate exotic germplasm. However, all of these exotics possibly traced back to original sources in the USA. Of tropical exotic sources, 0.9 per cent of the acreage contained from three per cent to 35 per cent exotic germplasm from the Caribbean, Mexico, Cuba and Argentina. Zero to 17 per cent of breeding lines contained exotic tropical or temperate germplasm. However, Seifert (cited by Brown, 1981) reported that the use of exotic maize germplasm had been instrumental in the improvement of hybrid maize adapted to the southern USA. Overall, there appeared to be no evidence that the acreage of US maize hybrids containing exotic germplasm would increase significantly, at least during the next 10 to 15 years (Goodman, 1985).

Wheat, sorghum and soyabean are more likely candidates for the inclusion of exotic germplasm since their sites of domestication and subsequent spread of landraces preclude the availability of germplasm that is locally adapted to the USA. In contrast, US maize breeders have available a broad range of landrace open-pollinated varieties that were bred from the Corn Belt dent racial complex which, together with its progenitor races, the northern flints and southern dents, evolved in North America.

Extensive use of exotic sorghum has been made since the conversion and evaluation programme which began in the mid-1960s at the International Crops Research Institute for the Semi-Arid Tropics (ICRISAT), Hyderabad, India, and at Texas A. & M. University, USA (Miller, 1968; Quinby, 1968; Dalton, 1970; Eberhart, 1970; Webster, 1975).

The soft red winter wheat breeding programmes in the USA had an infusion of exotic germplasm during 1954–69. This followed a 35-year period during which a small number of closely related varieties formed the germplasm base (Cox *et al.*, 1986).

Exotic germplasm from *Solanum* spp., in large part emanating from the International Potato Center (CIP), Peru, has been used mainly as a source of pest and disease resistance in potatoes. It has also contributed to increases of yield *per se* (Peloquin, 1984). In the US, about 33 per cent of released potato cultivars have exotic germplasm in their backgrounds. The comparable figure for European cultivars is greater than 50 per cent.

Tomato breeders have made extensive and critical use of exotic germplasm, thanks largely to the collection, evaluation and breeding work of C.M. Rick (Rick, 1979; 1984). Additional examples of exotic use for these and other crops are given by Chang (1984; 1985) and by Hawkes (1985*a*).

Future usage

Plant breeding and gene resource management

Duvick (1984*a*) surveyed 67 public and 34 private US plant breeding programmes, including cotton, soyabean, wheat, sorghum and maize. The majority of respondents for each crop reported that elite materials were generally more useful than landraces or related species as sources of pest resistance and stress tolerance.

Therefore, in spite of the contributions made by exotic germplasm, there is no evidence that exotic germplasm is now, or will be in the near future, the major germplasm source of elite varieties of the major US field crops.

This seeming paradox exists because of the goals of plant breeding and the nature of the materials with which plant breeders work. Plant breeders seek to provide cultivars of increased genotypic adaptation to agricultural environments. Genetic, cultural and environmental changes have provided increased productivity by optimising the fit of cultivars to increasingly uniform and optimum cultural conditions. The result, as well discussed by Simmonds (1962), Marshall (1977), and Timothy & Goodman (1979) is that there has been a simultaneous reduction in extant sources of adaptability available from diverse cultivated varieties, populations, and wild and weedy relatives. Such a trend is fundamentally linked to the whole organisation and development of society (Rindos, 1984). Its potential disadvantages can only be countered by first, the development of comprehensive, well documented, and accessible germplasm collections and secondly, by a plant breeding community that is willing to invest resources in long-term programmes of germplasm diversification.

'Exotic' germplasm and plant breeding

Several popular misconceptions exist with regard to exotic germplasm and its use. Differences in opinions as to what constitutes 'exotic' germplasm impair any objective measurement of the contribution of exotic collections to elite cultivated varieties.

A broad interpretation of exotic would include all untried germplasm. A more useful definition is that of Hallauer & Miranda (1981), 'all germplasm that does not have immediate usefulness without selection for adaptation to a given area'.

Thus, both landrace ('unimproved') germplasm and foreign elite germplasm can be exotic. Exotic germplasm can, therefore, flow reciprocally between temperate and tropical regions and between breeding programmes of 'northern' developed countries and less developed 'southern' or Third World countries.

Since the term exotic refers to adaptation rather than to genetic variability *per se*, a lack of what might be considered as exotic germplasm in elite cultivated varieties does not necessarily indicate a lack of genetic diversity in those materials. The amount of genetic diversity in elite maize breeding stocks has been sufficient to allow significant and still continuing increases in yield, resistance to pests and diseases, and tolerance to environmental stress. It has also led to cultivars making more efficient use of nutrients (Russell, 1974; 1985; Duvick, 1977; 1984*b*; Castleberry *et al.*, 1984). Likewise, consistent increases in genetic yield potential, mostly through the use of elite germplasm, have been shown for sorghum (Miller & Kebede, 1984), soyabean (Specht & Williams, 1984), cotton (Meredith & Bridge, 1984) and wheat (Schmidt, 1984).

In view of this steady but highly focused progress, nearly always based on progressive improvement of a few highly selected germplasm families, it is hardly surprising that the array of elite adapted germplasm in a given species should be less than the total array of adapted and unadapted genes. Uninformed critics of plant breeders' use of genetic diversity can fail to appreciate this elementary fact. However, it is also often forgotten that plant breeders do create new genetic variability (Brim, 1973; Fehr, 1976; Specht & Williams, 1984). For example, a significant increase in genetic yield potential of soyabean occurred in the 1940s when breeders released novel genetic variants by recombining genes from a relatively few plant introductions (Specht & Williams, 1984).

Plant breeders' future needs

Dwelling on past successes, however, is both short-sighted and dangerous. Plant breeders can never afford to be satisfied with their achievements, particularly because they are working in a biological environment where a new pest or disease can quickly reveal the risks resulting from complacency. It is for this reason that plant breeders included in their list of important goals the need to increase the amount of genetic diversity in breeding programmes (Kleese & Duvick, 1980).

A further reason for increasing use of exotic germplasm is the belief that the relatively few plant introductions and races that are the foundation of developed agriculture can hardly be expected to contain all genes of agronomic worth of any one cultivated species. A contrary opinion has been expressed by some breeders who believe that exotic plant introductions can make little or no additional contribution, at least to yield, beyond that which is already available within elite germplasm. If this should be true, then the effort to bring in useful new germplasm through exotics is futile. However, there is clear evidence to suggest that this is indeed not the case.

Jennings (pers. comm.) has analysed the pedigrees of high yielding soyabeans and found that, in most instances, such lines have pedigree backgrounds that trace back to different sources. Thus, there may be additional genes for yield in different new sources of soyabean. Similar data have been found for exotic races of maize (Gerrish, 1983; Salhuana, pers. comm.) Indeed, the continued success of maize breeders rests in large part upon the bringing together (in the nineteenth century) of two genetically distinct races to form the Corn Belt dent racial complex (Anderson & Brown, 1952; Doebley *et al.*, 1986).

The notion that additional genes of agronomic worth are available in foreign or exotic sources is well supported (Chang, 1984, 1985). Plant breeders have initially used exotic collections as sources of disease and pest resistance. Genes controlling these traits are often few in number and can, therefore, be readily observed, screened and incorporated through routine breeding procedures. Traits under multigenic control, useful in the improvement of more complex agronomic characteristics such as environmental adaptation and yield, have also been found among exotic materials. Numerous examples citing the contributions of exotic germplasm to the improvement of traits that are under both simple and complex genetic control are given by Zeven & van Harten (1979), Hallauer & Miranda (1981), Burton (1984), Peloquin (1984), Rick (1984), Chang (1985), and Goodman (1985). The future availability of exotic germplasm has been cited as a critical factor that will determine continued progress in raising genetic yield potential of sorghum and soyabean (Miller & Kebede, 1984; Specht & Williams, 1984).

However, building a broader germplasm base *per se* is pointless as a practical goal (Sprague, 1982). Genetic diversity, in itself, is practically worthless unless it encompasses genes that are useful, either in themselves, or in combination with other previously evaluated germplasm. This fact is particularly important to private plant breeders whose support for research is generated by sales of products meeting the demands of farmers, processors and consumers.

The use of exotic germplasm: difficulties and needs

Major hindrances to the use of exotic germplasm by private plant breeding companies are the uncertainties and difficulties of progress in a reasonable time-frame. Often, needed selections of exotic germplasm cannot be effectively identified since, for most crops, much of the available exotic germplasm in collections (or *in situ*) is poorly documented. Secondly, even when germplasm enhancement (pre-breeding) has been attempted, it may have been through breeding strategies that have proved successful with adapted germplasm but which may be inappropriate for unadapted germplasm. These difficulties only exacerbate the many problems of poor agronomic performance that can be expressed by exotic germplasm (Hallauer, 1978; Goodman, 1984, 1985).

The need for cataloguing and maintenance

Systematic strategies and adequate resource allocations are needed for the cataloguing, maintenance and evaluation of germplasm (Goodman, 1984). Lists describing the identity and provenance data (passport data) *must* be complete and easily accessible.

The need for evaluation

From the plant breeder's perspective, evaluation could be considered a two-step procedure of preliminary evaluation applicable to numerous accessions followed by more detailed evaluation on selected accessions. Preliminary evaluation data would include information on basic agronomic characters such as maturation time, plant height, per cent germination and disease resistance. These data are of critical importance and should be available for all accessions. Alternatively, for taxonomically well-understood cultivated species (as, for example, the races of maize) further agronomic data for traits such as grain yield and lodging resistance could be acquired for typical accessions (Goodman, 1985).

Plant breeders can only invest intensive breeding efforts upon a relatively small number of accessions at any one time. Thus, some rational basis is needed to identify the accessions of greatest potential use. Taxonomic and provenance data alone are probably insufficient for this purpose. Curators of gene banks will not be able to achieve the objective of usefully describing collections unless preliminary evaluation data are available. However, it is private and public plant breeders who will be among the primary users of germplasm resources. The plant breeding and germplasm resource communities must, therefore, act in a systematic and co-operative fashion to provide major financial, organisational, political and practical stimuli for the acquisition of these data (Hawkes, 1985*b*).

More detailed evaluation data could be collected following the choice of potentially useful accessions on basis of preliminary evaluations. These data would include information from replicated trials on a greater number of relatively simply inherited traits such as grain quality, rate of grain maturation, reaction to a wide range of disease and insect pests and the more complexly inherited traits such as tolerance to heat and drought stress, combining ability and yield. Detailed evaluations should be performed on all typical accessions of taxonomically well-understood cultivated species. Availability of these data will encourage the use of exotic germplasm.

The need for pre-breeding

Pre-breeding, or germplasm enhancement, is often necessary because even simply inherited traits can be masked by other genetic effects such as maturity response. Furthermore, several cycles of recombination and mild selection may be required to develop gene pools that are sufficiently adapted to allow further progress to be made using exotic germplasm (Hawkes, 1985*a*). For maize, Lonnquist (1974) has recommended the incorporation of exotic accessions into germplasm pools with mild selection occurring for a minimum of five generations of random mating before useful recombinants can be selected. The elite maize inbred lines of today are themselves the product of 100 years of pre-breeding in the form of mass selection by farmers followed by 50 years of recurrent selection by plant breeders (Anderson & Brown, 1952).

The importance of pre-breeding is well demonstrated by the experiences of sorghum breeders. The world collection of sorghum is represented by 20,000 to 25,000 accessions. About half of these could not have been evaluated or used in the USA and other temperate regions except for a programme that altered their tropically adapted maturity response. Since the instigation of several conversion programme (by Texas A. & M. University, ICRISAT, and a few private plant breeders), significant advances in sorghum breeding have been made through the incorporation of exotic germplasm. These include resistance to greenbug, downy mildew, anthracnose and smut (from Chinese collections); increased yield (from Feterita, Kaura, and Zera Zera germplasm); grain quality (from Indian accessions); better performance under stress of heat and drought (from Sudanese accessions); enhanced standability and good beer making quality (from Ugandan collections). Further advances are expected.

Clearly pre-breeding to allow the recombination and evaluation of characters under complex genetic control is a time-consuming and difficult process. However, willingness to invest in longer term breeding programmes incorporating unfamiliar germplasm will depend in a large part on the success of relatively shorter term pre-breeding programmes to identify potentially useful collections that can repay long-term breeding effort.

Private plant breeders, often more isolated and with pressure to fulfill short-term goals, will be more likely to use exotics if pre-breeding is performed first. However, private plant breeders must also, when possible, perform pre-breeding or at least contribute resources to regional or inter-regional programmes. Private/public co-operation can be especially useful in pre-breeding programmes.

The need for research

The need for certain fundamental and long-term applied research programmes is also obvious. Basic taxonomic and evolutionary studies of crop plants are prerequisites (Hawkes, 1985*a*). Appropriate breeding strategies that will allow the effective incorporation and use of exotic germplasm must be identified. Questions which must be addressed are: which exotic sources should be used; what degree of recombination is necessary; how much exotic germplasm should be incorporated into elite germplasm; which specific elite by exotic combinations show good results?

Long-term research in these fields should continue to be undertaken by public institutions since the time scale is such that it is unlikely to be done by private plant breeders. Public institutions also have a larger concentration of necessary resources and trained personnel to undertake such programmes, compared to private companies. However, private breeders have the obligation to communicate with public breeders, to help to provide funds for such research, and to lobby for public support. These programmes will not only produce useful information and germplasm; they will also provide a training ground for future generations of plant breeders, both public and private.

Procedures for identifying useful genetic diversity would be invaluable to breeders working with unfamiliar germplasm. Laboratory-derived data establishing germplasm relationships will help breeders construct and understand the outcome of possible genetic associations. This becomes increasingly important as more 'fresh' germplasm with no pedigree breeding information becomes available for incorporation into breeding programmes.

Allozymic and chromatographic data have identified specific genes of agronomic worth through their mutual association (Burnouf & Bouriquet, 1980; Tanksley & Rick, 1980; Tanksley *et al.*, 1981; Burnouf & Beitz, 1984). However, advances in DNA technology, specifically restriction fragment length polymorphisms (RFLPs), and gel electrophoresis using two-dimensional gels (Anderson *et al.*, 1985) offer much greater promise to tag genes of interest and to quantify genetic variation (Beckman & Soller, 1983; Helentjaris *et al.*, 1985; Robertson, 1985; Evola *et al.*, 1986; Vallejos *et al.*, 1986; see also Bernatsky & Tanksley this volume).

Potential contributions of private plant breeders to the use of collections

Generate commitment and support for germplasm conservation and use

The private plant breeder can help to document and publicise the wealth of available diversity and articulate its potential benefits. In so doing, the breeder can garner public and private support for long-term conservation and use efforts essential to maintain and apply genetic resources for the general good. Verbal support from private plant breeders will, however, need to be underpinned by a long-term commitment to action by the management of plant breeding companies. Such commitment will best be sought from those in management who understand the biological basis of genetics, plant breeding and biology. This commitment can then be demonstrated by private plant breeders, through their actions in the description, evaluation, maintenance, use, transfer of data and interchange of germplasm collections.

Maintenance and evaluation of collections by private breeders

Two examples of direct involvement by private plant breeders are the Latin American Maize Project (LAMP) and the United States and Canadian Co-operative Soybean Improvement Project.

Ten Latin American countries (Argentina, Brazil, Colombia, Chile, Guatemala, Mexico, Paraguay, Peru, Uruguay, and Venezuela) and the USA are involved in the maize project. This programme seeks to evaluate all maize collections held in each country and to make the data freely available to plant breeders. The programme is supported by both private and government funds and personnel and was instigated by a private US plant breeding company.

The soyabean programme is an outstanding example of what can be achieved by private plant breeders working in concert with public (state and federal) breeders. The germplasm base of US soyabeans is notoriously narrow (Delannay *et al.*, 1983). In the early 1970s it became apparent to Clark Jennings, a private soyabean breeder, that definite action to broaden the germplasm base should be taken, even though improvements in the yield and performance of elite cultivars were continuing. Jennings felt that if a programme to incorporate exotic soyabeans was delayed further, the ever-increasing performance gap between elite and exotic soyabeans would make the use of exotics a decreasingly attractive proposition for breeders seeking elite products.

Therefore, 80 public and private breeders were surveyed to determine their willingness to participate in a programme of evaluation, pre-breeding and exchange of germplasm and evaluation data. The programme was initiated in 1977.

Initially, each of 20 private and 20 public breeders were sent 50 different strains that had been shown to be promising, primarily for yield and disease resistance, in preliminary evaluations by the United States Department of Agriculture (USDA). These strains were scored visually for agronomic features. The most promising 25 lines were then yield tested in co-operation with another breeder at a second location. Out of these 25, the best six were chosen for yield testing across several locations. This programme continues with fresh infusions of 50 lines per year at each location so that at any one time many lines are in various stages of testing.

Currently, about 7,500 of the 10,000 USDA Plant Inventory accessions are working through the programme. At the end of each testing sequence, the most promising lines are then crossed on to adapted varieties, thus beginning the introgression or pre-breeding stage. The most promising introgressed populations can then be included in the co-operative strain test where each line must have a minimum of 25 per cent exotic germplasm and three replications of performance data. All germplasm in this programme is available to all co-operators as breeding material. These materials are also available to non-co-operators at the behest of the individual company to which a request is made.

The programme requires a relatively small share – typically less than five per cent – of the total soyabean breeding effort of participating breeders (Jennings, pers. comm.). To date, very high yielding derivatives are coming from the first cycle of breeding and even more promising materials are expected in later cycles.

A third example of private efforts in support of exotic germplasm is work included as part of the maize population improvement programme of Pioneer Hi-Bred International. It involves the annual increase of about 350 accessions of the original collections of maize races held by CIMMYT in Mexico for which routine maintenance is now a vital need. Evaluation for basic agronomic characters or traits such as maturity, plant height and ear height are recorded and are publicly available. Increased seed is returned to CIMMYT in Mexico with duplicate collections sent to Fort Collins, Colorado for long-term genetic conservation.

The contribution to collections by private plant breeders

Private plant breeders can themselves add to the reservoir of collections of cultivars and breeding stocks and so greatly increase their value. Modern cultivars and breeding populations (both public and private) contain a wealth of gene combinations resulting from 50–70 years of intense breeding effort directed towards improving pest, disease and stress resistance, greater efficiency of nutrient usage, and large increases in yield potential (Russell, 1974; 1985; Duvick, 1977; 1984*b*; Castleberry *et al.*, 1984; Fehr, 1984). The diversity of germplasm, environments and selection schemes among private breeders has led to a wealth of increasingly useful genetic diversity both in private breeding stocks and in cultivars.

Private plant breeders can contribute basic broad-based breeding populations and collections of original farmers' varieties or landraces to germplasm collections. They also can release inbred lines (parents of hybrid varieties such as maize and sorghum) upon which their operation no longer relies. While valuable genetic resources are always available for breeding and collection purposes in released self-pollinated varieties and in F_1 hybrids, germplasm in currently used parental inbred lines *per se* is not available to the public. However, plant variety protection laws protect the use of inbred lines for a limited time, after which these proprietary lines then become available to the public. Plant variety protections laws thus simultaneously stimulate increased breeding effort and make numerous collections of well documented elite germplasm widely available.

The flow of useful exotic germplasm can be reciprocal. Plant breeders in the 'south' or 'Third World' can benefit by incorporating some of the elite (but, to them, exotic) materials from temperate breeding programmes. For example, some Brazilian maize breeders consider southern US lines to be exotic and have devoted much of their breeding effort toward adapting them to the Brazilian environment. These elite US lines provide high yield potential, standability and other desirably agronomic traits for the improvement of locally bred materials (J. Anderson, pers. comm.). In India, maize breeders also can successfully use inbred lines of maize from the southern US Corn Belt, often without changing the lines in any way (Rai, pers. comm.).

Conclusions

Private plant breeders can clearly play a major role in broadening the genetic base of cultivated varieties. The achievement of this goal requires a system of well maintained and appropriately described collections. Classification, characterisation and pre-breeding of materials in collections will greatly stimulate their use by private plant breeders; in fact, without the information to be gained from these activities, private breeders will, in general, ignore the collections. A real commitment to diversify the germplasm base is needed. Private plant breeders should actively participate in the evaluation and pre-breeding of germplasm. Private plant breeders can also help to enrich collections by making available certain lines, varieties and populations that are the result of many years of intensive breeding efforts.

One of the major considerations for plant breeders is genetic diversity. The world's population depends upon the continued success of plant breeders in developing cultivars that will continue to sustain and nourish cultural and scientific progress. While the world is changing, it also remains the same. The original plant breeders who domesticated crop plants 10,000 years ago transformed totally exotic germplasm and thus transformed society. Today, most examples of the effective use of exotic plant genetic resources come from relatively few breeding programmes, headed by hard-working and dedicated scientists whose efforts are often insufficiently recognised and rewarded. We owe it to them, and to future generations, to bring more of our resources, and more effective organisation, to bear on the responsible management of one of the world's key resources, plant germplasm.

Acknowledgements

We wish to thank the following plant breeders for their enthusiastic support and help in writing this paper: Drs J. Anderson, J. Baker, B. McBratney, G. Dalton, J. Gerloff, C. Hayward, M. Iwig, C. Jennings, K. Porter, G. S. Rai, W. Salhuana, H. Schmidt, and R. Ward.

References

Anderson, E. & Brown, W. L. (1952). Origin of corn belt maize and its genetic significance. In *Heterosis*, pp. 124–48, Gowen, J. W. (ed.) Iowa State University Press, Ames.

Anderson, N. G., Tollaksen, S. L., Pascoe, F. H., & Anderson, L. (1985). Two-dimensional electrophoretic analysis of wheat seed proteins. *Crop Science* **25**, 667–74.

Beckman, J. S. & Soller, M. (1983). Restriction fragment length polymorphisms in genetic improvement: methodologies, mapping, and costs. *Theoretical Applied Genetics*, **67**, 35–43.

Brim, C. A. (1973). Quantitative genetics and breeding. In *Soybeans: Improvement, Production, and Uses*, pp. 155–86. Caldwell, B. E. (ed.). American Society of Agronomy, Inc., Madison, Wis.

Brown, W. L. (1981). Exotic germplasm in cereal crop improvement. *In Plant Improvement and Somatic cell Genetics*, pp. 29–42, Vasil, I. K., Scowcroft, W. R., & Frey, K. J. (eds.). Academic Press, New York.

Burnouf, T. & Bietz, J. A. (1984). Reversed-phase high-performance liquid chromatography of durum wheat gliadins: relationships to durum wheat quality. *Journal of Cereal Science*, **2**, 3–14.

Burnouf, T. & Bouriquet, R. (1980). Glutenin subunits of genetically related

European hexaploid wheat cultivars: Their relation to bread-making quality. *Theoretical and Applied Genetics*, **58**, 107–11.

Burton, G. W. (1984). Conservation and use of exotic germplasm to improve Bermudagrass and pearl millet. In *Conservation and utilization of Exotic Germplasm to Improve Varieties*, pp. 93–117. 1983 Plant Breeding Res. Forum. Pioneer Hi-Bred Int., Inc., Des Moines, IA.

Castleberry, R. M., Crum, C. W. & Krull, C. F. (1984). Genetic yield improvement of U.S. maize cultivars under varying fertility and climatic environments. *Crop Science*, **24**, 33–6.

Chang, T. T. (1984). Conservation of rice genetic resources: luxury or necessity? *Science*, **224**, 251–6.

Chang, T. T. (1985). Germplasm enhancement and utilization. *Iowa State Journal of Research*, **59**, 399–424.

Cox, T. S., Murphy, J. P. & Rodgers, D. M. (1986). Changes in genetic diversity in the red winter wheat regions of the United States. *Proceedings National Academy of Science, USA*, **83**, 5583–6.

Dalton, L. G. (1970). The use of tropical germplasm in sorghum improvement. In *Proceedings 25th Annual Corn and Sorghum Research Conference*, pp. 21–7. ASTA, Washington, DC.

Delannay, X., Rogers, D. M. & Palmer, R. G. (1983). Relative genetic contributions among ancestral lines to North American soybean cultivars. *Crop Science*, **23**, 944–9.

Doebley, J. F., Goodman, M. M. & Stuber, C. W. (1986). Exceptional genetic divergence of northern flint corn. *American Journal of Botany*, **73**, 64–9.

Duvick, D. N. (1977). Genetic rates of gain in hybrid maize yields during the past 40 years. *Maydica*, **22**, 187–96.

Duvick, D. N. (1984*a*). Genetic diversity in major farm crops on the farm and in reserve. *Economic Botany*, **38**, 161–78.

Duvick, D. N. (1984*b*). Genetic contributions to yield gains of U.S. hybrid maize, 1930–1980. In *Genetic Contributions to Yield Grains of Five Major Crop Plants*, pp. 15–47, Fehr, W. R. (ed.) Crop Science Society of America Special Publications. Madison, WI.

Eberhart, S. A. (1970). Progress report on the sorghum conversion program in Puerto Rico and plans for the future. In *Proceedings 25th Annual Corn and Sorghum Research Conference*, ASTA, Washington, DC.

Evola, S. V., Burr, F. A. & Burr, B. (1986). The suitability of restriction fragment length polymorphisms as genetic markers in maize. *Theoretical and Applied Genetics*, **71**, 765–71.

Fehr, W. R. (1976). Description and evaluation of possible new breeding methods for soybeans. In *World Soybean Research*, pp. 268–75, L. D. Hill (ed.). The Interstate Printers and Publishers, Inc., Danville, IL.

Fehr, W. R. (1984). Genetic contributions to yield gains of five major crop plants. *Crop Science Society of American Special Publications* No. 7. Madison, WI.

Gerrish, E. E. (1983). Indications from a diallel study for interracial hybridization in the corn belts. *Crop Science*, **23**, 1082–4.

Goodman, M. M. (1984). An evaluation and critique of current germplasm programme. In *Conservation and Utilization of Exotic Germplasm to Improve Varieties*, 1983 Plant Breeding Res. Forum. Pioneer Hi-Bred Intl., Inc., Des Moines, IA.

Goodman, M. M. (1985). Exotic maize germplasm: status, prospects, and remedies. *Iowa State Journal of Research*, **59**, 497–527.

Hallauer, A. R. (1978). Potential of exotic germplasm for maize improvement. *Maize Breeding and Genetics*, pp 229 – 48, Walden, D. B. (ed.) John Wiley and Sons, New York.

Hallauer, A. R. & Miranda, J. B. (1981). Quantitative genetics in maize breeding. Iowa State Univ. Press, Ames.
Hawkes, J. G. (1983). *The Diversity of Crop Plants.* Harvard University Press, Cambridge, Mass.
Hawkes, J. G. (1985*a*) *Plant Genetic Resources; the Impact of the International Research Centres.* Study paper No. 3 CGIAR, The World Bank, Washington, DC.
Hawkes, J. G. (1985*b*). Genetic resources evaluation; an overview. In *Proceedings EUCARPIA Genetic Resources Section International Symposium.*, pp. 69–77. Prague.
Helentjaris, T., King, G., Slocum, M., Siedenstrang, C., & Wegman, S. (1985). Restriction fragment polymorphisms as probes for plant diversity and their development as tools for applied plant breeding. *Plant Molecular Biology*, **5**, 109–18.
Kleese, R. A. & Duvick, D. N. (1980). Genetic needs of plant breeders. In *Genetic Improvement of Crops, Emergent Techniques*, pp. 24–43. Rubenstein, I., Gengenbach, B., Phillips, R. L., and Green, C. E. (eds.). University of Minnesota Press, Minneapolis.
Lonnquist, J. M. (1974). Consideration and experiences with recombinations of exotic and corn belt maize germplasm. In *Proceedings 29th Annual Corn and Sorghum Research Conference*, pp. 102–17. ASTA, Washington, DC.
Marshall, D. R. (1977). The advantages and hazards of genetic homogeneity. *Annuals New York Academy of Sciences*, **287**, 1–20.
Meredith, W. R. & Bridge, R. R. (1984). Genetic contributions to yield changes in upland cotton. In *Genetic contributions to Yield Gains of Five Major Crop Plants*, pp. 75–87, Fehr, W. R. (ed.). Crop Science Society of America. Madison, WI.
Miller, F. R. (1968). Some diversity in the world sorghum collection. In *Proceedings 23rd Annual Corn and Sorghum Research Conference*, pp. 120–8. ASTA, Washington, DC.
Miller, F. R. & Kebede, Y. (1968). Genetic contributions to yield gains in sorghum, 1950 to 1980. In *Genetic Contributions to Yield Gains of Five Major Crop Plants*, pp. 1–14, Fehr, W. R. (ed.) Crop Science Society of America, Madison, WI.
Nelson, H. G. (1972). The use of exotic germplasm in practical corn breeding programs. In *Proceedings 27th Annual Corn and Sorghum Research Conference*, pp. 115–18. ASTA, Washington, DC.
Peloquin, S. J. (1984). Utilization of exotic germplasm in potato breeding; germplasm transfer with haploids in 2n gametes. In *Conservation and Utilization of Exotic Germplasm to Improve Varieties*, pp. 145–67. Plant Breeding Research Forum. Pioneer Hi-Bred Intl., Inc., Des Moines, IA.
Quinby, J. R. (1968). Opportunities for sorghum improvement. In *Proceedings 23rd Annual Corn and Sorghum Research Conference*, pp. 170–6. ASTA, Washington, DC.
Rick, C. M. (1979). Potential improvement of tomatoes by controlled introgression of genes from wild species. In *Proceedings Conference on Broadening the Genetic Base of Crops*, pp. 167–73. Zeven, A. C., and Var Harten, A. M. (eds.) Pudoc, Wageningen.
Rick, C. M. (1984). Conservation and use of exotic tomato germplasm. In *Conservation and Utilization of Exotic Germplasm to Improve Varieties*, Plant Breeding Research Forum. Pioneer Hi-Bred Intl., Inc., Des Moines. IA.
Rindos, D. (1984). *The Origins of Agriculture, an Evolutionary Perspective.* Academic Press, Orlando.

Robertson, D. S. (1985). A possible technique for isolating genic DNA for quantitative traits in plants. *Journal of Theoretical Biology*, **117**, 1–10.
Ross, H. (1979). Wild species and primitive cultivars as ancestors of potato varieties. In *Proceedings Conference on Broadening the Genetic Base of Crops*, pp. 237–45. Zeven, A. C. and van Harten, A. M. (eds.) Pudoc, Wageningen.
Russell, W. A. (1974). Comparative performance for maize hybrids representing different eras of maize breeding. In *Proceeding, 29th Annual Corn and Sorghum Research Conference*, pp. 81–101. ASTA, Washington, DC.
Russell, W. A. (1985). Comparison of the hybrid performance of maize inbred lines developed from the original and improved cycles of BSSS. *Maydica*, **30**, 407–19.
Schmidt, J. W. (1984). Genetic contributions to yield gains in wheat. In *Genetic Contribution to Yield Gains of Five Major Crop Plants*, pp. 89–101, Fehr, W. R. (ed.) Crop Science Society of America, Madison, WI.
Simmonds, N. W. (1962). Variabilty in crop plants, its use and conservation. *Biological Reviews*, **37**, 422–65.
Simmonds, N. W. (1976). *Evolution of Crop Plants*. Longman, London.
Specht, J. E. & Williams, J. H. (1984). Contribution of genetic technology to soybean productivity – retrospect and prospect. In *Genetic Contributions to Yield Gains of Five Major Crop Plants*, pp. 49–74, Fehr, W. R. (ed.) Crop Science Society of America Special Publications, Madison, WI.
Sprague, G. F. (1982). The adequacy of current plant breeding procedures, evaluation techniques, and germplasm resources relative to future food, feed, and fiber needs. In *The 1982 Plant Breeding Research Forum*, pp. 205–33. Pioneer Hi-Bred Intl., Inc., Des Moines, IA.
Tanksley, S. D. & Rick C. M. (1980). Isozyme gene linkage map of the tomato; applications in genetics and breeding. *Theoretical and Applied Genetics*, **57**, 161–70.
Tanksley, S. D., Medino-Filho, H. & Rick, C. M. (1981). The effect of isozyme selection on metric characters in an interspecific backcross of tomato: basis for an early screening procedure. *Theoretical and Applied Genetics*, **60**, 291–6.
Timothy, D. H. & Goodman, M. M. (1979). Germplasm preservation: The basis of future feast or famine, genetic resources of maize – an example. In *The Plant Seed: Development, Preservation, and Germination*, pp. 171–200, Rubenstein, I., Phillips, R. L., Green, C. E. & Gengenbach, B. G. (eds.). Academic Press, New York.
Vallejos, C. E., Tanksley, S. D. & Bernatzky, R. (1986). Localization in the tomato genome of DNA restriction fragments continuing sequences homologous to the rRNA (45S), the major chlorophyll A/B binding polypeptide and the ribulose biphosphate carboxylase genes. *Genetics*, **112**, 93–105.
Webster, O. J. (1975). Use of tropical germplasm in a sorghum breeding program for both tropical and temperate areas. In *Proceedings 30th Annual Corn and Sorghum Research Conference*, pp. 1–12. ASTA, Washington, DC.
Zeven, A. C. & van Harten, A. M. (1979). Proceedings Conference on Broadening the Genetic Base of Crops. Pudoc, Wageningen.

3
Germplasm collections and the experimental biologist[1]

R.G. PALMER

Introduction

Although plant breeders are usually regarded as the primary managers and users of germplasm collections, scientists in many disciplines use genetic resources. These include researchers engaged in genetics, ecology, taxonomy, evolution, speciation, physiology, tissue culture, molecular biology, and social sciences. With respect to the use of collections in research, three questions will be considered: 1) What do experimental biologists need in germplasm collections? 2) How do experimental biologists use these germplasm collections? and, 3) Why do experimental biologists fail to use existing germplasm collections more fully? Examples from soyabean research will then be given to highlight how the findings of one discipline augment those of other disciplines in the use of germplasm.

Experimental biologists and germplasm collections

When experimental biologists turn to germplasm collections for material, they are seeking a rich source of genetic variation. Obviously their objectives differ among the various disciplines as well as among individual scientists within a discipline. Large collections are useful to nematologists, pathologists, and entomologists, for instance, who survey accessions to identify sources of pest resistance. In contrast, genetic stock collections, which may or may not be in gene banks, are specialised germplasm collections valuable to geneticists and others who use, for example, aneuploid lines.

[1] Joint contribution USDA ARS, and Journal Paper No. J-12410 of the Iowa Agriculture and Home Economics Experiment Station, Ames, Iowa 50011, USA. Project 2763.

What do experimental biologists need in germplasm collection?

Germplasm collections must represent as many as possible of the geographic and ecological habitats of a species, whether wild or cultivated. Collection sites should be representative of an area so that valid generalisations can be drawn. Germplasm accessions should be well documented, well maintained and readily available for distribution. Regrettably, political decisions may dictate germplasm collection expeditions and distribution of seeds. For example, the wild annual soyabean, *Glycine soja* Sieb. & Zucc., is considered a national treasure in China, its country of origin, and as a result germplasm exchanges are restricted. Recent developments suggest that future plant explorations and dissemination of germplasm may be more strongly influenced by political dictums (Mooney, 1980). Germplasm conservation should be a global effort.

For germplasm resources to be used effectively, a directory of germplasm sources is necessary. The International Board for Plant Genetic Resources (IBPGR) has issued a number of these and has sponsored workshops on several crops. The IBPGR goals are to formulate a comprehensive plan to unite the various aspects of germplasm acquisition, identification, cataloguing, evaluation and preservation. For soyabeans, the first phase was the preparation of an international directory of germplasm collections. This directory has been compiled from the responses of 87 institutions in 43 countries (Juvik *et al.*, 1985).

Specialised collections, such as genetic stock centres, are often of great use to experimental biologists. Genetic stock centres perform the indispensable function of disseminating mutant and wild type strains of genetically important organisms to a wide variety of users in both the public and private sectors. For several reasons these centres must be under the direction of investigators trained in the biology and genetics of given organism. First, specific mutant stocks may undergo changes with time due to reversion, suppression and secondary mutations. Stocks carrying cytological markers must be characterised microscopically to ensure that the variant of interest has not been lost. A second important function of genetic stock centres involves documentation. Specific alleles must be traced precisely through the stocks carrying them. Origins of both wild type and mutant strains must be determined and linkage data based on two or three factor crosses collected. Stock centres frequently serve as repositories for genetic mapping data on a given species and stock centre personnel are responsible for updating genetic maps of a given organism (Palmer & Shoemaker, 1987). A third function that curators perform is advisory. They are continually called upon to provide advice on the best genetic variant to be used for given experiments. Finally, stock centre curators often involve themselves in the construction of strains carrying multiple markers, rearrangements, deletions, etc, that will be useful to the scientific community as a whole.

The Genetics Stocks Committee of the Genetics Society of America comprises curators of collections of barley, cotton, cucurbits, *Datura*, *Brassica*, maize, pea, sorghum, soyabean, tomato and wheat. At a meeting of this committee in June 1986, the problem of funding and the possibility of charging user fees was discussed. Since so much research with genetically important organisms in the USA is supported by federal grants, the consensus was that it made little sense to change user fees since the money went from one federal pocket (the grant) into another (the stock centre). Administrative complications on some university campuses make user fees inadvisable; additional administrative expenses would be greater than the funds generated by user fees. Use of germplasm collections by many investigators with limited funds would be severely limited. Donations are often made to genetic stock collection centres by private users (Rick, pers. comm.)

Maize (*Zea mays* L.) is one of the best understood higher plant species, genetically and agronomically. In addition to germplasm collections, the Maize Genetic Cooperation Stock Center was established.

The Maize Genetics Cooperation Newsletter acts as a clearing house for genetic nomenclature and describes and lists new stocks available for distribution from the centre. Documentation of the use of genetic stocks by experimental biologists is given in *Maize for Biological Research* (Sheridan, 1982).

Another prerequisite for genetic resource collections to be universally useful would be a common descriptor list. The IBPGR has developed these for most crops, and in the USA there are about 38 crop advisory committees developing descriptor lists. These lists help form a common link among curators from different countries working with the same crop.

How do experimental biologists use these germplasm collections?

Experimental biologists have requirements of germplasm collections that are as varied as the investigators themselves. Geneticists and molecular biologists may be interested primarily in individual loci (or alleles). Germplasm is examined until the particular allele or set of alleles is identified. These accessions then are subjected to detailed examination. Systematists, however, view phenotypes (eventually they may describe genotypes) rather than individual loci as their primary interest. Typical questions asked by systematists concern where, how and why a particular combination of genes arose. A germplasm collection should represent a living record of the genotypic variation found in the species or genus. Intra- and inter-population genotypic variability needs to be maintained for systematic studies. Similarly, molecular biologists are beginning to recognise the variability inherent in plant species. Rapid DNA isolations from small amounts of tissue are now done so that individual plants (genotypes) can be sampled.

Frequently, the experimental biologist wants to make comparisons in expressions between the allele under study and its dominant or recessive counterpart. It is usually desirable to have different alleles in an isogenic background to eliminate the confounding effects of different genetic backgrounds. Isolines are defined as two or more lines differing from each other genetically at one locus only. Few germplasm collections have isoline stocks. More than 300 isoline combinations of one or several genes are available in soyabean. These lines are developed by backcrossing and are selected in the BC_5 generation. This phase of germplasm development has been essential for genetic stocks that are used in paired comparisons by biologists. Seed stocks from the isoline collection are one of the most requested genetic materials. The single most requested genotypes are the non-nodulating isolines (Bernard, pers. comm.).

As molecular biologists become aware of germplasm collections, they have the dilemma of selecting a 'representative sample' of genotypes for analysis. Their methods are often expensive and time-consuming, and their research areas are highly competitive. If the descriptor lists for a crop are complete, the molecular biologist can select only those accessions that meet pre-selected criteria, e.g. certain endosperm pigmentation patterns.

Why do experimental biologists fail to use existing germplasm collections more fully?

The most consistent response to this question was that most experimental biologists do not know that such collections exist. Even scientists at agricultural inversities, colleges or academies may be uninformed about plant germplasm. For many collections, the curator is a plant breeder. The germplasm has been managed by plant breeders primarily for plant breeders. The traits measured on accessions often reflect the interest of the plant breeder.

Information transfer from the curators of germplasm collections to plant breeders, experimental biologists and others has varied from almost non-existent to excellent. The reasons for this disparity are numerous and complex. Some curators say that their main responsibility is to maintain rather than to evaluate for agronomic traits. Others cite inadequate funding for personnel and facilities as one of the main reasons for the lag in transfer of information. The time required from collection to evaluation to dissemination of results is longer than most scientists realise. The phenotype of a plant may not be a good indicator of genotype. Thus large numbers of accessions may need to be examined before the desired genotype is found. This is a costly and time-consuming process.

In the USA many curators are research scientists and teachers first, curators second. Professional advancement for most curators depends on refereed publications and not experiment station or USDA reports on germplasm evaluation. This obviously must be changed if we are to make germplasm collections more widely used. In other countries, collections are maintained by full-time professional curators.

The majority of descriptors, of necessity for the most part, are easily assessed. They tend to be qualitative traits and/or ones that are of interest to curators or plant breeders, e.g. seed colour, plant height, maturity and reaction to pests. Traits that require more detailed expertise and equipment to measure are often not included in descriptions of accessions. These traits include chromosome aberrations, restriction fragment-length polymorphisms and protein profiles. As information is gained on such traits, it should be added to the data for each accession, such as lipoxygenase and Kunitz trypsin inhibitor in soyabeans. The need for speciality collections increases as more non-agronomic traits are described.

Many times curators have identified traits of interest, such as insect, nematode, disease or chemical resistance. The transfer of the trait(s) of interest to adapted germplasm is called pre-breeding, or germplasm enhancement. The development of isogenic lines in soyabeans, as mentioned above, represents a type of pre-breeding. Germplasm resources may be used more efficiently by experimental biologists if pre-breeding efforts have occurred. Such efforts may or may not be the responsibility of germplasm curators. Sometimes pre-breeding is done by the experimental biologist, for example, McCoy (1985), Bingham & Saunders (1974), and Peloquin (1981). With the sorghum conversion programme, a special pre-breeding programme was developed (Gerick & Miller, 1984). The question of who does pre-breeding will depend upon the crop and the country or agency interested in using the crop.

In some interspecific crosses, the incompatibility problems can be resolved through the use of embryo rescue techniques (McCoy, 1985). Photoperiod manipulation is often necessary to permit the flowering of both parents synchronously so that hybridisations are possible; for example, the sorghum conversion programme at Texas A. & M. University by Gerick & Miller (1984). Manipulation of the ploidy level of alfalfa and potato germplasm has been used to bring the chromosome number to the desired level for plant breeding (Bingham & Saunders, 1974; Peloquin, 1981).

Curators and users of germplasm collections need to publicise the available literature on the collections. One method would be to give more complete information on the source of the plant materials in the materials and methods sections of scientific publications.

Soyabean germplasm

The United States Department of Agriculture (USDA) maintains soyabean germplasm, based on a maturity group classification from 000 (early) to X (Late maturity), in two large collections – one at Urbana, Illinois, and the other at Stoneville, Mississippi. The cytological collection is maintained at Ames, Iowa. Efforts have been made to include accessions and public varieties of the cultivated species, the wild annual species and the perennial species, as well as specialised germplasm such as genetic types, cytological types and genetic isolines (Table 3.1). More than 11,000 strains are maintained in the germplasm collections, and about 30,000 packets of seeds are distributed annually (Kilen, pers. comm.).

Soybean scientists have several avenues available to publicise genetic resources. USDA, university, private industry and Canadian soyabean scientists meet annually at the Soybean Breeders Workshop. Concurrently, the Soybean Genetics Committee meets to discuss items on genetic interpretation and gene symbols concerning the genus *Glycine*. The editor

Table 3.1. *USDA Soyabean Germplasm Collection*

Description*	Number of strains	Maturity group
Introduced and pre-1945 domestic varieties	140	000 to IV
Introduced and pre-1960 domestic varieties	75	V to X
Post-1974 public varieties	160	000 to IV
FC and PI strains (to PI 494182)	7,027	000 to IV
Genetic types (to T284)	104	
Cytological types	101	
Genetic isolines	300	
Wild soyabeans (to PI 487431)	675	000 to X
Perennial *Glycine* species (to PI 429809)	66	

* FC = Forage Crop number; PI = Plant Inventory number; T = Genetic Type Collection number.

of the Soybean Genetics Newsletter and the curator of the soyabean genetics collection are ex-officio members and six members are elected.

At the Soybean Breeders Workshop the Soybean Germplasm Crop Advisory Committee holds its annual meeting. One of its major responsibilities is to develop and provide a strategic overview of the total national scientific effort in the study and use of soyabean germplasm and to recommend co-operative approaches for improvement in the germplasm management system where needed.

The Advisory Committee consists of 14 members. Included are the curators of the northern and southern divisions of the germplasm collection, the research geneticists working with the germplasm curators at each location, and a representative of USDA. The remaining nine committee members are elected to the committee to represent various research areas and/or various geographic areas: there is a private breeder from both the north and south, a public breeder from the north and the south, a public pathologist or nematologist from the north and the south, a public entomologist, a public physiologist or biochemist, and a public cytogeneticist or molecular geneticist.

Examples of the use of germplasm collections by experimental biologists

The use of germplasm collections by experimental biologists from many disciplines has occurred. Lectins in common bean, *Phaseolus vulgaris* L. (Osborn *et al.*, 1985), seed storage proteins in soyabean (Ladin *et al.*, 1984; Davies *et al.*, 1985), oil quality in soyabean (Kitamura *et al.*, 1985; Kitamura *et al.*, 1983), and nodulation response in soyabean (Devine & Breithaupt, 1981) are representative examples of germplasm collections being used by experimental biologists. Details from three additional examples in soyabean are presented.

Soyabean: sterility mutants

The Soybean Genetic Type Collection contains both spontaneously occurring and mutagen-induced mutants (Palmer & Kilen, 1987). Male-sterile, female-fertile mutants have provided genetic material for comparative studies of microsporogenesis and microgametogenesis between normal development and abnormal development (Albertsen & Palmer, 1979; Graybosch & Palmer, 1985*a*; 1985*b*). These light- and electron-microscope studies have provided detailed information on reproductive processes in soyabean that is relevant for many legumes. These mutants have been used in genetic linkage studies (Palmer, 1985).

Some male-sterile mutants have some female sterility. The *ms1* mutant may have aneuploids, haploids, and polyploids among its progeny (Kenworthy *et al.*, 1973; Beversdorf & Bingham, 1977; Chen *et al.*, 1985). Haploids have been used in tissue culture mutagenesis research (Weber & Lark, 1980). Other mutants are male sterile and almost completely female sterile, but these meiotic mutants are a consistent source of aneuploids (Palmer, 1985). Aneuploids are useful in chromosome mapping studies.

Application of soyabean male-sterile mutants to recurrent selection schemes in breeding programmes has been described (Brim & Stuber, 1973). Male-sterile lines may also be used to facilitate the introgression of genes from soyabean accessions into adapted modern cultivars (May & Wilcox, 1986).

Soyabean: seed lectin

Lectins or phytohaemaglutinins are present in the seeds of many plant species. Although quantitative and qualitative variations among accessions and cultivars of soyabean and related species are known, the adaptive and/or biological significance of soyabean lectin (SBL) has not been established.

Seed from the USDA soyabean collection was screened by polyacrylamide gel electrophoresis (PAGE) for the presence or absence of electrophoretic forms of SBL (Stahlhut & Hymowitz, 1980). Two electrophoretic forms of SBL were identified among the 2,784 accessions examined and 18 accessions were identified as lacking the SBL (Pull *et al.* 1978). Orf *et al.* (1978), using PAGE, demonstrated that the presence of SBL is controlled by a single dominant allele, *Le*, and that the homozygous genotype, *le le*, results in the absence of SBL.

Seed from 559 accessions of wild annual soyabean were screened by PAGE for the presence or absence of SBL. A total of 272 accessions was identified as lacking the SBL (Stahlhut *et al.*, 1981). The seed lectin from *Glycine soja* and the cultivated soyabean exhibited antigenic homology, migrated to identical positions in PAGE and had identical subunit compositions, sensitivity to haemaglutinating activities to sugar haptens, neutral carbohydrate content and isolectin profiles (Pueppke, 1981; Stahlhut *et al.*, 1981; Pueppke *et al.*, 1982). These results indicate that the SBL molecule is conserved structurally and functionally and may be produced characteristically by both *G. soja* and *G. max*.

Glycine soja is the presumed wild ancestor of the cultivated soyabean. The results of these PAGE studies suggest that perhaps the soyabean was domesticated from a narrow germplasm pool of wild soyabean that happened to contain seed lectin.

Fifty-six accessions from six species of the subgenus *Glycine* were examined for the presence of absence of SBL. All accessions were lacking SBL (Pueppke & Hymowitz, 1982). The species were *G. canescens* F.J. Herm., *G. clandestina* Wendl., *G. falcata* Benth., *G. latifolia* (Benth.) Newell & Hymowitz, *G. tabacina* (Labill.) Benth., and *G. tomentella* Hayaka.

Molecular comparisons have been conducted between soyabean germplasm containing lectin and germplasm lacking lectin. Vodkin (1981) reported that lectin-specific polysomes could not be obtained from the *le le* genotype. It was suspected that this genotype was deficient or substantially reduced in functional lectin mRNA. This reduction was either at the level of mRNA synthesis or at the level of processing (Goldberg *et al.*, 1983). They found a 3.4 kb DNA segment, consisting of repeated sequences, that had been inserted into the coding region of the lectinless genotype. Vodkin *et al.* (1983) concluded that this 3.4 kb insertion reduced transcription in the *le le* genotype and resulted in the lectinless phenotype.

The insertion sequence in the *le le* genotype has structural features of a mobile genetic element (Rhodes & Vodkin, 1985), although it has not been shown to be unstable genetically. The soyabean lectin gene insertion has been named Tgm1 (transposon *Glycine max*).

Three lectinless accessions of *G. max* (Goldberg *et al.*, 1983) and six lectinless accessions of *G. soja* (Vodkin & Rhodes, 1986) were analysed for the presence or absence of a unique DNA insertion sequence. This insertion was located in similar positions in the DNA in all nine accessions. These data could not discriminate between the hypotheses that the lectin element transposed into the seed lectin locus on more than one occasion or that all accessions were derived from a common progenitor.

The soyabean germplasm contained different electrophoretic forms of SBL that were of interest to experimental biologists from different disciplines. The original accessions were used and no pre-breeding was needed.

Soyabean: Kunitz trypsin inhibitor

The trypsin inhibitor proteins in raw mature soyabean seed have been suggested to be one of the major factors responsible for the poor nutritional value of the unheated meal. Several different trypsin inhibitors have been reported in soyabean. Much of the inhibition in soyabean is due to the Kunitz trypsin inhibitor (SBTI-A_2).

Seed from the USDA soyabean collection was screened by PAGE for electrophoretic variants of SBTI-A_2. Four electrophoretic forms were identified among the 2,944 accessions examined (Hymowitz, 1973; Orf & Hymowitz, 1979). One form of SBTI-A_2 does not exhibit a protein band in the gels. These accessions have 30 to 50 per cent less trypsin activity per gram of protein than do crude protein extracts from seeds of SBTI-A_2 containing cultivar Amsoy 71.

Nutritional trials with chicks, rats and swine have been conducted with Kunitz trypsin-inhibitor variants. Soyabean accessions without Kunitz trypsin inhibitor were more nutritious than the ones with it (Hymowitz, 1983).

The four electrophoretic forms of SBTI-A_2 have been released as isolines (Bernard & Hymowitz, 1986*a*; 1986*b*). These lines are similar to their respective recurrent parents in field appearance and performance and may be of interest in feeding studies to evaluate for possible economic merit. The electrophoretic null genotype may be useful as a parent in breeding programmes. Seed protein extracts from European soyabean germplasm introduced into the USA were analysed electrophoretically for the frequency of the Kunitz trypsin inhibitor alleles. In all, 434 accessions were studied (Skorupska & Hymowitz, 1977). The most frequent allele, *Ti-a*, represented 93.3 per cent of the total. This germplasm traced its origin to germplasm from the USSR (Asia) and north-eastern China. The *Ti-b* allele represented 6.7 per cent of the total germplasm, and these lines were derived from northern Japan. None of the accessions examined had the *Ti-c* or the *ti* allele.

In a similar study, seed protein extracts from 477 Japanese soyabean cultivars were analysed by PAGE to determine the distribution of alleles at the *Ti* locus with respect to district of adaptation (Hymowitz & Kaizuma, 1979). About 60 per cent of the cultivars examined had the *Ti-a* allele. The *Ti-b* allele was identified in 38.7 per cent and the *Ti-c* allele in 1.3 per cent of the cultivars, respectively. The *ti* allele was not found among these cultivars.

In a survey of the distribution of electrophoretic forms of SBTI-A_2, Hymowitz and Kaizuma (1981) analyzed 1,603 accessions from 15 countries or regions in Asia. The allele distribution was *Ti-a* (93.9 per cent), *Ti-b* (5.9), *Ti-c* (0.1), and *ti* (0.1). The *Ti-c* allele was reported in one accession from Korea and one from Pakistan. Two accessions from Korea had the *ti* allele.

Hymowitz & Kaizuma (1979; 1981) combined their seed protein data with agronomic, biogeographical and historical literature to form a hypothesis regarding dissemination of soyabean germplasm from China to other Asia areas. Seven germplasm pools were identified.

The SBTI-A_2 and the Kunitz trypsin inhibitor-like proteins (SBTIL-

A_2) were compared chemically from a diverse germplasm pool (Nakamura *et al.*, 1984). *Glycine max*, *G. soja*, *G. tomentella*, *G. tabacina*, *G. clandestina*, *G. canescens*, and *Neonotonia wightii* (Arnott) Lackey (syn. *G. wightii* (Wight & Arn.) Verdc.) were the source materials. Chromatographic and inhibitory profiles of SBTI-A_2 from the perennial *Glycine* species were similar to those of SBTI-A_2. Immunological tests indicated that SBTIL-A_2 to SBTI-A_2 antiserum was positive with all plants of the *Glycine* species but negative with *N. wightii*. In addition, two forms of SBTIL-A_2 were identified among the four perennial species, but *G. max*, *G. soja*, and *N. wightii* had a single form. *Glycine max* and *G. soja* had an identical form that differed from the single form found in *N. wightii*. These results suggest that the Kunitz trypsin inhibitor can be used to provide phylogenetic relationships among species related to the cultivated soyabean.

The SBTI research represents an integrated approach by experimental biologists to go from evaluation of accessions to isoline development. Further development may result in the release of SBTI-lacking cultivars.

Conclusions

The experimental biologist's greatest need of the germplasm collections, either general collections or speciality collections, is as a reservoir of genetic diversity. The expansion of speciality collections is seen as a result of the increased participation by experimental biologists in germplasm characterisation. The existence of large working germplasm collections and speciality collections needs to be publicised. The most consistent response concerning the failure of experimental biologists to use collections is that they did not know about the collections. Germplasm has been the foundation of plant improvement and provides the building blocks for future progress. Co-operation among scientists across disciplines augurs well for more efficient uses of germplasm collections. An integrated approach is one step to ensure that a continuous array of improved plant materials will be available.

References

Albertsen, M. C. & Palmer, R. G. (1979). A comparative light- and electron-microscope study of microsporogenesis in male-sterile (*ms1*) and male-fertile soybeans (*Glycine max* (L.) Merr.). *American Journal of Botany*, **66**, 253–65.

Bernard, R. L. & Hymowitz, T. (1986*a*). Registration of L81-4590, L81-4871, and L83-4387 soybean germplasm lines lacking the Kunitz trypsin inhibitor. *Crop Science*, **26**, 650–1.

Bernard, R. L. & Hymowitz, T. (1986*b*). Registration of L82–2024 and L82–2051 soybean germplasm lines with Kunitz trypsin inhibitor variants. *Crop Science*, **16**, 651.

Beversdorf, W. D. & Bingham, E. T. (1977). Male-sterility as a source of haploids and polyploids of *Glycine max*. *Canadian Journal of Genetics and Cytology*, **19**, 283–7.

Bingham, E.T. & Saunders, J. W. (1974). Chromosome manipulations in alfalfa. Scaling the cultivated tetraploid to seven ploidy levels. *Crop Science*, **14**, 474–7.

Brim, C. A. & Stuber, C. W. (1973). Application of genetic-male sterility to recurrent selection schemes in soybeans. *Crop Science*, **13**, 528–30.

Chen, L.-F. O., Heer, H. E. & Palmer, R. G. (1985). The frequency of polyembryonic seedlings and polyploids from *ms1* soybean. *Theoretical and Applied Genetics*, **69**, 272–7.

Davies, C. S., Coates, J. B. & Nielsen, N. C. (1985). Inheritance and biochemical analysis of four electrophoretic variants of β-conglycinin from soybean. *Theoretical and Applied Genetics*, **71**, 351–8.

Devine, T. E. & Breithaupt, B. H. (1981). Frequencies of nodulation response alleles, *Rj2* and *Rj4*, in soybean plant introduction and breeding lines. United States Department of Agriculture, Technical Bulletin 1628.

Gerick, T. J. & Miller, F. R. (1984). Photoperiod and temperature effects on tropically and temperately adapted sorghum. *Field Crops Research Journal*, **9**, 29–40.

Goldberg, R. G., Hoschek, G. & Vodkin, L. O. (1983). An insertion sequence blocks the expression of a soybean lectin gene. *Cell*, **33**, 465–75.

Graybosch, R. A. & Palmer, R. G. (1985). Male sterility in soybean (*Glycine max*). II. Phenotypic expression of the *ms4* mutant. *American Journal of Botany*, **72**, 1751–64.

Hymowitz, T. (1973). Electrophoretic analysis of SBTI-A_2 in the USDA soybean germplasm collection. *Crop Science*, **13**, 420–1.

Hymowitz, T. (1983). Variation in and genetics of certain antinutritional and biologically active components of soybean seed. In *Better Crops for Food*, pp. 449–60. CIBA Foundation Symposium 97, Pitman Books, London.

Hymowitz, T. & Kaizuma, N. (1979). Dissemination of soybeans (*Glycine max*): Seed protein electrophoresis profiles among Japanese cultivars. *Economic Botany*, **33**, 311–19.

Hymowitz, T. & Kaizuma, N. (1981). Soybean seed protein electrophoresis profiles from 15 Asian countries or regions: Hypothesis on paths of dissemination of soybeans from China. *Economic Botany*, **35**, 10–23.

Juvik, G. A., Bernard, R. L. & Kauffman, H. E. (1985). *Directory of Germplasm Collections. 1. II. Food Legumes (Soyabean)*. International Board for Plant Genetic Resources, Rome, Italy.

Kenworthy, W. J., Brim, C. A. & Wernsman, E. A. (1973). Polyembryony in soybeans. *Crop Science*, **13**, 637–9.

Kitamura, K., Davies, C. S., Kaizuma, N. & Nielsen, N. C. (1983). Genetic analysis of a null-allele for lipoxygenase-3 in soybean seeds. *Crop Science*, **23**, 924–7.

Kitamura, K., Kumagai, T. & Kikuchi, A. (1985). Inheritance of lipoxygenase-1, -2 and -3 isozymes in soybean seeds. *Japanese Journal of Breeding*, **35**, 413–20.

Ladin, B. F., Doyle, J. J. & Beachy, R. N. (1984). Molecular characterization of a deletion mutant affecting the α1-subunit of β-conglycinin of soybean. *Journal of Molecular and Applied Genetics*, **2**, 372–80.

May, M. L. & Wilcox, J. R. (1986). Pollinator density effects on frequency and randomness of male-sterile soybean pollinations. *Crop Science*, **26**, 96–9.

McCoy, T. J. (1985). Interspecific hybridization of *Medicago sativa* L. and *M. rupestris* M.B. using ovule–embryo culture. *Canadian Journal of Genetics and Cytology*, **27**, 238–45.

Mooney, P. R. (1980). *Seeds of the Earth – a private or public resource*. Mutual Press, Ottawa.

Nakamura, I., Kitamura, K., Kaizuma, N. & Futsuhara, Y. (1984). Speciation of the genus *Glycine*. I. Electrophoretic properties of Kunitz trypsin inhibitor-like proteins. *Japanese Journal of Breeding*, **24**, 468–77.

Orf, J. H. & Hymowitz, T., Pull, S. P. & Pueppke, S. G. (1978). Inheritance of a soybean seed lectin. *Crop Science*, **18**, 899–900.

Orf, J. H., & Hymowitz, T. (1979). Inheritance of the absence of the Kunitz trypsin inhibitor in seed protein of soybeans. *Crop Science*, **19**, 107–9.

Osborn, T. C., Brown, J. W. S. & Bliss, F. A. (1985). Bean lectins. 5. Quantitative genetic variation in seed lectins of *Phaseolus vulgaris* L. and its relationship to qualitative lectin variation. *Theoretical and Applied Genetics*, **70**, 22–31.

Palmer, R. G. (1985). Soybean Cytogenetics. In *World Soybean Research Conference – III: Proceedings*, pp. 337–44. Shibles, R. S. (ed.). Westview Press, Boulder, Colorado.

Palmer, R. G. & Kilen, T. C. (1987). Qualitative Genetics and Cytogenetics (pp. 135–209). In *Soybeans: Improvement, Production and Uses*. J. R. Wilcox (ed.). Agronomy Monograph 16, American Society of Agronomy, Madison WI 53711, USA.

Palmer, R. G., & Shoemaker, R. C. (1987). Linkage map of soybean (*Glycine max* L. Merr.). In *Genetic Maps IV*. O'Brien, S. J (ed.). Cold Spring Harbor Laboratory, Cold Spring Harbor, N.Y.

Peloquin, S. J. (1981). Meiotic mutants in potato breeding. *Stadler Genetic Symposium*, **14**, 1–11.

Pueppke, S. G. (1981). Comparison of subunit compositions and profiles of the seed lectins purified from *Glycine max* and *G. soja*. *Soybean Genetics Newsletter*, **8**, 48–50.

Pueppke, S. G., Benny, U. K. & Hymowitz, T. (1982). Soybean lectin from seeds of the wild soybean, *Glycine soja* Sieb. & Zucc. *Plant Science Letters*, **26**, 191–7.

Pueppke, S. G. & Hymowitz, T. (1982). Screening the genus *Glycine* subgenus *Glycine* for the 120,000 dalton seed lectin. *Crop Science*. **22**, 558–60.

Pull, S. P., Pueppke, S. G., Hymowitz, T. & Ord, J. H. (1978). Soybean lines lacking the 120,000 dalton seed lectin. *Science*, **200**, 1277–9.

Rhodes, P. R. & Vodkin, L. O. (1985). Highly structured sequence homology between an insertion element and the gene in which it resides. *Proceedings of the National Academy of Sciences (USA)*, **82**, 493–7.

Sheridan, W. F. (1982). *Maize for Biological Research*, pp. 1–434. University Press, University of North Dakota, Grand Forks, North Dakota.

Skorupska, H. & Hymowitz, T. (1977). On the frequency of alleles of two seed proteins in European soybean (*Glycine max* (L.) Merr.) germplasm: Implications on the origin of European soybean germplasm. *Genetica Polonica*, **18**, 217–24.

Stahlhut, R. W. & Hymowitz, T. (1980). Screening the USDA soybean germplasm collection for lines lacking the 120,000 dalton seed lectin. *Soybean Genetics Newsletter*. **7**, 41–3.

Stahlhut, R. W., Hymowitz, T., & Ord, J. H. (1981). Screening the USDA *Glycine max* collection for the presence or absence of a seed lectin. *Crop Science*, **21**, 110–12.

Vodkin, L. O. & Rhodes, P. R. (1986). Common structural patterns between the soybean lectin insertion and transposable elements in other plant species, pp. 97–105. In *Molecular Biology of Seed Storage Proteins and Lectins*, Shannon, L. M. & Chrispeels, M. J. (eds.). The American Society of Plant Physiologists.

Vodkin, L. O. (1981). Isolation and characterization of messenger RNAs for seed lectin and Kinitz trypsin inhibitor in soybeans. *Plant Physiology*, **68**, 766–71.

Vodkin, L. O., Rhodes, P. R. & Goldberg, R. B. (1983). A lectin gene insertion has the structural features of a transposable element. *Cell*, **34**, 1023–31.

Weber, G. & Lark, K. G. (1980). Quantitative measurement of the ability of different mutagens to induce an inherited change in phenotype to allow maltose utilization in suspension cultures of the soybean, *Glycine max* (L.) Merr. *Genetics*, **96**, 213–22.

Part II

Use of collections

4
International use of a sorghum germplasm collection

K. E. PRASADA RAO, M. H. MENGESHA,
AND V. G. REDDY

Introduction

Sorghum (*Sorghum bicolor* (L.) Moench) is one of the most important cereals of the semi-arid tropics. It was probably domesticated in the north-east quadrant of Africa, an area that extends from the Ethiopia–Sudan border westward to Chad (Doggett, 1970; de Wet *et al.* 1976*b*). From here it spread to India, China, the Middle East and Europe soon after its domestication (Doggett, 1965). Collection and conservation of sorghum germplasm attracted the attention of breeders and botanists about three decades ago, when the vulnerability of landraces following the release of new varieties and hybrids was realized. Experience gained from germplasm collection missions shows that we are now in a critical transitional stage; there is an urgent need to collect and conserve the traditional landraces and their wild relatives.

The range of genetic diversity available among the cultivated sorghums and their wild relatives assembled at ICRISAT is truly amazing. Extreme types are so different as to appear to be separate species. Much of this genetic diversity is still available in areas of early cultivation in Africa and regions of early introduction in Asia. In Africa, genetic variability is available both in the cultivated races and the wild progenitors of the crop. De Wet & Harlan (1971) have reported the distribution of both the wild varieties and the major cultivated complexes of *S. bicolor*, and classified them into races based mainly on spikelet morphology (Harlan de Wet, 1972). This classification is simple to use, and helps to elucidate the variation patterns and reveal the paths of evolutionary history.

However, it is difficult to categorise the variation in cultivated sorghums for economic purposes, merely by examining preserved panicles in herbaria. A knowledge of the useful genes in each accession is important in the use of germplasm for crop improvement. In the 1960s and 1970s, extensive work was done to elucidate the taxonomic and evolutionary relationships between subspecies and races of *S. bicolor*, but little attention was paid to the possible use of this natural variability to broaden the genetic base for sorghum improvement.

Past research on sorghum improvement can be summarized as follows:

1. Early work on pure-line selection among cultivated landrace populations in Africa and India that resulted in somewhat improved cultivars (Doggett, 1970).
2. Selection among dwarf populations, and subsequent utilisation of cytoplasmic male sterility to develop commercial hybrids (Quinby & Martin, 1954).
3. Conversion programmes initiated to introduce diverse alleles from tropical germplasm into dwarf, photoperiod-insensitive breeding lines from USA (Stephen, Miller & Rosenow, 1967).
4. Expansion of variation by crossing and/or backcrossing between adapted introductions and local types followed by selection. This provided valuable source material from which new lines are being developed for direct release or as parents for hybrids. Part of the collection of yellow-endosperm types has been extensively used (House, 1985).

It is obvious from past progress in sorghum improvement that much work has yet to be done in germplasm utilisation. Only a small fraction of the total available collection can be fully utilised by breeders at any one time. Crop improvement programmes are often interested in portions of the collection that carry desirable traits. As a prerequisite to efficient use of germplasm, accessions must be properly evaluated, characterised and documented with a workable retrieval system so that well-defined sets of samples with specific combinations of desirable traits can easily be retrieved and used in breeding programmes. The world collection of sorghum needs to be evaluated, characterised and catalogued on the basis of useful genetic characters so that it can be effectively used (Brhane, 1982).

This paper deals with the status of the sorghum world collection maintained by the Genetic Resources Unit (GRU) at ICRISAT, its evaluation, documentation and utilisation in the ICRISAT Sorghum Improvement Program and by other institutions.

Sorghum germplasm collection status at ICRISAT

The first major efforts in the assembly of a sorghum world collection was made in the 1960s by the Rockefeller Foundation in the

Indian Agricultural Research Programme (House, 1985; Murty *et al.*, 1967; Rockefeller Foundation, 1970). A total of 16,138 accessions were assembled from different countries and IS (International Sorghum) numbers were assigned. Of these, only 8,961 could be transferred to the ICRISAT gene bank in 1974 because the rest had already lost their viability. Special efforts were made by ICRISAT to fill the gaps by obtaining duplicate sets from the USA (Purdue University, the National Seed Storage Laboratory, Fort Collins, and from Mayaguez, Puerto Rico). This yielded 3,000 of the missing accessions, leaving a permanent gap of about 4,000 accessions in the world collection presently conserved in the ICRISAT gene bank (Mengesha & Prasada Rao, 1982).

The addition of sorghum germplasm to the world collection became the responsibility of ICRISAT in 1974. This is in accordance with the recommendation made by the Advisory Committee on Sorghum and Millets Germplasm sponsored by the International Board for Plant Genetic Resources (IBPGR, 1976). At present, ICRISAT is the major repository for the world sorghum germplasm collection, with a total collection of 26,564 accessions. Organisations such as IBPGR and ORSTOM, in collaboration with national and international institutions and several individuals, have played key roles in the assembly of this collection. The accessions are listed according to their country of origin in Table 4.1. Among donors, the most important are the Ethiopian Sorghum Improvement Project, Ethiopia; Gezira Agricultural Research Station, Sudan; the All India Co-ordinated Sorghum Improvement Project (AICSIP), and several Indian Agricultural Universities. Almost 80 per cent of the total sorghum collection has come from the developing countries in the semi-arid tropics.

Types of collection

Several types of collections are maintained at ICRISAT Center. These were established on the recommendations of various sorghum workers (Harlan, 1972).

Accessions collection

The available world collection and new accessions assembled by ICRISAT. Seed samples of about 500 g of each accession.

Spontaneous collection

The wild and weedy races (Table 4.2) maintained separately.

Table 4.1. *Sorghum germplasm collection status as ICRISAT up to June 1986*

Origin	Assembled by Rockefeller Foundation	Assembled by ICRISAT up to June 1986	Total
Africa			
Angola	23	6	29
Benin	1	196	197
Botswana	28	162	190
Burkina Faso	160	399	559
Burundi	—	70	70
Cameroon	1,753	183	1,936
Cape Verde Islands	1	1	1
Central African Rep.	37	2	39
Chad	125	13	138
Egypt	15	7	22
Ethiopia	1,446	3,018	4,464
Ghana	11	137	148
Ivory Coast	1	—	1
Kenya	313	559	872
Lesotho	—	8	8
Malagasy Rep.	—	1	1
Malawi	58	385	443
Mali	95	566	661
Morocco	—	8	8
Mozambique	—	42	42
Namibia	—	1	1
Niger	25	383	408
Nigeria	897	473	1,370
Rwanda	—	70	70
Senegambia	12	282	294
Sierra Leone	—	100	100
Somalia	5	120	125
South Africa	483	420	903
Sudan	855	1,494	2,349
Swaziland	18	1	19
Tanzania	31	401	432
Togo	—	258	258
Uganda	471	141	612
Zaire	24	—	24
Zambia	3	288	291
Zimbabwe	123	289	412
Asia			
Afghanistan	5	1	6
Bangladesh	—	9	9
Burma	2	6	8
China	24	58	82
India	2,732	1,406	4,138
Indonesia	6	26	32
Iran	6	1	7

Table 4.1. (*cont.*)

Origin	Assembled by Rockefeller Foundation	Assembled by ICRISAT up to June 1986	Total
Iraq	2	2	4
Israel	22	—	22
Japan	106	5	111
Lebanon	—	360	360
Nepal	7	1	8
Pakistan	1	11	29
Philippines	1	4	5
Saudi Arabia	—	1	1
South Korea	2	—	2
Sri Lanka	—	25	25
Syria	—	4	4
Taiwan	12	1	13
Thailand	5	—	5
Turkey	1	50	51
Yemen Arab Republic	—	1,306	1,306
Yemen, People's Democratic Republic	—	1	1
USSR	5	64	69
Europe			
Belgium	—	1	1
Cyprus	1	—	1
France	5	—	5
German Democratic Republic	—	4	4
Greece	1	—	1
Hungary	—	88	88
Italy	8	—	8
Portugal	—	6	6
UK	—	77	77
America			
Argentina	2	14	16
Cuba	1	2	3
El Salvador	—	1	1
Gautemala	—	6	6
Honduras	—	1	1
Mexico	207	27	234
Nicaragua	—	1	1
Spain	—	3	3
Uruguay	—	1	1
USA	1,208	674	1,882
Venezuela	—	1	1
West Indies	—	3	3
Australia and Oceania			
Australia	6	22	28
New Guinea	—	1	1
Unknown	370	27	397
Total	11,778	14,786	26,564

Table 4.2. *Wild relatives of sorghum assembled at ICRISAT Center up to June 1986*

Genus	Section	*Species*	*Subspecies*	Race	Subrace	Number of accessions
Sorghastrum		*Sorghastrum rigidifolium*	—	—		7
Sorghum	Para sorghum	*Sorghum versicolor*	—	—		17
		Sorghum purpureosericeum	*deccanense*	—		5
			dimidiatum	—		3
		Sorghum nitidum	—	—		3
		Sorghum australiense	—	—		3
	Chaeto sorghum	*Sorghum macrospermum*	—	—		1
	Stipo sorghum	*Sorghum intrans*	—	—		5
		Sorghum brevicallosum	—	—		1
		Sorghum stipodeum	—	—		9
		Sorghum plumosum	—	—		4
		Sorghum matarankense	—	—		3
	Sorghum	*Sorghum halepense*	—	Halepense	Halepense	15
					Johnson grass	5
					Almum	5
				Miliaceum		4
				Controversum		4
		Sorghum propinquum	—	—		3
		Sorghum bicolor	*drummondii*	—		86
		Sorghum bicolor	*Verticilliflorum*	Verticilliflorum		97
				Arundinaaceum		36
				Virgatum		16
				Aethiopicum		13
Total						345

Named cultivar collection

237 named cultivars released by private and public institutions in different countries. Seed samples of 2 kg of each accession are maintained to meet seed requests.

Genetic stock collection

Genotypes resistant to diseases and pests, stocks with identified genes, and cytoplasmic-genic male steriles. Seed samples of 1 kg are maintained by selfing, except in the case of male-sterile lines that are maintained by hand pollination (Table 4.3).

Conversion Collection

176 IS conversion lines obtained from USA. Seed samples of about 500 g of each accession are maintained. A tropical conversion programme has been initiated at ICRISAT and, after conversion, established lines will be added.

Basic collection

A basic collection consisting of 1,275 accessions selected from the world collection and stratified taxonomically, geographically and on the

Table 4.3. *Range of variation in selected characters*

Descriptors	Range of variation	
Days to 50% flowering (number of days)	36	199
Plant height (cm)	55	655
Pigmentation	Tan	Pigmented
Midrib colour	White	Brown
Peduncle exsertion (cm)	0	55.0
Head length (cm)	2.5	71.0
Head width (cm)	1.0	29.0
Head compactness and shape	Very loose stiff branches	Compact oval
Glume colour	Straw	Black
Glume covering	Fully covered	Uncovered
Grain colour	White	Dark brown
Grain size (mm)	1.0	7.5
100-seed mass (g)	0.58	8.56
Endosperm texture	Completely starchy	Completely corneous
Threshability	Freely threshable	Difficult to thresh
Luster	Lustrous	Non-lustrous
Subcoat	Present	Absent

basis of their ecological adaptation at ICRISAT. Seed samples of about 500 g of each accession are maintained. This exercise needs to be repeated at other locations to select comprehensive basic collections for different regions.

Characterisation and range of variation

The entire sorghum germplasm collection, except for very recent acquisitions, has been characterised for important morpho-agronomic characters at the ICRISAT Center during both the rainy and post-rainy seasons. The observed range of variation in morpho-agronomic characters is summarised in Table 4.3.

Documentation and computerisation

Morphological and agronomic data, with passport information from IS 1 to IS 16676 have been documented using the ICRISAT Data Management Retrieval System (IDMRS) program. Computer print-outs are available on request. Additional data are added to this computer file as they become available. This program has the facility to retrieve the information in full or in part, as well as to retrieve sets of accessions with specific combinations of desirable characters, it is especially useful for identifying accessions to fit specific requirements of breeders.

Computer printouts of evaluation data have already been supplied to sorghum scientists in India, Cameroon, Chad, China, Ethiopia, Federal Republic of Germany, UK, USA and Mexico.

Utilising genetic diversity

Screening for sources of resistance

Traditional landraces and their wild relatives, through centuries of natural and human selection, can be expected to have acquired resistance to specific pests, diseases and environmental stresses, and can therefore be used as sources of resistance. Screening the world collection for insect and disease resistance was started soon after is assembly by the Rockefeller Foundation. Significant progress has been made in India in identifying sources of resistance, and a catalogue of sorghum genetic stocks with resistance to pests and diseases was published by the Indian Council of Agriculture Research (ICAR) and the Indian Agricultural Research Institute (IARI), New Delhi (Gupta & Rachie, 1961).

Sorghum germplasm is being screened for resistance traits at the ICRISAT Center by scientists from various disciplines under artificially infested conditions (Taneja & Leuschner, 1985*a*, *b*, Sharma, 1985*a*, *b*; ICRISAT, 1985*a*). The results of screening, indicating the number of promising lines identified, are summarised in Table 4.4.

Screening for drought resistance

It would be extremely difficult to systematically screen all the 26,000 accessions assembled to date for this reaction to drought. Thus, a representative sample based on morphological and physiological traits was selected and screened by essentially observing desiccation tolerance, recovery resistance and agronomic score (Peacock *et al.*, 1985). Of the 26 lines sown in 1974, four resistant lines were selected for further testing, to examine the physiological basis of resistance to midseason stresses due to heat and lack of water. Laboratory techniques will eventually be developed to screen germplasm rapidly for drought tolerance, especially those collections from the drier areas of the tropics.

Screening for crop establishment

Using a field technique developed at the ICRISAT Center (Soman *et al.*, 1984) good progress has been made in screening 814

Table 4.4. *Genetic stocks collection maintained at ICRISAT Center as of June 1986*

Type	Accessions
Promising lines for pest resistance	
Shoot fly (*Atherigona soccata*)	60
Stem borer (*Chilo partellus*)	70
Midge (*Contarinia sorghicola*)	14
Headbug (*Calocoris angustatus*)	6
Promising lines for disease resistance	
Grain mold	156
Anthracnose (*Colletotrichum graminicola*)	15
Rust (*Puccinea purpurea*)	31
Downy mildew (*Peronosclerospora sorghi*)	155
Striga low stimulant lines (Lab screening)	645
Striga resistant lines (Field screening)	24
Other characters	
Glossy lines	501
Pop sorghum lines	36
Sweet-stalk sorghum lines	76
Scented sorghum lines	17
Twin-seeded lines	131
Large-glume lines	71
Bloomless sorghum lines	207
Broomcorn sorghum lines	52
Cytoplasmic A&B lines	240

germplasm accessions for seedling emergence though soil crust, a major problem in the semi-arid tropics.

Over 50 promising lines have been identified using this technique. High soil temperatures reduce seedling emergence in sorghum. In a preliminary study using a laboratory technique developed at ICRISAT Center (Soman & Peacock, 1985) 52 selected germplasm lines were screened for temperature stress. Seven accessions were found to be promising. They could emerge at higher soil temperature.

Diversifying the cytoplasm

The association of T-cytoplasm with southern leaf blight (*Helminthosporium maydis*) in maize became matter of grave concern among plant breeders worldwide. If such a disease should become associated with milo-type serile cytoplasm in sorghum that is at present used in the production of almost all the hybrids in the world, the entire hybrid-sorghum seed industry would be doomed. To avoid such possible hazards associated with using narrow cytoplasm, it is necessary to diversify the male-sterile source in sorghum.

The diversity in the world collection provided an excellent opportunity to advance in this direction. Work is in progress in India and the USA, and several potentially useful diverse cytoplasms have been isolated. Scientists have identified cytoplasms with different sterility responses from those of milo cytoplasm. Among the new male-sterile lines isolated, M 35–1 A, M 31–2 A, VZM2 A, and GI A from India (source yet to be identified) are important. Sterility in IS 12662C cytoplasm was identified by Schertz & Ritchey (1977) as probably different from milo (A1), and released as A2 (Schertz, 1977; Rosenow *et al.*, 1980; Schertz *et al.*, 1981). The sterility mechanism in IS 1112C was probably that of a different cytoplasm and was designated as A3 by Quinby (1980). To identify the sources of cytoplasm, other characteristics including mitochondria, chloroplasts and polypeptides are being studied in India, the USA, and Scotland, U.K. These studies are providing information and useful lines to diversify germplasm in sorghum breeding programmes (Schertz & Pring, 1982).

Germplasm enhancement

Conversion programme

The US conversion programme (Stephen *et al.*, 1967) provided breeders in temperate areas with the great genetic diversity available in tropical sorghum, especially the sources of resistances to diseases and pests. However, the conversion programme can handle only a small portion of the available germplasm, and the converted lines so developed are only adapted to temperate areas.

A major portion of the world collection consists of tall, photoperiod-sensitive landraces that are of limited value in crop improvement. To augment the use of tropical germplasm in breeding programmes and to broaden the genetic base, we began a tropical conversion programme using the long-day rainy season and the short-day post-rainy season at the ICRISAT Center. The technique originally developed at Texas A & M University, Texas, USA, was adopted, except that the female parent used in the first cross is the landrace and contributes the cytoplasm. Moreover, unlike the converted genotypes from USA, the converted material developed in India, or the partially segregating material selected by breeders during the process of conversion, is adapted to tropical countries, especially those in the semi-arid tropics. Over the past few years, eight Zerazera landraces from Ethiopia and Sudan have been converted into photoperiod-insensitive lines. It took six years to convert the Sudanese and Ethiopian Zerazeras to photoperiod-insensitivity (ICRISAT, 1985*b*), and the final converted lines are in three maturity and three plant-height backgrounds. All these lies are being assigned ICRISAT Sorghum Conversion (ISC) numbers.

We are also in the process of converting three sorghum landraces from Nigeria, Guineense, Kaura, and Farafara. The material is currently in BC_1F_1 generation. The tropical landraces were selected for conversion mainly on the basis of our own original notes made at the time of collection. During the past three years, thousands of selections from the partially converted Zerazera populations have been made by breeders from ICRISAT, AICSIP, agricultural universities and private seed companies in India and other sorghum breeders from some 20 countries.

Introgression

Interest in the use of exotic germplasm for cereal improvement has markedly increased. Exotic germplasm varies greatly in its capacity to contribute positively to crop improvement. For this reason, the choice of the appropriate exotic genotype is as important as the choice of the correct adapted cultivar as parent. With the possible exception of oat, the cultivated pools of major cereals, including sorghum, contain sufficient genetic variability to permit further genetic gains in yield and other agronomic traits. There appears to be no well-organised introgression programme anywhere that aims to transfer agronomically useful genes from wild to cultivated taxa. This task must be given more emphasis.

Attempts have been made to transfer shoot fly (*Atherigona soccata*) resistance from sugarcane (*Saccharum officinarum* L.) to sorghum (de Wet *et al.*, 1976*a*). Modified sorghums carrying sugarcane genes have been recovered from such crosses whether or not these sorghums have any shoot fly resistance has not been reported (Brhane, 1982).

At the ICRISAT Center, the available wild relatives of sorghum have been screened for resistance to sorghum shoot fly and sorghum downy mildew (*Peronosclerospora sorghii*) and sources of resistance have been identified.

Since appreciable levels of resistance to shoot fly are not available in any cultivated sorghums, it is necessary to search for resistance in wild species. Crosses were made between resistant wild sorghum species and adapted cultivars during 1985. Presumed hybrids are being grown in a greenhouse. Because there are differences in chromosome number it is unlikely that hybrids will be fertile. All possible techniques, including embryo rescue, are being used in our attempts to produce successful hybrids.

As there are several cultivated landraces with fairly good resistance to sorghum downy mildew, we have not yet initiated any work for transferring downy mildew resistance from wild sorghums to cultivated genotypes.

Alternate uses of sorghum

Sorghum growing areas in some developing countries are diminishing (FAO, 1985). However its alternative uses, e.g. as forage and for making beer, alcohol, syrup, etc, can slow down or even reverse this trend. There is a need for sorghum scientists to develop new agricultural and industrial uses for sorghum.

Sorghum for forage

In recent years the search for forage-sorghum genotypes has intensified. The world collection at ICRISAT has been screened by several forage breeders in India and promising lines identified. IS 4133, IS 4866 and IS 11148 were identified as suitable forage plant types possessing such desirable attributes as plant height, profuse leafiness, high seed productivity and comparable quality characteristics for protein and dry-matter digestibility components. A wild genetic base for forage attributes was reported by Tripathi & Ahluwalia, 1984.

We are collaborating with the National Bureau of Plant Genetic Resources (NBPGR), New Delhi, in a systematic evaluation of forage sorghum. Evaluation of 1,500 forage-type accessions at Hisar, New Delhi, Jhansi, and Akola is in progress. Results from this multilocational evaluation should identify the most promising lines for use in forage breeding programmes.

Sorghum for beer

Large-scale urbanisation in Africa has resulted in a shift in sorghum beer production from what used to be a family brew, or at best a community affair, to an industrialised process. Sorghum beer production is a highly specialised industry in South Africa. It is distinct from beer made from barley malt in that the usual process of starch conversion and alcohol fermentation is preceded by a souring process which influences the body, stability and alcohol content of the final product (Doggett *et al.*, 1970). The most important characteristic of a sorghum for brewing is its ability to produce amylase upon germination (Novellie, 1982). Most sorghums collected from Rwanda and Burundi, and some from Ethiopia, are used for making beer; some are also used for making porridge and traditional bread. These collections belong to the race Caudatum or Durra-Caudatum (Prasada Rao & Mengesha, 1982). Substantial research remains to be done to establish quality criteria for making beer. The available germplasm collections from eastern and southern African countries provide an excellent base material for this purpose.

Sweet-stalk sorghums for syrup and alcohol production

Sorghum landraces that have sweet stalks are sparingly distributed across sorghum-growing areas of Africa and India. The green and tender stalks are chewed like sugarcane. In Ethiopia, sweet-stalk sorghums are also used for a confection (Damon, 1962). Schaffert & Gourley (1982) reported that sweet-stalk sorghums can also be used to produce alcohol by adopting the technology applied to sugarcane.

In view of the growing importance for sweet-stalk sorghums, a part of the world collection maintained at ICRISAT was screened for stalk-sugar content. Calculated on a dry-weight basis of the 78 lines tested, the sugar content ranged from 16.2 to 38.1 per cent (Subramaniam & Prasada Rao, unpublished). The lines were identified among collections from Botswana, Cameroon, Chad, Ethiopia, India, Kenya, Malawi, Niger, Nigeria, Somalia, South Africa, Sudan, Thailand, Uganda, USA, Zambia and Zimbabwe (Prasada Rao & Murty, 1982).

Seed samples of sweet-stalk sorghums have been supplied to scientists in many countries for use in their research programme.

Pop sorghums

A broad survey of the geographical distribution of pop sorghums in the world collection showed that a majority of them originated in India

(Prasada Rao & Murty, 1982) where popped sorghum grains are consumed as a snack food and as a delicacy. Most of the pop sorghums in India are identified by colloquial names. Of the 3,682 accessions screened for popping quality, 36 showed good popping qualities, and could be useful in breeding programmes that aim to improve this quality (Murty *et al.*, 1982).

Broomcorn sorghums

Broomcorns are characterised by a very short rachis and very long panicle branches. The kernels are small and often enclosed in long ellipsoid glumes belonging to the taxonomic race Bicolor. The stalks are dry, non-sweet, and have a hard rind. Inflorescences of broomcorn, with their long panicle branches, are used to make household, warehouse, whisk, and toy or hearth brooms (Weibel, 1970). Broomcorn types apparently developed in the Mediterranean region from material coming from India or Africa via the Middle East. In Italy they were grown before AD, 1596 and their cultivation and the manufacture of brooms spread to Spain, France, Austria and southern Germany. Considerable development of broomcorn subsequently took place in the USA, where it is said to have been introduced by Benjamin Franklin (Doggett, 1970). Spikelets in recently developed cultivars disarticulate before the inflorescences are harvested, eliminating the need for threshing.

In the world collection maintained at ICRISAT, there are 52 accessions belonging to the broomcorn group. Seed samples of these types have been supplied to scientists in Bolivia, Ethiopia, Mexico, Yemen PDR and Yugoslavia for use in their breeding programmes.

Taxonomic systems and germplasm utilisation

Of the several classification systems proposed for cultivated sorghums, the most important ones are those of Snowden (1936), Murty *et al.* (1967), and Harlan & de Wet (1972). Among these, the classification system proposed by Harlan & de Wet has gained rapid popularity. The Advisory Committee on Sorghum and Millets has accepted and recommended this system for classifying germplasm at ICRISAT (IBPGR & ICRISAT, 1980).

Harlan & de Wet's classification system recognises five basic races, i.e. Bicolor, Guinea, Caudatum, Kafir, and Durra, and all possible combinations of these races, referred to as intermediate races. It is easy to identify races on the basis of spikelet morphology. This classification system tends to be over-simplified, and is of limited use, particularly from the utilisation point of view. After studying the world collection, and specimens filed at the Royal Botanic Gardens, Kew, UK, we are working out a 'subrace' classification system which is essentially an extension of the Harlan & de Wet classification. For example, the major cultivars under race Caudatum can be classified thus:

Cultivar	Harlan	ICRISAT
Hegari sorghums from Sudan	Race: Caudatum	Race: Caudatum Subrace: Hegari
Beer sorghums from the highlands of Burundi	Race: Caudatum	Race: Caudatum Subrace: Nigricans

In this subrace classification, it will be possible to reflect the geographic origin as well as the potential of particular landraces to sorghum users. It is known to almost all sorghum scientists that Hegaris are grain sorghums originating from Sudan, and Nigricans are beer sorghums originating from eastern Africa. Information on the subrace will help users to visualise the type of material they are dealing with.

Availability of germplasm

If the world collection is to serve a useful purpose, it should be readily available to all research workers, institutions and universities. The supply of seed material to scientists worldwide is one of the major

Table 4.5. *Sorghum germplasm samples distributed up to June 1986*

Year	ICRISAT Center	Within India	Other countries	Total samples distributed
1973	—	—	3	3
1974	4,133	3	359	4,495
1975	6,574	2,102	1,090	9,766
1976	3,977	1,788	2,729	8,494
1977	11,691	2,812	3,446	17,949
1978	8,563	2,159	696	11,415
1979	7,870	3,720	5,785	11,375
1980	23,197	1,798	1,897	26,892
1981	36,322	3,571	9,718	49,611
1982	17,556	1,668	12,795	32,019
1983	15,967	3,415	18,080	37,462
1984	17,477	2,123	14,668	34,268
1985	20,979	1,816	12,376	35,171
1986	3,639	10,998	6,576	21,213
Total	177,945	37,973	90,215	306,133

responsibilities of ICRISAT and the promising accessions discussed in this paper are available on request. All exported seed material from ICRISAT must pass through the Indian Plant Quarantine Authority and this passage is facilitated by the Export Certification Quarantine Laboratory established at ICRISAT. Table 4.5 shows the number of sorghum germplasm samples distributed by ICRISAT since 1973.

Conclusions

Although germplasm collections are useful in elucidating the taxonomic and evolutionary relationships between different species and races, their principal justification is to assemble natural variability that can be used to broaden the genetic base for present and future sorghum improvement.

Some germplasm accessions may be directly recommended for cultivation. For example, E 35–1 (a selection from a Zerazera landrace from Ethiopia) was recommended for release in Burkina Faso (ICRISAT, 1984) and IS 9302 and IS 9323 (Kafir landraces from South Africa) were released into intermediate-altitude areas of Ethiopia (Abebe & Yilma, 1984). Germplasm accessions are, however, more commonly used as source material for transferring useful genes into adapted types. Perhaps the most extensive use of primitive and wild material has been in the breeding for resistances to diseases and pests. The search for various kinds of resistance may be intensified in the future. The transfer of desirable genetic traits from wild to cultivated material depends on cytogenetic and genetic relationships between the respective species. In the case of wide crosses, special techniques including biotechnology may have to be adopted in the future.

Much germplasm utilisation work has yet to be done. There is far more variability in the present collection than that used. As a prerequisite to efficient use, germplasm must be properly evaluated to identify the potential of accessions for use in breeding programme.

Regional evaluation of germplasm at or close, to the place of origin is vital to exploit the true potential of a genotype. Few countries have the desire or the resources to satisfy these requirements. Therefore, international collaboration in evaluation is not only desirable but essential. Such efforts will not only strengthen and enhance the utilisation of sorghum germplasm, but could also create new uses for the crop throughout the world.

References

Abebe, M. & Yilma, K. (1984). High rainfall intermediate altitude sorghum (HRIAS). *Sorghum Newsletter*, **27**, 3–4.

Brhane, G. (1982). Utilization of germplasm in sorghum improvement. In *Sorghum in the Eighties*: Proceedings of the International Symposium on Sorghum, 2–7 Nov. 1981, ICRISAT, India. Vol. 1. pp. 335–45, International Crops Research Institute for the Semi-Arid Tropics, India.

Damon, E. G. (1962). The cultivated sorghums of Ethiopia. *Experiment Station Bulletin* No. 6. Imperial Ethiopian College of Agriculture and Mechanical Arts.

De Wet, J. M. J., Gupta, S. C., Harlan, J. R. & Grassl, C. O. (1976). Cytogenetics of introgression from *Saccharum* into *Sorghum*. Crop *Science*, **16**, 568–72.

De Wet, J. M. J., & Harlan, J. R. (1971). The origin and domestication of *Sorghum bicolor*. *Economic Botany*, **25**, 128–35.

De Wet, J. M. J., Harlan, J. R. & Price, E. G. (1976*b*). Variability in *Sorghum bicolor*. In *Origins of African Plant Domestication* pp. 493–63, Harlan, J. R., de Wet, J. M. J. & Stemler, A. B. L. (eds.), The Mountain Press, The Hague.

Doggett, H. (1965). The development of cultivated sorghums. In *Crop Plant Evolution*, pp. 50–69. Hutchinson, J. (ed.), Cambridge University Press, Cambridge.

Doggett, H. (1970). *Sorghum*. Longmans, Green, London.

Doggett, H., Curtis, D. L., Laubscher, F. X. & Webster, O. J. (1970). Sorghum in Africa. In *Sorghum Production and Utilization*, pp. 288–327, Wall, J. S., & Ross, W. M. (eds.), Avi Publishing, Westport, Conn.

FAO (1985). *1984 Production Year Book*. Food and Agriculture Organization, Rome.

Gupta, V. P. & Rachie, K. O. (1961). *Catalogue of Sorghum Genetic Stocks Resistant to Pests-1*. Indian Council of Agricultural Research, New Delhi.

Harlan, J. R. (1972). Genetic resources in sorghum. In *Sorghum in the Seventies*, pp. 1–13. Rao, N. G. P. & House, L. R. (eds.), Oxford and IBH Publishing Co., New Delhi.

Harlan, J. R., & de Wet, J. M. J. (1972). A simplified classification of cultivated sorghum. *Crop Science*, **12**, 172–6.

House, L. R. (1985). *A Guide to Sorghum Breeding*. 2nd Edn. India: International Crops Research Institute for the Semi-Arid Tropics, India. International Board for Plant Genetic Resources (1976). *Proceedings of the Meeting of the Advisory Committee on Sorghum and Millets Germplasm*, Rome. (mimeographed).

IBPGR & ICRISAT (1980). *Sorghum Descriptors*. IBPGR, Rome.

ICRISAT (1984). Sorghum. In *Annual Report 1983*, pp. 15–64. ICRISAT.

ICRISAT (1985*a*). Sorghum. In *Annual Report 1984*, pp. 17–80. ICRISAT, India.

ICRISAT (1985*b*). Augmenting genetic resources. In *ICRISAT Research Highlights 1984*, pp. 28–9. ICRISAT, India.

Mengesha, M. H. (1984). International germplasm collection, conservation, and exchange at ICRISAT. In *Conservation of Crop Germplasm An International Perspective*, pp. 47–54. Crop Science Society of America, Madison.

Mengesha, M. H. & Prasada Rao, K. E. (1982). Current situation and future of sorghum germplasm. In *Sorghum in the Eighties*: Proceedings of the International Symposium on Sorghum, 2–7 Nov. 1981, Vol. 1, p. 323, ICRISAT India.

Murty, B.R., Arunachalam, V. & Saxena, M. B. L. (1967). Classification and catalogue of a world collection of sorghum. *Indian Journal of Genetics and Plant Breeding*, 27 (Spl. Number), 1–312.

Murty, D. S., Patil, H. D., Prasada Rao, K. E. & House, L. R. (1982). A note on screening the Indian sorghum collection for popping quality. *Journal of Food Science and Technology*, **19**, 79–80.

Novellie, L. (1982). Fermented Beverages. In *Proceedings of the International Symposium on Sorghum Grain Quality*. 28–31 October 1981, pp. 113–20, ICRISAT, India.

Peacock, J. M., Azam-Ali, S. N. & Matthews. (1985). An approach to screening for resistance to water and heat stress in sorghum (Sorghum bicolor (L.) Moench). In *Arid Lands: Today and Tomorrow*. Proceedings of the International Arid Lands Research and Development Conference. 20–25 October 1985, Tucson, Arizona, USA.

Prasada Rao, K. E. & Mengesha, M. H. (1982). Sorghum germplasm collection in Rwanda. *Genetic Resources Unit ICRISAT Progress Report*, **46** (limited distribution).

Prasada Rao, K. E. & Murty, D. S. (1982). Sorghum for special uses. In *Proceedings of the International Symposium on Sorghum Grain Quality*, pp. 129–34. ICRISAT, India.

Quinby, J. R. (1980). Interaction of genes and cytoplasms in male sterility in sorghum. In *Proceedings 35th Corn and Sorghum Research Conference*, pp. 175–84. American Seed Trade Association, Chicago.

Quinby, J. R. & Martin, J. H. (1954). Sorghum improvement. *Advances in Agronomy*, **6**, 305–59.

Rockefeller Foundation (1970). *World Collection of Sorghums. List of Pedigrees and Origins*. Rockefeller Foundation, Indian Agricultural Program, New York.

Rosenow, D. T., Schertz, K. F. & Sotomayor, A. (1980). Germplasm release of three pairs (A and B) of sorghum lines with A2 cytoplasmic-genetic sterility system. *Texas Agricultural Experiment Station*, MP-1448, 1–2.

Schaffert, R. W. & Gourley, L. M. (1982). Sorghum as an energy source. In *Sorghum in the Eighties*, Vol. 2. pp. 605–23. ICRISAT, India.

Schertz, K. F. (1977). Registration of A2 TX 2753 and BTX 2753 sorghum germplasm (Reg. No. GP 30 and 31). *Crop Science*, **17**, 983.

Schertz, K. F. & Pring, D. R. (1982). Cytoplasmic sterility systems in sorghum. In *Sorghum in the Eighties*. Vol. 2, pp. 373–83. ICRISAT, India.

Schertz, K. F. & Ritchey, J. R. (1977). Cytoplasmic-genic male-sterility systems in sorghum. *Crop Science*, **18**, 890–3.

Schertz, K. F., Rosenow, D. T. & Sotomayor-Rios, A. (1981). Registration of three pairs (A and B) of sorghum germplasm with A2 cytoplasmic-genic sterility system. *Crop Science*, **21**, 148.

Sharma, H. C. (1985*a*). Screening for sorghum midge resistance and resistance mechanisms. In *Proceedings of the International Sorghum Entomology Workshop*, pp. 275–92, ICRISAT, India.

Sharma, H. C. (1985*b*). Screening of Host-Plant Resistance to Midrid Head Bugs in Sorghum. In *Proceedings of the International Sorghum Entomology Workshop*, pp. 317–35, ICRISAT, India.

Snowden, J. D. (1936). *The Cultivated Races of Sorghum*. Adlard. London.

Soman, P. & Peacock, J. M. (1985). A laboratory technique to screen seedling emergence of sorghum and pearl millet at high soil temperature. *Experimental Agriculture*, **21**, 335–41.

Soman, P., Peacock, J. M. & Bidinger, F. R. (1984). A field technique to screen seedling emergence of pearl millet and sorghum through soil crust. *Experimental Agriculture*, **20**, 327–34.
Stephen, J. C., Miller, F. R. & Rosenow, D. J. (1976). Conversion of alien sorghums to early combine types. *Crop Science*, **7**, 396.
Subramanian, A. (1956). Jowar pops are a real delicacy. *Indian Farming*, **6**, 25–6.
Taneja, S. L. & Leuschner, K. (1985*a*). Resistance screening and mechanisms of resistance in sorghum to shoot fly. In *Proceedings of the International Sorghum Entomology Workshop*, pp. 115–29, ICRISAT, India.
Taneja, S. L. & Leuschner, K. (1985*b*). Methods of rearing, infestation and evaluation for *Chilo partellus* resistance in sorghum. In *Proceedings of the International Sorghum Entomology Workshop*, pp. 175–88. ICRISAT, India.
Tripathi, D. P. & Ahluwalia, M. (1984). Promising forage sorghum sources in sorghum germplasm. *Sorghum Newsletter*, **27**, 14.
Weibel, D. E. (1970). Broomcorn. In *Sorghum Production and Utilization*, pp. 441–68, Wall, J. S. & Ross, W. M. (eds.), Avi Publishing, Westport, Conn.

5
Current use of potato collections

J. G. Th. HERMSEN

Genetic variation for use in potato research and breeding

From the time potato breeding began in the nineteenth century, it has been hampered by insufficient genetic variation despite the existence of many varieties. After the first few casual introductions of potato in the sixteenth century, little genetic variation was added to the gene pool until the twentieth century. Indeed, since the introduction of potato, large gene losses must have occurred in Europe mainly due to adaptive selective, virus and other diseases. When, in the early twentieth century, government-supported breeding programmes replaced (USA, UK) or supplemented (The Netherlands, Germany) private breeding, various local gene pools with specific valuable characters were exchanged between countries. This resulted in some new varieties with a good breeding value and these have been used extensively in subsequent breeding.

Financial support from governments from the 1920s onwards and more recently from private companies, enabled professional scientists to undertake a series of expeditions to the potato gene centres from the USA southwards to Chile aimed at obtaining the maximum amount of genetic diversity.

Although there is probably little left of the early materials, a large amount of collected material is still available in the living state in the major germplasm collections listed in Table 5.1 as well as in many smaller working collections.

The codes in the first column of Table 5.1 refer to the following centres:

CIP = Centro Internacional de la Papa, Peru.
IR-1 = Interregional Potato Introduction Station, USA.
BGRC = Braunschweig Genetic Resources Center;
collaborative German/Netherlands potato collection.

G-LKS = Gross-Lüsewitz Kartoffel Sammlung, DDR.
CPC = Commonwealth Potato Collection, Scotland.
INTA = Potato collection of Instituto Nacional de Tecnologia Agropecuaria, Argentina.
VIR = Vavilov Institute of Plant Industry, USSR.

It should be noted that the reduction of cultivated accessions at CIP from 1976–84 in Table 5.1 was mainly brought about by discarding duplicate clones. True seeds of all duplicates are still being stored.

Every gene bank has a core of unique material collected by its own expeditions. However, many accessions occur in two or more gene banks mainly because of exchange of material. The occurrence of duplications may also be due to collecting identical genotypes, especially of cultivated material. The number of accessions available per species varies greatly.

Potato germplasm can be maintained and distributed as tubers, as *in vitro* plantlets, or as true seeds. Each method has advantages and limitations. The method of maintenance and distribution influences the use of wild and primitive *Solanum* species. Most private breeders prefer to receive healthy tubers of well-known genotypes for evaluation and use in crosses. Public breeders and research institutes are more prepared to handle seed samples of accessions. Indeed, they often prefer the genetic variation present in a true seed progeny to the availability of only one genotype per accession.

Table 5.1. *Number of accessions of wild species and primitive forms in six major germplasm banks in the years indicated. Codes in capitals explained in the text*

Code germplasm bank	Year	Number of accessions		Reference
		Wild spp.	Cultiv. spp.	
CIP	1972	3,000		Ochoa (1976)
	1976		12,400	Ochoa (1976)
	1984	1,500	5,000	International Potato Center (1984)
IR-1	1969	760	770	Ross & Rowe (1969)
	1976	4,000[1]		Hanneman (1976)
BGRC	1984	2,163	1,060	Van Soest *et al.* (1984)
G-LKS	1965	719	1,717	Rothacker (1966)
CPC	1969	460	839	Anon. (1969)
INTA	1976	718	41[2]	Okada (1976)
VIR	No data available about present size			

[1] Wild + cultivated with an approximate 2 to 1 majority for the wild
[2] Probably under-rated because of Okada's collecting after 1976

The use of germplasm in research and breeding can be promoted by systematic evaluation and the publication of evaluation data at regular intervals.

Ideally, evaluation should include the following activities.

1. Systematic evaluation of all accessions of a gene bank for all relevant characters.
2. Computer-aided storage and retrieval of all collected data.
3. Publishing an inventory of available seed stocks with evaluation data.
4. Updating the inventory at regular time intervals by means of supplements to, or new editions of, the inventory.

Although every gene bank curator would agree with the desirability of these objectives, the real policy is different because it is dependent on the available facilities and personnel. As a result the policy of evaluation differs greatly among potato germplasm centres. Systematic evaluation is an integral part of the potato breeding programme, focussed primarily on the needs of developing countries, only at CIP. In principle, CIP's material and data are available to other countries as well. As stated before, an *index seminum* with evaluation data on CIP's collection would greatly increase the accessibility and therefore also the ultilisation of this invaluable collection of primitive cultivated forms in other breeding programmes.

Also the BGRC collection is being systematically evaluated, but mainly outside the centre by Dutch and German scientists, and for a limited number of traits chosen on the basis of priorities set by the two participating countries. The amount of evaluation data on these traits is considerable and, furthermore, easily accessible because inventories and detailed publications are available. Screening of the collection for seed-borne diseases is in progress, but such diseases may still hamper distribution and use of the material for some time.

The IR-1 and CPC collections have been established for much longer than the previous ones. As a result, many accessions have been distributed to many countries. Although there is no practice of systematic evaluation, much data is available, particularly about the IR-1 material. The present policy of CPC is: no direct involvement in evaluation activities nor in the collection and publication of evaluation data from elsewhere. The IR-1, however, has collected and published evaluation data from receivers of their material, although the most recent inventory dates back to 1969. Breeders are looking forward to the updated inventory. Screening for seed-borne diseases demands extra rejuvenation efforts and may hamper the use of IR-1 material at least temporarily.

The use of the potato collections in research and breeding

Introduction

The material from the centres of diversity can be used in research and breeding owing to the efforts of germplasm collectors and taxonomists. Breeders not only need the material as such, but also a framework of knowledge about its nature. Taxonomists have elucidated the pattern of variation and the boundaries of *Solanum* species enabling an adequate classification of collected material. They have mapped the geographic range of species enabling adequate sampling by expeditions (Bukasov, 1933; Hawkes, 1963; 1978; Ochoa, 1962; Correll, 1962; Hawkes & Hjerting, 1969). Taxonomists and geneticists have revealed relationships between species enabling the breeders to choose the most accessible species and to handle them in the most efficient way (Hawkes, 1958; Hermsen, 1986). There is also a relationship between maintenance and use of collections. The accessions should be available in a healthy state, adequately classified taxonomically and propagated in such a way that gene loss is minimised.

The importance of evaluation, of efficient storage and of the retrieval and publication of evaluation data, for promoting the use of germplasm collections has been pointed out earlier. Evaluation mostly includes preliminary screening for separate characters, most of them being associated with resistance to biotic and abiotic factors and some with quality components.

Materials thus selected for their potential value in utilisation are included predominantly in various programmes of basic and practical breeding research. The results can be found in scientific publications and in the release by potato breeders of varieties with one or more wild or primitive *Solanum* species in their pedigree.

The potato is unique in the use, since 1960, of large numbers of accessions of tetraploid and diploid primitive species for broadening the genetic base of potato breeding via recurrent selection for adaptation to climatic conditions (also long daylength) in temperate zones. In this section the pattern of use of the world potato collections will briefly be treated and illustrated with case histories.

Use in basic research on nature and inheritance of useful traits

Solanum species are being widely used for basic research into the nature and inheritance of useful characters detected in the species. A few divergent examples will briefly be presented. Most questions under investigation are related to resistance:

Is the resistance durable, and if so, is it based on one or few genes or polygenically controlled?
Is the lack of symptoms a matter of hypersensitivity based on gene-for-gene relations or of extreme durable resistance?
What is the nature of aphid-trapping ability of glandular hairs in *S. berthaultii*, *S. tarijense* and *S. polyadenium*? Can it be transferred into advanced breeding lines?
What is the nature of pest resistance, e.g. in *S. demissum*, *S. stoloniferum* and *S. chacoense* where glandular hairs are not involved? Is it related to content of glyco-alkaloids that make the potato unsuitable for human consumption?

The nature of stress tolerance and its relationship with the accumulation of sugars and amino acids has been studied in many plant species. In tuber-bearing Solanums large-scale genetic and breeding research, as well as physiological and biochemical research, has been carried out on frost tolerance (Richardson & Weiser, 1972; Van Swaaij, 1986).

Some tuber-bearing Solanums are a source of different types of cytoplasmic male sterility (CMS). *S. verrucosum* was studied by Grun *et al.* (1962) and Abdalla & Hermsen (1972). Some CMS-types from the *S. verrucosum* source are being investigated at the molecular level by the group of Galun (Weizmann Institute, Rehovot, Israel). Two types, viz tetrad sterility (Abdalla & Hermsen, 1972) and eclipse sterility (Ramanna & Hermsen, 1974) are being studied for their use in streamlining true seed production for the purpose of the new technology of growing potatoes from true seeds (TPS).

Solanum species are also a rich source of different kinds of mutants which can be used as breeding tools (e.g. embryo-spot and meiotic mutants for breeding at the diploid level; see later section) but also for mapping the chromosomes of potato. Gene mapping of potato, which is in its initial stages, is of great importance and should be promoted.

Basic research on interspecific barriers

There is another field of basic research on *Solanum* species in which public breeders are involved: the nature of breeding barriers and ways to overcome or circumvent such barriers. A general theory on interspecific breeding barriers in terms of 'barrier capacity' of stylar parents and 'penetration capacity' of the pollen parents has been put forward by Hogenboóm (1973). Most wild *Solanum* species can readily be crossed with the cultivated potato once their ploidy level is adjusted. This implies that an equal functional ploidy level should be realised either by doubling the chromosome number of diploid and allotetraploid (=

functionally diploid) species or by halving the chromosome number of cultivars (haploidisation). Routine methods are available for these ploidy manipulations. However, some promising wild species cannot be made accessible along this line, e.g. *S. etuberosum* and *S. brevidens* which are excellent sources of multiplication resistance to PLRV (leafroll virus). These species have been made accessible to breeding by Hermsen (1984*a*) according to the general breeding scheme in Fig. 5.1, which may have an

Table 5.2. *Successful applications of the scheme in Fig. 5.1*
S = source of resistance; B1 and B2 bridging species used as male (♂) *or female* (♀) *as indicated. Detailed data in Hermsen* (*1984*a)

S	B1	B2
S. bulbocastanum	*S. acaule*♀	*S. phureja*♂
S. pinnatisectum		
Ditto	*S. verrucosum*♂	*S. stoloniferum*♂
S. brevidens	*S. pinnatisectum*	*S. verrucosum*♀
S. etuberosum	or *S. jamesii*♂	
Ditto	Ditto	*S. acaule*♀

Fig. 5.1. General breeding scheme for recalcitrant *Solanum* species according to Hermsen (1984*a*). S = genome of species providing the resistance; B1 and B2 = genomes of two bridging species; T = genome of *S. tuberosum* cultivars 2×, 4×, 5×, 6× = indication of ploidy level. Successive steps indicated by (1), (2), (3), (4), (5).

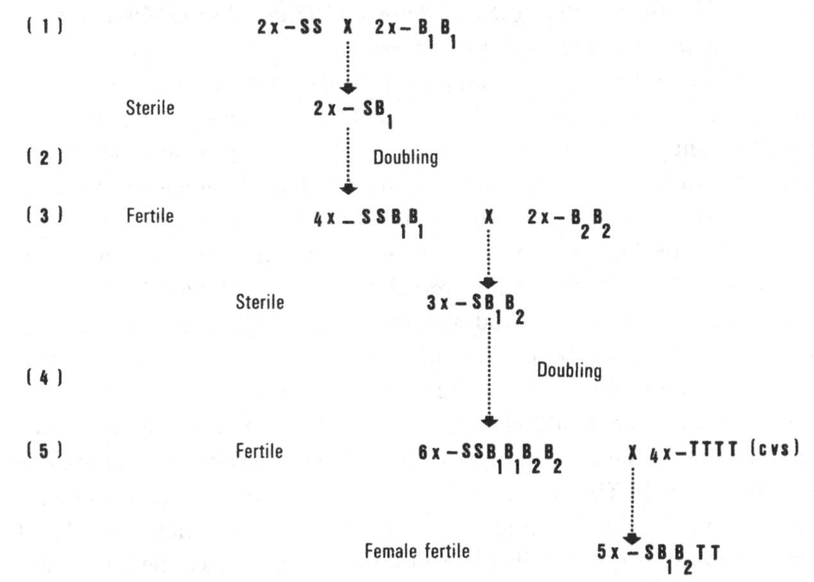

applicability for more recalcitrant species than those for which it has been used successfully up to now (Table 5.2).

The scheme has the following characteristics.

1. The use of two species, B_1 and B_2, which in addition to their bridging function have the potential to contribute desirable genes to the final population (steps 1 and 3).
2. Doubling the chromosome number at two stages where sterility is blocking further progress (steps 2 and 4).
3. Only conventional crosses, making use of previous experience and of optimal environmental conditions.
4. The occurrence of one narrow bottleneck at the stage, where *S. tuberosum* is being included for the first time (step 5).
5. Once the pentaploid has been obtained (step 5) there are no barriers to further enhancement through backcrossing with cultivars. This final stage can basically be attained in the third year from the start.

Some research groups, such as that of Dr J.C. Helgeson (Madison, USA), have successfully applied somatic hybridisation of *S. brevidens* and a diploid primitive *Solanum* species hybrid (Austin *et al.*, 1985).

Such hybridisation leads directly to step 3 and has the advantage that a cultivated clone can be used instead of a wild partner. It is not known to the author whether the allotetraploid fusion product can be crossed directly with *S. tuberosum* or with diploid cultivated potato.

The use of potato germplasm in variety breeding in developed countries; public v. *private breeders*

It depends on the country whether variety breeding is carried out predominantly by public breeders, by both public and private breeders, predominantly by private breeders or exclusively by private breeders. On average, public breeders are more inclined to include *Solanum* species in their breeding programmes. Most public breeders can afford to carry out long-term breeding programmes because they are working under the auspices of universities or other research institutes. Private breeders are more reluctant to initiate wild species breeding, although they feel the need of new genes when their varieties are in danger of succumbing to new races or pathotypes of diseases and pests. Private breeders, and also some public breeders, need 'link-men' (Hawkes, 1979) who are able to provide the necessary characters or genes of the wild species in manageable breeding lines. In The Netherlands the Plant Breeding Department of the Agricultural University and the Foundation for Agricultural Plant Breeding have performed this intermediate function since the late 1940s.

Fig. 5.2. Frequency distribution of 508 varieties among 20 European countries, subdivided per country into pure *S. tuberosum* varieties (□) and varieties with one or more wild species or primitive forms in their parentage (▦). Data calculated from Table 5.3 in Stegemann & Schnick (1986)

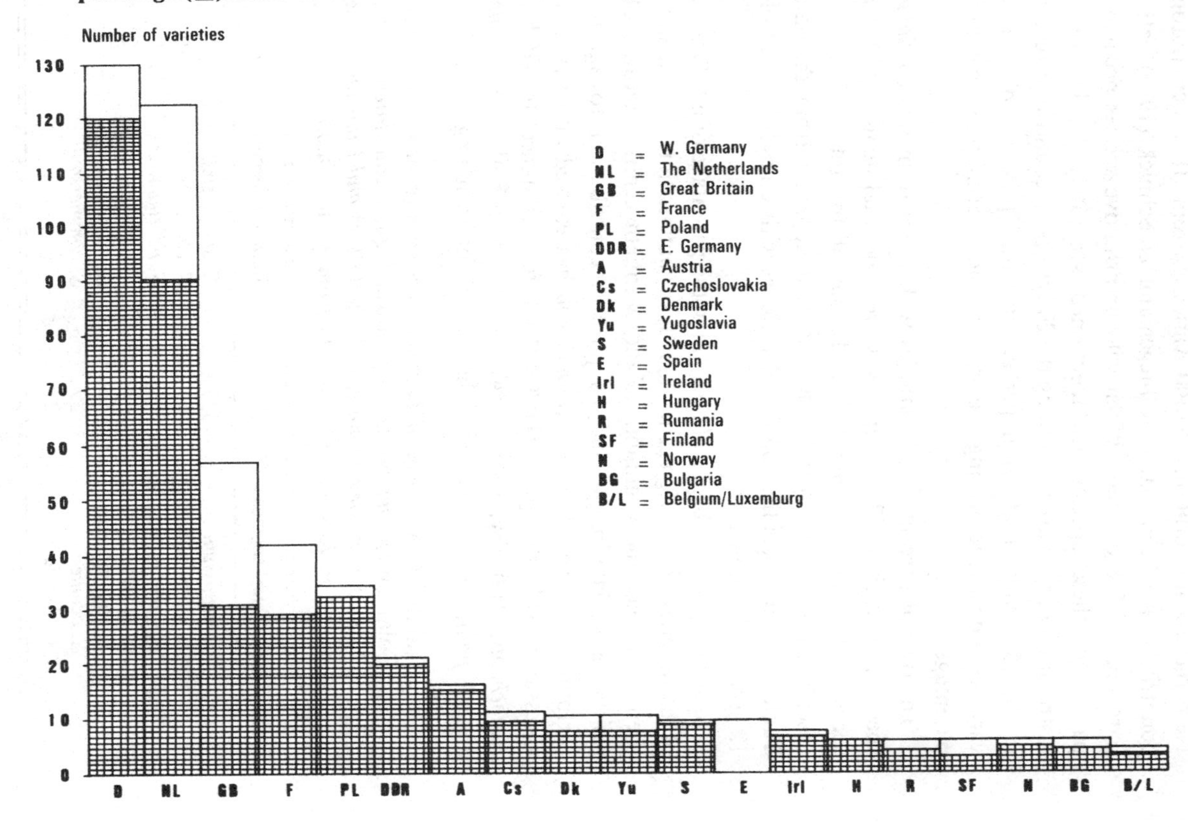

The most important intermediate in Germany has been until recently the Max Planck Institute for Plant Breeding Research at Cologne. These institutes have contributed greatly to the incorporation of genes from wild and primitive species into private breeding programmes, and thus indirectly into potato varieties.

About 300 varieties from the Federal Republic of Germany and The Netherlands, in addition to 300 varieties from 18 other European countries, have been listed by Stegemann & Schnick (1986) with the parentage of 508 of them. When wild or primitive species occur in the parentage, these species are mentioned separately in the list. The distribution of the varieties among the 20 European countries is presented in Fig. 5.2 with subdivision per country into pure subsp. *tuberosum* varieties and varieties having one or more wild or primitive species in their parentage.

The overall percentage of varieties with genes from related *Solanum* species is as high as 77 per cent. However, the total number of *Solanum* species involved is no more than 12 out of the total of about 175 described up to now. Another limitation is apparent from the data in Table 5.3, which lists the number of varieties with genes from each of the 12 species mentioned.

S. demissum and subsp. *andigena* are in the parentage of 60–70 per cent of the varieties with foreign genes, mostly in combination. *S. demissum* has been used since the beginning of the twentieth century because of its resistance to late blight and its positive contribution to agronomic performance of varieties. Subsp. *andigena* has entered into varieties via accessions with the gene H1 (= Ro1) for resistance to *Globodera rostochiensis* (Ellenby, 1948; 1954). Resistance to both *G. rostochiensis* and *G. pallida* pathotypes was the reason for including *S. vernei* and

Table 5.3. *Numbers of European varieties* (n) *with the given* Solanum *species in their parentage. Figures calculated from Table 5.3 in Stegemann & Schnick (1986) and from the Tables 4 and 15 in Ross (1986). No data from Soviet Union.*

Solanum species	*n*	*Solanum* species	*n*
S. demissum	335	*S. spegazzinii*	11
S. tuberosum		*S. chacoense*	10
ssp. andigena	298	*S. microdontum*	2
S. vernei	41	*S.verrucosum*	2
S. stoloniferum	41	*S. sparsipilum*	1
S. acaule	39	*S. commersonii*	1
S. phureja	27		

S. spegazzinii in breeding programmes, mainly in The Netherlands and the Federal Republic of Germany.

Ross (1986) listed 44 varieties deriving their cyst nematode resistance from these three species, either separately or in combination. The species *S. phureja* occurs together with *S. demissum* in the parentage of 25 out of 27 varieties, because the tetraploid F1 of these species is fertile and crosses well with subsp. *tuberosum*. The same holds true for *S. demissum* × *S. chacoense*. *S. stoloniferum* has been used mainly because of its extreme resistance to PVY and PVA. *S. acaule* was of interest mainly because of its resistance to frost and PVX (Ross, 1986).

The North American potato varieties show a similar overall pattern of occurrence of foreign material in their parentage as the European varieties (Chase, 1984; Ross, 1986). Only *S. vernei* and *S. spegazzinii* have hardly been used because there has not yet been a need of resistance to *Globodera pallida* in North America.

Overall, the data presented here suggest a considerable introduction of foreign genes into varieties in developed countries. However, the high percentage of varieties with foreign genes (77 per cent) is only due to the frequent use in breeding programmes of derivatives from *S. demissum* and subsp. *andigena*, of a few complicated species hybrids, and of varieties derived from these materials.

As pointed out by Glendinning (1979), traditional breeding was aimed at adding, by backcrossing procedures, genes required to control existing problems. Unlike the population breeding procedures (see later sections) these backcrossing procedures involve only a limited and incidental expansion of the gene pool.

Breeding at the diploid level, a stimulus to the use of diploid Solanum species

The cultivated *S. tuberosum* is an autotetraploid species. This implies a complicated inheritance pattern and hence the need of large breeding populations, and relatively slow progress in breeding. On the other hand, autotetraploidy allows it to exploit any extra performance gained from the potential for tetra-allelism (four different alleles per locus = maximal heterozygosity) at loci for polygenic traits including yield. So the conclusion is obvious that the best possible procedure for varietal improvement of potato is: breeding at the diploid level followed by intact transfer of selected superior diploid genotypes to the tetraploid level. The idea was already put forward by Chase (1963). The necessary tools are available:

1. Reducing the ploidy level from tetraploid to diploid through pseudogamy is a routine method particularly since superior prickle pollinators in *S. phureja* homozygous for a dominant seed marker are available (Hermsen & Verdenius, 1973).
2. Intact transfer of selected diploid genotypes to the tetraploid level (sexual polyploidisation) has become possible with the discovery of meiotic mutants which cause deviations from normal meiosis and lead to 'unreduced' or rather $2n$ gametes with the same ploidy level and largely the same genotype as the parent plant (Peloquin, 1982; Hermsen, 1984*b*).

Breeding at the diploid level offers a strong stimulus for including the wealth of diploid wild and primitive species directly in the breeding programme. The finding by Peloquin *et al.* (1984) that particular dihaploids from potato varieties are very effective in putting diploid wild species in a nobilised form, is highly important, because it may remove the breeders' reluctance to include wild diploid species as such in their breeding programme and thus promote their use in breeding.

The use of a single species as a source of various traits

The occurrence of various desirable traits in one species has been established for several well-known species, e.g. *S. demissum*, *S. acaule*, *S. chacoense*, *S. vernei*, *S. phureja*, and *S. tuberosum*. It is striking that only a few of the characters known to occur in this kind of species have been used and, if a species is used for different purposes, this has occurred nearly always in separate breeding programmes. The main reason is that desirable traits of a species are scattered among different accessions and even among different plants within an accession. This hampers any efficient use in breeding. An obvious solution would be, on the plant level, to apply combination breeding within a species aiming at superior multi-purpose single genotypes; or, on the population level, to apply population breeding within a species aiming at increasing the frequency of favourable genes and decreasing the frequency of undesirable genes. Probably the best approach would be population breeding followed by combination breeding in the enhanced population. In fact, this is, by and large, the procedure followed in the adaptation programmes with subsp. *S. andigena*, and with *S. phureja* and *S. stenotomum* as discussed in the following chapter.

Although such procedures are too laborious to be applied to many separate species, the approach would probably pay for carefully chosen promising wild species which are known to be genetically close to cultivated species. Peloquin *et al.* (1984) have shown a potential of rapid nobilisation of a number of wild species through crosses with certain diploid or tetraploid spp. *tuberosum* genotypes.

Modifications which could restrict time and labour are (i) taking two or three species which have complementary characteristics instead of one, and (ii) crossing the population of a wild species with advanced diploid subsp. *tuberosum* clones and applying the aforementioned breeding procedures to the F1 progeny.

Adaptation of primitive **Solanum** *species by recurrent selection*

It was Simmonds (1966) who in 1959 initiated an extensive programme of selection for adaptation of the short-day potato *S. tuberosum* subsp. *andigena* to the environment (mainly daylength) of temperate zones. This approach was based on two considerations: (i) the narrow genetic base of potato breeding at that time, and (ii) the rich genetic variation observed in subsp. *andigena*. Behind this approach there was also the scientific question about the immediate ancestry of subsp. *tuberosum*: either subsp. *andigena* or the subsp. *tuberosum*-like Chilean potatoes. If subsp. *andigena* were the immediate predecessor of subsp. *tuberosum*, it should be possible to recreate subsp. *tuberosum* from subsp. *andigena* via systematic recurrent selection and to perform perhaps in decades, what has taken centuries in nature. A practical objective was to obtain varieties more rapidly because the need for backcrossing would be restricted. Furthermore the occurrence of heterotic yields was a realistic expectation. The procedure applied by Simmonds was the introduction of about 300 accessions into the programmes, selection of plants with the best yields of tubers, isolated planting of these tubers next season and harvest of open-pollinated fruits from them. The seeds obtained were the starting material for the next selection cycle. This procedure was repeated several times, the selection criteria being modified as the populations improved. Already during the early cycles half the seedlings were grown in Cornwall, where late blight regularly occurs, and selected survivors were also included in the seed production plots. A similar adaptation programme involving broadly based diploid material of *S. phureja* and *S. stenotomum* was also initiated by Simmonds in the early 1960s and in the mid-1960s by Haynes (USA) and Maris (Netherlands).

Plaisted in 1963 started a programme to produce an adapted source of subsp. *andigena* germplasm in the USA using first cycle seeds from Simmonds after one cycle of selection in England, original subsp. *andigena* seed from South America, and from 1966 onwards, second cycle seed progeny from Simmonds. Finally in 1969 and 1970 he obtained more than 1,000 new accessions of subsp. *andigena* from South America and from the CPC and IR-1 collections.

In the 1960s, Simmonds' procedure was successfully continued leading to further enhancement of late blight resistance and to the discovery of PVY and PVX resistance in the material. However, two important shortcomings of the procedure had become apparent.

1. Fruits from open pollination contain predominantly self seeds (Glendinning, 1976) leading to gradual inbreeding depression which is counterproductive for effective selection (Plaisted, 1986).
2. Selection without regard to lineage seriously reduced the number of original females endangering the main goal of the programme: broadening the genetic base of breeding.

So, with the 1,000 new accessions, controlled pollinations of selected clones, with bulk pollen of these clones, were made for reproducing the cycles. Furthermore, every effort was made to select at least one clone from each of the accessions. Finally the programme was speeded up and improved by bulk pollinations with pollen from the more advanced populations selected in the 1960s from Simmonds' material having also genes for PVY immunity and resistance to late blight.

Several researchers have reported heterotic yields in hybrids between selected subsp. *andigena* material and cultivars of subsp. *tuberosum* (Paxman, 1966; Glendinning, 1969; Tarn & Tai, 1977; Cubillos & Plaisted, 1976; Muñoz & Plaisted, 1981). However, Cubillos & Plaisted (1976) and Muñoz & Plaisted (1981) demonstrated that hybrid advantage in yield was due to a larger set of tubers, the tuber weights being inferior to those of control crosses between subsp. *tuberosum* cultivars and even more so to tuber weight of subsp. *tuberosum* cultivars.

In the meantime the improved subsp. *andigena* or Neotuberosum has entered into private and public potato breeding programmes aimed at producing superior varieties. The first cultivar, ROSA, which is a subsp. *tuberosum* × subsp. *andigena* hybrid has been released in 1981 by Plaisted and his group at Cornell University.

The use of different source species for one trait

If there are arguments or evidence to suggest that the genetic basis for a desired trait is different in various species or groups of species, it is obvious to use these different sources in breeding for that trait. A good example is late blight resistance in Mexican species and in South American species. Both groups of species have evolved in separate gene centres, so both should be used for late blight resistance breeding and finally combined. Another example is resistance to the ubiquitous leaf roll virus. In potato cultivars, polygenic resistance to infection is known to occur. Recently a high level of resistance to multiplication of leaf roll virus was discovered by CIP researchers in the wild species *S. brevidens*, *S. etuberosum* and *S. acaule* (Jones, 1979; Brown, 1984). We might be dealing here with two components of resistance to leaf roll which, when combined, might provide an effective protection against that virus.

Special situation in developing countries and the breeding approach at the International Potato Center

According to Mendoza (1986) the potato plant may be affected by up to 266 diseases and pests. Some, such as late blight, PLRV and PVYO are distributed worldwide, while others are mainly restricted to tropical zones (bacterial wilt, root knot nematode, tuber moth) or localised within certain geographic areas. Mendoza (1986) also points to the frequent occurrence of climatic stresses and soil nutrient deficiencies and toxicities in developing countries.

Most developing countries have depended, and still depend, upon certified seed of imported varieties bred in the temperate zones (Europe, N. America), because they had no national breeding programmes to develop cultivars adapted to their specific agricultural conditions. In the meantime CIP, according to its mandate, has developed a breeding strategy aimed at meeting the specific requirements of developing countries. This strategy is population improvement based on phenotypic recurrent selection complemented in later stages by progeny testing. It was developed and recently described and discussed by Mendoza (1986), head of the breeding and Genetics Department of CIP.

In the starting material from the Potato Collection, the major source of CIP's breeding programme, the frequency of desirable genes is extremely low and occurs only in small parts of it. The results summarised by Mendoza (1986) clearly demonstrate that the population breeding strategy has led to a dramatic increase of the frequency of desirable genes. The case history on Neotuberosum (earlier section) describes analogous results.

Besides increasing the frequency of genes controlling yield, resistance or tolerance to biotic stresses and adaptation to heat, cold, daylength, etc, the breeding strategy maintains a wide genetic diversity with ample opportunity for germplasm prescreening in the regional distribution centres founded by CIP and for the selection of locally adapted outstanding varieties by national programmes. The first varieties from CIP material have already been released in a number of developing countries. From this and the preceding section it will be clear that owing to CIP's breeding activities, which are largely based on the use of the potato collection, and to the distribution of enhanced populations to regional centres and national programmes, the proportion of local varieties with genes from wild species and primitive forms will rapidly increase.

Factors limiting the use of collections and ways to remove them

Limiting factors related to the composition of collections

The large differences in the number of accessions available per wild species, have been mentioned earlier. Attention has also been given to the occurrence of duplicates of accessions of wild species and of primitive cultivars.

In the context of this chapter it should be emphasised that much more effort should be put into collecting genetically broader samples of those potentially important wild species, of which only one or few accessions are available. The geographically unbalanced composition of CIP's collection of primitive cultivars (87 per cent from Peru) need not be a limiting factor, because of the tremendous concentration of variation in CIP's 'home country'. However, a more extensive exploration in neighbouring countries, particularly Bolivia, might still reveal some new cultivars not detected before.

Limiting factors in maintenance of collections

Many accessions, especially in the early days of collecting, have been lost due to maintenance problems, like virus diseases with tuber maintenance and sterility, or poor flowering with true seed increase, which leads either to loss of the accession or to genetic drift, because too few plants contribute to the progeny. Vegetative maintenance by tissue culture methods would as such be ideal but also has limitations.

First it requires a tremendous initial investment of time and efforts, because:

1. All clones (e.g. 5,000 at CIP) have to be (made) free from viruses;
2. All clones have to be tested for presence of seed-borne diseases and infected clones removed; and
3. All healthy clones have to be put into *in vitro* culture.

Second maintenance *in vitro* of 5,000 clones which at regular intervals have to be rejuvenated, requires an enormous investment of time and facilities for maintenance only. This is not feasible unless the rejuvenation intervals can be extended considerably via methodological improvements (Schilde, 1979).

Actually CIP is bound to the vegetative maintenance of its collection, because the extensive evaluation data refer to the clones, which would be lost in any switch to true seed maintenance.

All other collections use predominantly true seed preservation. The rejuvenation period of true seeds is presently about 25 years. True seed propagation is laborious and genetic drift may occur. This sets a limit to the number of accessions that can be handled adequately in one germplasm centre. There are two ways to increase this limit. First, by basic research aimed at improving long-term storage of true seeds, which may reduce the rejuvenation frequency, and secondly, by the identification and elimination of duplicate accessions within and between germplasm centres.

Factors limiting accessibility and distribution of **Solanum** *material and evaluation data*

In principle all collected materials and data are available to all scientists. However, there are quite a few limitations. The pivotal problem of accessibility is the lack of acquaintance with material and data. In this respect inventories, although inherently inadequate, are very useful. An *Index Seminum* comprises only passport data, but no evaluation data (CPC, GLKS, INTA collections). Inventories containing also evaluation data (BGRC, IR-1) are useful, but have a number of limitations. The BGRC inventory is nearly up to date (1984), has been evaluated systematically and almost completely, but for few characters only. The IR-1 inventory, though outdated (1969) comprises evaluation data for many resistances; especially those for ring rot, bacterial wilt, golden nematode, and aphids are nearly complete for some species. However, more than 60 per cent of the present accessions (those obtained after 1969) have not yet been listed. An updated inventory is in preparation.

At CIP, large-scale evaluation has been carried out and a wealth of data, mainly of primitive cultivars, are stored. However, they are largely unknown to potato breeders and scientists, because they have not yet been listed. Inventories of germplasm collections which are rapidly expanding and being evaluated extensively, soon get outdated. Preparing up-to-date printed inventories is very laborious. One may wonder if there is no alternative way of providing information. The large collections have the facilities for computer-aided storage and retrieval of data, which can be printed and communicated in nearly any form required (BGRC, CIP, IR-1). The possibilities of this system should be better known to potential users. They ought to know how to formulate their requests for information or material; in other words, which kinds of request can be met and which cannot.

There is another limitation to the use of genetic resources in potato improvement. The exchange of material between collections and the dispatch of material to users is greatly hampered for reasons of health of tubers (tuber-borne diseases) and true seeds (seed-borne diseases, like PSTV and a few viruses) from the collections. In particular, the viroid PSTV has severely hampered the distribution of material, because screening for PSTV infection is troublesome and freedom of a true seed collection from PSTV can only be warranted if it has been rejuvenated completely from healthy parent plants (Glendinning, 1976). PSTV has also been a limiting factor for progress in some breeding programmes involving *Solanum* species, because all material had to be screened, and infected material to be eliminated.

Perspectives of the use of *Solanum* germplasm in breeding

The use of *Solanum* germplasm in potato breeding will steadily increase both in developed and developing countries. The wealth of variation among and within species is getting better known to breeders. Several wild species have been found to be genetically close to cultivated species. This has become apparent from easy crossability, rapid nobilisation after initial crosses and unexpected extra contributions of wild species to agronomic performance. Several recalcitrant, but valuable, species have been made accessible and put into a cultivated background. The large-scale adaptation programmes involving subsp. *andigena* (Neotuberosum) and *S. phureja* and *S. stenotomum* have been included into practical breeding in developed countries and into CIP's breeding programme. These adaptation programmes will contribute greatly to the broadening of the genetic base of potato breeding, much more so than traditional breeding aimed at introducing, by back-crossing procedures, new specific characters from *Solanum* species into existing varieties.

The breeders in the developed countries and at CIP have become acquainted with some new tools in breeding (haploidisation and sexual polyploidisation) and are becoming more familiar with breeding at the diploid level. As explained before, this offers excellent opportunities for directly including diploid *Solanum* species and catching their treasures.

The active involvement of CIP in the collection, evaluation and use of wild species, and even more so of primitive forms, will lead to the introduction of a broadly based *Solanum* gene pool into developing countries.

References

Abdalla, M. M. F. & Hermsen, J. G. Th. (1972). Plasmons and male sterility types in *Solanum verrucosum* and its interspecific hybrid derivatives. *Euphytica*, **1**, 209–20.

Anon. (1969). *Commonwealth Potato Collection*. Inventory of seed stocks 1968. Scottish Plant Breeding Station, Pentlandfield.

Austin, S., Baer, U. A. & Helgeson, J. P. (1985). Transfer of resistance to potato leaf roll virus from *Solanum brevidens* into *Solanum tuberosum* by somatic fusion. *Plant Science*, **39**, 75–82.

Brown, C. R. (1984). Genetic studies and breeding for resistance to PLRV, PVY and PVX. In rept. 26th Planning Conf. CIP on *Present and Future Strategies for Potato Breeding and Improvement*, pp. 17–44. International Potato Center, Lima, Peru.

Bukasov, S. M. (1933). (The potatoes of South America and their breeding possibilities). Suppl. 58, Bulletin of Applied Botany, Genetics and Plant Breeding. Leningrad Academy of Agricultural Sciences, Leningrad (Russian, with English summary).

Chase, R. W. (1984). *North America Potato Variety Inventory*. Certification Section PAA.

Chase S. S. (1963). Analytic breeding in *Solanum tuberosum* L. A scheme utilizing parthenotes and other diploid stocks. *Canadian Journal of Genetics and Cytology*, **5**, 359–63.

Correll, D. S. (1962). *The Potato and Its Wild Relatives*. Texas Research Foundation Botanical Studies 4. Texas Research Foundation, Renner.

Cubillos, A. G. & Plaisted, R. L. (1976). Heterosis for yield in hybrids between *S. tuberosum* ssp. *tuberosum* and *S. tuberosum* ssp. *andigena*. *American Potato Journal*, **53**, 143–50.

Ellenby, C. (1948). Resistance to the potato root eelworm. *Nature*, **162**, 704.

Ellenby, C. (1954). Tuber forming species and varieties of the genus *Solanum* tested for resistance to the potato root eelworm *Heterodera rostochiensis* Woll. *Euphytica*, **3**, 195–202.

Glendinning, D. R. (1969). The performance of progenies obtained by crossing Groups Andigena and Tuberosum of *Solanum tuberosum*. *European Potato Journal*, **12**, 13–19.

Glendinning D. R. (1976). Neo-Tuberosum: new potato breeding material. 4. The breeding system of Neo-Tuberosum, and the structure and composition of the Neo-Tuberosum gene pool. *Potato Research*, **19**, 27–36.

Glendinning, D. R. (1979). The potato gene pool, and benefits deriving from its supplementation. In *Broadening the Genetic Base of Crops*, pp. 187–94, Zeven, A. C. & Van Harten, A. M. (eds.), PUDOC, Wageningen.

Grun, P., Aubertin, M. & Radlow, A. (1962). Multiple differentiation of plasmons of diploid species of *Solanum*. *Genetics*, **47**, 1321–33.

Hanneman, R. E. (1976). The Inter-regional Potato Introduction Project (IR-1). *Exploration and Maintenance of Germplasm Resources*, pp. 85–7, International Potato Center, Lima, Peru.

Hawkes, J. G. (1958). Taxonomy, cytology and crossability. In *Handbuch der Pflanzenzüchtung III Züchtung der Knollen- und Wurzelfruchtarten*, 2nd edn, pp. 1–43, Kappert, H. & Rudorf, W. (ed.), Paul Parey, Berlin.

Hawkes, J. G. (1963). A revision of the tuber-bearing Solanums (second edition). *Report of the Scottish Plant Breeding Station*, 76–181.

Hawkes, J. G. (1978). Biosystematics of the potato. In *The Potato Crop: the Scientific Basis for Improvement*, pp. 1–14, Harris, P. M. (ed.), Chapman & Hall, London.

Hawkes, J. G. (1979). Genetic poverty of the potato in Europe. In *Broadening the Genetic Base of Crops*, pp. 19–27, Zeven, A. C. & van Harten, A. M. (eds.), PUDOC, Wageningen.
Hawkes, J. G. & Hjerting, J. P. (1969). *The Potatoes of Argentina, Brazil, Paraguay and Uruguay – A biosystematic study*. University Press, Oxford.
Hermsen, J. G. Th. (1984*a*). Utilization of wide crosses in potato breeding. In *Present and Future strategies for Potato Breeding and Improvement*, pp. 115–32, International Potato Center, Lima, Peru.
Hermsen, J. G. Th. (1984*b*). Mechanisms and genetic implications of 2n-gamete formation. *Iowa State Journal of Research*, **58**, 421–34.
Hermsen, J. G. Th. (1986). Efficient utilization of wild and primitive species in potato breeding. In *The Production of New Potato Varieties*: *Technological Advances*, Jellis, G. J. & Richardson, D. E. (eds.), Cambridge University Press, Cambridge.
Hermsen, J. G. Th. & Verdenius, J. (1973). Selection from *Solanum tuberosum* Group Phureja of genotypes combining high frequency haploid induction with homozygosity for embryo-spot. *Euphytica*, **22**, 244–59.
Hogenboom, N. G. (1973). A model for incongruity in intimate partner relationships. *Eurphytica*, **22**, 219–33.
International Potato Center (1984). *Potatoes for the Developing World*. Lima, Peru.
Jones, R. A. C. (1979). Resistance to potato leafroll virus in *Solanum brevidens*. *Potato Research*, **2**, 149–52.
Mendoza, H. A. (1986). Advances in population breeding and its potential impact on the efficiency of breeding potatoes for developing countries. In *The Production of New Potato Varieties*: *Technological Advances*, Jellis, G. J. & Richardson, D. E. (eds.), Cambridge University Press, Cambridge.
Muñoz, F. J. & Plaisted, R. L. (1981). Yield and combining abilities in andigena potatoes after six cycles of recurrent phenotypic selection for adaptation to long day conditions. *American Potato Journal*, **58**, 469–79.
Ochoa, C. M. (1962). *Los Solanum tuberiferos silvestres del Peru* (*Secc. Tuberarium, sub-secc. Hyperbasarthrum*). International Potato Center, Lima, Peru.
Ochoa, C. M. (1976). Review of progress in explorations 1973–1975; cultivated potatoes. In *Exploration and Maintenance of Germplasm Resources*, pp. 19–26, International Potato Center, Lima, Peru.
Okada, K. A. (1976). Argentine collection. In *Exploration and Maintenance of Germplasm Resources*, pp. 99–108, International Potato Center, Lima, Peru.
Paxman, G. J. (1966). Heterosis in Andigena hybrids. *Report of the John Innes Institute for 1965*, 51–3.
Peloquin, S. J. (1982). Meiotic mutants in potato breeding. In *Stadler Genetics Symposium 14*, Redei, G. W. (ed.), University of Missouri, Columbia.
Peloquin, S. J., Hermunstad, S. A., Okwuagwu, C. O. & Chujoy, J. E. (1984). Haploid-species hybrids in germplasm transfer and breeding. In *Abstracts of Conference Papers of the 9th Triennial Conf. of the European Association for Potato Research*, p. 54, Winiger, F. A. & Stöckli, A. (eds.), Juris Druck, Zürich.
Plaisted, R. L. (1986). Advances and limitations in the utilization of Neotuberosum in potato breeding. In *The Production of New Potato Varieties*: *Technical Advances*, Jellis, G. J. & Richardson, D. E. (eds.), Cambridge University Press, Cambridge.
Ramanna, M. S. & Hermsen, J. G. Th. (1974). Unilateral 'eclipse sterility' in reciprocal crosses between *Solanum verrucosum* Schlechtd and diploid *S. tuberosum L. Euphytica*, **23**, 417–21.

Richardson, D. G. & Weiser, C. J. (1972). Foliage frost resistance in tuber-bearing *Solanum*. *HortScience*, **7**, 19–22.
Ross, H. (1986). Potato Breeding – Problems and Perspectives. Advances in Plant Breeding, Suppl. 13 to *Zeitschrift Pflanzenzüchtung*, Paul Parey, Berlin.
Ross, R. W. & Rowe, P. R. (1969). *Inventory of Tuber-bearing Solanum Species*. Bull. 533 Wisconsin Agricultural Experiment Station, Madison.
Rothacker, D. (1966). *Sortiment Wilder und Kultivierter Kartoffelspecies des Instituts für Pflanzenzüchtung Gross-Lüsewitz* (G-LKS). Teil 1, Deutsche Akademie der Landwirtschaftswissenschaften, Berlin.
Schilde, L. (1979). *In vitro* maintenance of valuable *Solanum* resources at CIP. In *Exploration, Taxonomy and Maintenance of Potato Germplasm*, pp. 182–9, International Potato Center, Lima, Peru.
Simmonds, N. W. (1966). Studies of the tetraploid potatoes. III. Progress in the experimental re-creation of the Tuberosum Group. *Journal of the Linnean Society* (*Botany*), **59**, 279–88.
Stegemann, H. & Schnick, D. (1985). *Index 1985 Europäischer Kartoffelsorten*. Mitt. aus der Biol. Bundesanstalt für Land- und Forstwirtsch. Berlin-Dahlem 227, Paul Parey, Berlin.
Tarn, T. R. & Tai, G. C. C. (1977). Heterosis and variation of yield components in F1 hybrids between Group Tuberosum and Group Andigena potatoes. *Crop Science*, **17**, 517–21.
Van Soest, L. J. M., Dambroth, M. & Lamberts, H. (1984). The German–Netherlands potato collection. *Plant Genetic Resources Newsletter*, **58**, 11–15.
Van Swaaij, A. C. (1986). *Frost Tolerance in Potato. Use of hydroxyproline Resistant Cell Cultures*. PhD Thesis, University of Groningen, Department of Genetics.

6

Use of collections in cereal improvement in semi-arid areas

J. P. SRIVASTAVA AND A. B. DAMANIA

Within the system of the International Agricultural Research Centres (IARCs), the International Center for Agricultural Research in the Dry Areas (ICARDA) has a global responsibility for barley improvement and a joint mandate with the Centro International de Mejoramiento de Maiz y Trigo (CIMMYT) for wheat improvement in West Asia and North Africa. In ICARDA's region, about 97 per cent of durum wheat, 65 per cent of bread wheat and essentially all of barley is grown without irrigation under rainfed conditions. In co-operation with national and international institutions, the Cereal Improvement Programme at ICARDA endeavours to increase the productivity, as well as the stability of production, of barley and wheat in rainfed areas.

Common strategies of the three breeding projects, i.e. barley, durum wheat and bread wheat, are targeted crosses and multi-location testing as well as the selection of early segregating populations as a means to cope with the erratic and unpredictable climatic conditions typical of the dry areas. A modified bulk method is being used in the selection of early segregating populations. Additional strategies involve the use of landraces, unimproved locally adapted varieties and the wild relatives.

The region has diverse and variable climatic conditions. Rainfall and temperature regimes can differ considerably over short distances, as well as being quite different among years. Broadly speaking, there are two major agroclimatic zones in which cereals are grown. One zone has a Mediterranean climate characterized by hot, dry summers and cool, but not frigid, winters in which rainfall is concentrated. Crops are grown mostly during the winter. The second zone consists of various highland areas over 1,000 metres and areas of continental climate where winters are cold and summers are hot. Crops are usually planted in the autumn, though spring plantings are important in parts of Turkey, Afghanistan and Pakistan. Precipitation throughout the region is low. Most of the arable land gets only 200 to 600 mm of annual rainfall.

Eight of the countries in the ICARDA region, Afghanistan, Algeria, Iran, Iraq, Morocco, Pakistan, Turkey and the Yemen Arab Republic, have substantial crop growing land masses over 1,000 metres. Except in Turkey, little research has been done to improve cereal productivity in these elevated regions. Traditional methods of cultivation using landraces or unimproved local cultivars are common. Winter habit wheat and barley varieties are frequently grown to survive the cold seasons.

Barley, durum wheat and bread wheat together cover 70 per cent of the land devoted to food crops. Within the zone of 300 to 600 mm of annual precipitation, farmers in wetter areas (and farmers having irrigation) favour bread wheats, while farmers with less rainfall tend to grow durums. In more arid zones, 200 to 300 mm of rain, farmers usually grow barley. The unpredictable rainfall and hot, dry summers typical of the region have important implications for agricultural researchers. One of these, obviously, is that drought tolerance is a high priority need for farmers. The ability to survive during dry periods and to grow and ripen rapidly when rainfall resumes is essential for the success of cereal varieties.

Germplasm collections

The landraces form the backbone of crop genetic improvement. They are genetically diverse populations which are well adapted to their surroundings and pathogens as a result of which certain genotypes in the population may have advantages in one season whereas others do better in other seasons.

Populations of landraces collected in recent years from various countries within the primary, as well as secondary, centres of diversity have not been fully evaluated and the variability displayed for disease resistance and agronomic traits has yet to be fully described. Studies are needed on the evolutionary pathways, the distribution of characters and their frequencies, as well as associations among characters in different areas (Damania, Jackson & Porceddu, 1985) to allow better use of collected germplasm for the benefit of crop improvement programmes.

Many regions of the world have succeeded in improving crop production through the introduction of high yielding varieties. However, these varieties have not met with great success in the ICARDA region. In order to improve and stabilise cereal production in the semi-arid areas, characters such as earliness as well as tolerance to drought, heat and cold, low plant nutrients, diseases, etc need to be incorporated into the high yielding varieties. The genes for these characters can probably be found in the local landraces and wild relatives which are well adapted to such environments.

The global depletion of genetic diversity in crop plant species of economic importance has focussed attention on the nature and extent of genetic variability in the wild relatives of these species. Over the past decade or two the utilisation of wild germplasm of crop species in cereal breeding programmes has increased. However, information on genetic variability in the wild species and their centres of distribution is, on the whole, scanty.

The Genetic Resources Unit together with the Cereal Improvement Program at ICARDA has assembled a collection of more than 37,000 accessions of cereal landraces as well as wild germplasm of barley such as *Hordeum spontaneum*, and of wheat such as some *Triticum dicoccum* and *Aegilops* spp. The details are given in Table 6.1.

One of the major objectives of the Cereal Improvement Program at ICARDA is to provide the plant breeders and scientists with superior wheat and barley germplasm for utilisation in the national programmes. This is accomplished, to a great extent, through a system of international

Table 6.1. *Status and documentation of cereal genetic resources collection at ICARDA*

Crop	Total number of accessions in storage	Number of accessions evaluated and documented	Number of passport descriptors documented
Cultivated species			
Barley	15,942	12,138	22
Durum wheat	20,085	18,569	25
Bread wheat	8,219	1,540	10
Wild species			
H. bulbosum	205	—	—
H. spontaneum	1,308	1,308	4
Aegilops spp.	679	662	6
T. dicoccoides	919	855	19
Totals	47,357	35,072	—

Crop	*Major source*
Barley:	USDA; Germplasm Institute, Bari, Italy; ICARDA collections
Durum wheat:	USDA; Germplasm Institute, Bari, Italy; ICARDA collections
Bread wheat:	ICARDA Breeding lines; ICARDA collections
Wild spp.:	ICARDA collections; other sources

nurseries which operate not only in the countries of the region but are also distributed widely to other areas of the world on request. Essentially three diverse types of germplasm targeted for specific environments are made available through this system:

1. Regional crossing blocks: These are parental genotypes which scientists can use in their own crossing projects and are designed to complement and broaden the genetic base of the national improvement programmes. They are selected from germplasm collections following evaluation in response to the current needs of the national programmes. Entries in the crossing blocks are grouped by plant traits such as early maturity, tolerance to various stresses and diseases, high grain yield, good grain quality, and other characters of economic importance.
2. Segregating populations: These are F_2 populations, derived from crosses made at ICARDA, utilising selections from evaluated collections. National scientists use them as pools of genetic diversity and select promising genotypes adapted to their local environments.
3. Preliminary observation nurseries and Regional yield trials: These nurseries include a more finished product in the form of ICARDA's most promising wheat and barley lines. The best entries from the Preliminary observation nurseries are promoted to the regional yield trials for more rigorous testing and evaluation. Scientists use the lines in the national crossing programmes for specific breeding objectives or further select outstanding lines from these as candidates for local testing and possible release as varieties.

A network of testing sites utilised by ICARDA for screening local varieties as well as other germplasm maintained at the Genetic Resources Unit is fast developing in many mandated countries in the region. So far, evaluation and utilisation of germplasm had lagged behind the collecting and conservation activities. In practical terms, collections are only as good as the use which can be made of them. They must be evaluated for desired characters and documented in a manner to make it easy for breeders to use them in their improvement programmes.

Barley

Barley, *Hordeum vulgare* L., is one of the most dependable cereal crops in harsh environments. It is grown in semi-arid areas as well as in the cold, short-season ones. Local varieties and landraces of barleys occupy nearly 80 per cent of the cultivated areas in the ICARDA region. In many countries the entire barley growing zone is still under landraces and indigenous varieties (Table 6.2) and yields appear to have remained static for centuries (Srivastava, 1977). The present barley germplasm collection at ICARDA numbers 15,195 accessions. These accessions come from 51 different countries, however the major sources are as follows: Colombia, Cyprus, Ethiopia, Federal Republic of Germany, Morocco, South Africa, Syria, Tunisia, Turkey, USA, USSR and Yugoslavia (ICARDA, 1984). More collections are needed from countries in the Middle East and North Africa where a considerable amount of diversity can still be found.

The IBPGR had partially funded a project for the evaluation of barley germplasm at ICARDA. The results and data from the first 5,000 accessions has just been published in the form of a catalogue (Somaroo, Adham & Mekni, 1986). There was significant variability among the landraces in the collection for such characters as days to heading, plant height, 1000-grain weight, protein/lysine ratio and resistance to diseases.

The Cereal Improvement Program at ICARDA collected barley landraces from 33 different locations, mainly in the drier regions of Syria and Jordan during early summer of 1981 (Weltzien, 1982*b*). On subsequent evaluation it was found that some samples had a definite cold requirement). All such samples were from Syria indicating that under local conditions this trait may contribute to yield stability by preventing winter kill of early planted material. Information of this kind is valuable for a crop improvement programme, not only by showing which characteristics contribute to adaptation but also by indicating the degree of plasticity present in the landraces of the region for which varietal improvement is undertaken. These landraces were multiplied during 1981–2 at ICARDA and further tested for salt and drought tolerance. Eighty lines were identified at performing well at a saline and drought affected site near

Table 6.2. *Type of barley under cultivation (% of area) in regions of the world*

Region	Indigenous varieties or landraces	Improved cultivars
Middle East	80	20
South and Far East	80	20
North Africa	80	20
East Africa	90	10
Mediterranean Europe	30	70

Aleppo (Weltzien, pers. comm.) and have been distributed widely to breeders requesting lines with these traits.

Landraces of barley have good adaptation to constraints of drier areas and harbour a largely untapped reservoir of useful genes for adaptation to the arid and semi-arid regions of the world. During 1984–5 about 1,500 accessions of landraces from the collection were tested at Bouider, a very low rainfall site (i.e. less than 200 mm), for drought tolerance. The evaluation of landraces was further emphasised in 1985 with the main objective of exploring the usefulness of locally adapted germplasm to improve barley yields under harsh conditions. Barley improvement projects for dry areas are based both on conventional germplasm, that is, breeders modern lines, as well as on less conventional germplasm, the landraces and wild relatives. However, barley breeding for low rainfall areas (less than 250 mm per year) has proven to be difficult and challenging (Ceccarelli, pers. comm.).

Scientists at ICARDA postulate that the identification and assessment of the relative importance of morphological and physical traits present in naturally adapted genotypes may help the efficiency of selection for barley improvement (E. Acevedo, pers. comm.). The linkage of selected morphological and physiological traits to dry matter production, crop performance under drought, and cold, heat or salinity tolerance, will provide barley breeders with additional selection criteria for parents to be used in crosses for stress environments.

An example of the successful utilisation of the genetic variability available in landraces is offered by 'Tadmor', a pure line selected from a barley germplasm sample collected at Zaibe (near Palmyra) in Syria. This

Table 6.3. *Mean grain yield (kg/ha) in on-farm trials at six locations with less than 250 mm rainfall during 1985–6 in Syria*

	Means from six locations	
Variety	Yield	Rank
A. Black	1768	5
ACSAD 179	1802	2
Furat 171	1793	3
WT 2291	1791	4
Tadmor	1920	1
ACSAD 68	1696	7
A. White	1754	6
Furat 1911	1675	8
Doc-7	1333	9
Jlo-95	1189	10
L.S.D. 0.05	142	

line outyielded both the two local cultivars 'Arabic white' and 'Arabic black', especially under dry conditions (Table 6.3). 'Tadmor' is also characterised by a protein content of 10.7 per cent and a lysine content of 0.43 per cent. Another sample collected at Ain El Arab in Syria was also found to outyield significantly both local cultivars under dry conditions.

Landraces have also been used in the barley breeding programme at the national level. For example, two barley cultivars recently released in Turkey are 'Obruk 86' and 'Anadolu 86'. They are selections from crosses between Anatolian landraces collected earlier by the national programme and improved Turkish cultivars. The first cultivar is an excellent malting barley with a high kernel weight and winter hardiness. Also, its exceptionally long (approx. 8 cm) coleoptile allows the plant to regenerate even after winter kill above the soil level. The second cultivar is a feeding barley with a high protein content (13.5 per cent) and some tolerance to salinity.

In a few instances variability has been exploited in a simple mass selection exercise. For example in Iraq, lines selected from a landrace population increased grain yield by 88 per cent over unselected landraces under low rainfall conditions (ICARDA 1983).

As the wild progenitor of barley, *Hordeum spontaneum*, was probably domesticated within the ICARDA region, the centre is in a unique position to assess its potential for improvement of barley yields in the dry areas. In recent years the barley project has begun evaluating *H. spontaneum* collected from the region where it can be found growing under diverse ecological conditions. Because it also occurs in very harsh environments it is considered to be a probable source of resistance to drought.

From past observations we are aware that *H. spontaneum* can contribute genes towards powdery mildew resistance (Fischbeck *et al.*, 1976) and probably other diseases such as rust and scald. It can also provide useful genes for improving grain yield, harvest index, earliness, root development, drought and cold tolerance, etc. Collections made during 1981–2 seem to indicate that local barley landraces possess a high component of *H. spontaneum* as a result of intermediate forms gaining ground after, *H. vulgare* × *H. spontaneum* gene exchange which occurs in nature (Weltzien, 1982*a*).

It is also proposed to evaluate the wild barley accessions collected by Witcombe (1978), Jana (1982), Witcombe, Bourgois & Rifie (1982) and Jaradat *et al.* (1986) for agronomic and morphological characters as well as reactions to diseases. The utilisation strategies of *H. spontaneum* at ICARDA are as follows: a) screen for tolerance to drought; b) evaluate the extent of diversity within the species for agronomic traits; c) select a number of accessions to initiate a crossing project with cultivated barley; and d) evaluate a number of early generation families between the wild progenitor and cultivated barley plants with non-brittle rachis.

Evaluations carried out so far have revealed considerable variability in the collections for growth habit, cold resistance and days to heading. Accessions with a relatively good potential for crosses with improved cultivated barleys were identified (Ceccarelli *et al.*, 1985). Crosses will be made with selected cultivated lines (used as maternal parents) adapted to arid environments. Early generations of the hybrid material will be grown at several sites representing the actual barley growing environments.

Durum wheat

Durum wheat *Triticum durum* Desf., is the most important food crop in West Asia. It is not only extensively grown, but also constitutes a major part of the region's food imports. Despite its importance in the human diet, the progress made to improve the yield and nutritional qualities of this crop in West Asia is slow. This is mainly because, historically, durum wheat has not received sufficient and sustained attention by the plant breeders in the region. The need for improving the crop and widening its adaptation to a broad range of agroclimatic conditions that prevail in the region cannot be over-emphasised.

West Asia is known to be an important centre of diversity for durum wheat. Cultivated landrace populations of this species, as well as natural populations of its ancestors, constitute a major reservoir of genetic variability (Jana, Srivastava & Gautam, 1983). ICARDA's proximity to the centre of diversity is favourable for collection, evaluation and utilisation of available genetic resources of durum wheat. A major objective of ICARDA's cereal research is to develop a comprehensive germplasm bank for the present and long-term use in durum wheat breeding. Consistent with this objective the durum wheat project has assembled a collection of 19,438 accessions from all the durum growing areas of the world. A systematic evaluation of these accessions for adaptation to adverse conditions, such as drought and soil salinity, is an integral part of ICARDA's work. As in barley, local durum wheat varieties and landraces such as 'Haurani' predominate in most countries of the region, particularly in the moderate to low rainfall conditions. The proportion of the areas under local varieties, as well as improved cultivars for durum wheat, is given in Table 6.4.

Attempts to introduce high yielding semi-dwarf cultivars directly in Jordan, to increase yields by a single stroke, have met with little success due to poor adaptation and grain shrivelling as a result of a very short wet period. Recently attempts have been made by the national programme in that country to develop well-adapted high yielding cultivars utilising local landraces. Duwayri, Tell & Shqaidef (1985) crossed 'Stork', a semi-dwarf high yielding cultivar under optimum conditions, with 'Haurani', the local well-adapted landrace in Jordan and Syria which produces reasonable yields under stress conditions. A number of lines resulting from such crosses appear promising in low as well as moderate rainfall zones.

In an experiment at ICARDA, 144 accessions of durum wheat comprising landraces, crosses between local cultivars and improved varieties, and improved varieties were planted at two rainfed sites in Syria. The preliminary results are interesting. At Tel Hadya site (316 mm rainfall) a cross between a *T. turgidum* landrace and 'Anhinga', an improved variety from Mexico, gave the highest yield. The second highest yielder was a landrace collected from low rainfall area in Jordan. At Breda site (207 mm rainfall) the highest yielding line was another landrace collected from a similar area in Jordan. This indicates that there is considerable potential for utilisation of primitive forms and landraces in crosses to improve yields in low as well as moderate rainfall areas in the region.

Table 6.4. *Percentage distribution of* durum *wheat varieties in the countries of the ICARDA region*

Country	Indigenous varieties or landraces (%)	Improved cultivars (%)	New high yielding Varieties (%)
Afghanistan	90	0	10
Algeria	85	0	15
Cyprus	0	90	10
Ethiopia	100	0	0
Iran	100	0	0
Iraq	25	65	20
Jordan	20	80	0
Lebanon	30	65	5
Libya	0	55	45
Morocco	50	50	0
Saudi Arabia	100	0	0
Syria	64	29	7
Tunisia	0	65	35
Turkey	10	40	50
Pakistan	88	0	12

In addition to landraces and wild relatives, primitive forms in present collections can also provide valuable genes for disease resistance, protein content, tillering, drought tolerance and other desirable attributes. For example, *Triticum polonicum* accessions were crossed with 'Crane', a durum cultivar with high fertiliser input requirements, to produce 'Sebou' which was especially adapted to harsher environments where limited fertilisers are used. 'Sebou' out-performed 'Haurani' over three years of testing at on-farm trails in dry areas, demonstrating the favourable attributes of both parents in the cross (Nachit & Ketata, 1986). Similarly, a new durum variety 'Sahl' was released in Algeria in 1985 which came from a cross where one of the parents was *T. dicoccum*.

Salinisation of irrigated land is a widespread problem in West Asia due to the progressive accumulation of salts. Landraces of durum wheat have been cultivated on large tracts of these lands for a long time. Thus, it is reasonable to expect that genetic adaptation for salinity tolerance might be detectable in landrace populations from the region. A saline drought-affected field nursery site as Hegla (average rainfall less than 200 mm) in North-west Syria, is being used for screening landraces for salt tolerance. Srivastava & Jana (1983) first used the site to evaluate durum wheat lines from diverse origins. Later Jana *et al.* (1983) further tested 3,000 durum wheat accessions from various countries and ten lines were found to be highly tolerant to combined stresses of salinity and drought. So far, more than 80 salt-tolerant lines have been identified (Weltzien & Winslow, 1984) and distributed to breeders.

The entire durum wheat collection at ICARDA is being annually evaluated for 25 characters (Table 6.5) in collaboration with the University of Tuscia at Viterbo, Italy. This preliminary evaluation not only considers the current selection criteria of the breeders but may also provide information suited to their future breeding goals. The objectives of this project are: a) to document and disseminate information on these genetic resources so that they may be useful to breeders, b) to multiply the germplasm for distribution; and c) to test selected germplasm in co-operation with national programmes in other countries to confirm results. The aim is to produce a comprehensive catalogue of passport information and evaluation data with relevant statistical analysis.

Further evaluation of selected germplasm from the above collection at Swift Current, Saskatchewan, Canada, has shown that a large number of durum lines can be systematically screened for excised-leaf water retention capability, a trait which appears to be yield positive for certain patterns of drought stress. It is felt that further effort, in conjunction with the studies at Swift Current in screening for physiological traits across a wide spectrum of environments, could provide together with normal agronomic data, a very comprehensive information base for effective use by plant breeders. A similar project exists between ICARDA and the University of London's Birkbeck College for the study of physiological and biochemical responses of landraces in general to drought.

During 1984–5 a mission to Turkey from the University of Saskatchewan, Canada, and another expedition from the University of

Table 6.5. *Descriptors of cereals for which information is available in the ICARDA database*

Character	Barley	Durum Wheat
1. Botanical species	—	x
2. Growth habit	x	—
3. Growth class	—	x
4. Flag leaf size and shape	—	x
5. Flag leaf attitude	—	x
6. Stem colour	x	—
7. Stem solidness	—	x
8. Waxiness	—	x
9. Kernel row number	x	—
10. Hoodedness	x	—
11. Awnedness	x	x
12. Awn colour	x	x
13. Awn roughness	x	—
14. Kernel covering	x	—
15. Rachila hair length	x	—
16. Lemma colour	x	—
17. Grain colour	x	—
18. Glume colour	x	x
19. Days to heading	x	x
20. Days to maturity	x	x
21. Plant height	x	x
22. Spikelets per spike	x	x
23. Spike length	—	x
24. Spike density	x	x
25. Productive tillering	x	x
26. 1000-kernel weight	x	x
27. Percentage protein content	x	x
28. Lysine/protein ratio	x	—
29. Resistance to lodging	x	x
30. Agronomic score	x	x
31. Low temperature damage	—	x
32. Salinity tolerance	—	x
33. Drought tolerance	—	x
34. Electrophoretic analysis	—	x
35. Reaction to disease	x	x

Yarmouk, Jordan, collected 463 accessions of cultivated and wild barley (*H. spontaneum*) as well as cultivated durum and wild emmer wheat (*T. dicoccoides*) from Turkey and Jordan respectively. These collections from the primary centre of diversity are expected to contain substantial variability of value in plant breeding as well as basic research (Jaradat, Jana & Pietrzak, 1986). The complete evaluation of this germplasm is in progress. Earlier studies (Sharma, Waines & Foster, 1981) have indicated that *T. dicoccoides* could be a good source for high protein content, disease resistance and stress (cold and drought) tolerance.

In another study at ICARDA (Tahir, 1986), 15 samples of *T. dicoccoides* collected from high altitudes in Syria and Iraq, where the wheat crop always suffers due to frost, drought and yellow rust (*Puccinia striiformis*), were evaluated for various agronomic traits. The material was grown at two sites: at Sarghaya, a cold high elevation site in Syria (1,450 m) and at Tel Hadya, the principal ICARDA research station (285 m). The variability for all characters, except frost tolerance, is presented in Table 6.6. The versatility of this wild species is such that it has several desirable traits and therefore its usefulness as a germplasm resource is appreciated. However, utilisation of wild genetic resources by hybridisation with cultivated forms takes more time and effort because of genetical barriers and undesirable linkages.

Subsequently 75 crosses between *T. durum* and *T. dicoccoides* were made at ICARDA. The accessions and their crosses were artificially inoculated for yellow rust with indigenous isolates at Tel Hadya and only resistant lines were selected for further evaluation. There was a considerable variability for the agronomic characters recorded. Despite temperatures ranging from −1 °C to −12 °C at Tel Hadya and Sarghaya, respectively, the *dicoccoides* accessions did not suffer frost damage except accession numbers SY 20184 and IQ 55132 (the only accession from Iraq) where slight frost damage was recorded similar to the control variety

Table 6.6. *Evaluation of 15* T. dicoccoides *lines collected from Syria (figures in the Table represent minimum and maximum values).*

Protein %	Plant height (cm)	Days to heading	Yellow rust resistance	Spike length (cm)	Frost tolerance	1,000 Kernel weight (g)
20.4–25.0	80–120	159–179	S–R	7.6–14.0	All lines tolerant	10.5–26.2

S = Susceptible
R = Resistant

'Bezostaya'. In addition, four accessions, SY 20010, SY 20017, SY 20021 and SY 20089 were found to be highly resistant to yellow rust. This is perhaps due to the fact that they were collected from yellow rust prone mountainous areas.

The *T. durum* × *T. dicoccoides* cross also transfers some disease resistance qualities from the wild species to the cultivated. These crosses have also shown that selection for high protein content as well as yield can be transferred to the cultivated form (Table 6.7). Whereas the durums have a relatively low percentage of protein and the *dicoccoides* a higher figure, in the resulting progenies high percentage protein is maintained. Similarly, the 1,000-kernel weight of the *durum* varieties is much higher than that of the *dicoccoides*. Nevertheless, in the progenies high 1,000-kernel weight is retained. In a simple *durum* × *dicoccoides* cross, characters from the wild species, such as brittle rachis, glume hairiness, profuse unsynchronized tillering, hybrid necrosis, grass clumping and loose crown, persist in subsequent generations but rapid progress can be made by making a top cross of this material with durum wheat.

Bread wheat

Bread wheat *Triticum aestivum* L., ranks high among cereals grown in the ICARDA region. It is grown largely under rainfed conditions and it is estimated that 50 per cent of the crop is grown in areas receiving less than 400 mm of rainfall. In these areas, besides the erratic and unpredictable rainfall, both in amount and distribution, and temperature (frost and heat), as well as other biotic stresses such as

Table 6.7. *Protein percentage and 1,000 kernel weight of F_4 seeds from interspecific crosses* (Triticum durum × T. dicoccoides)

Parents and Crosses	Protein (%)	1,000 kw (g)
BD 272 (*T. durum*)	15.2	38.0
BD 1658 (*T. durum*)	15.8	37.5
T. dicoccoides S.Y. 20101	20.5	20.0
T. dicoccoides S.Y. 20189	22.1	26.0
BD 272 × *T. dicoccoides* S.Y. 20101	21.6	37.6
BD 272 × *T. dicoccoides* S.Y. 20101	18.1	37.6
BD 272 × *T. dicoccoides* S.Y. 20101	20.7	37.6
BD 272 × *T. dicoccoides* S.Y. 20101	18.3	41.4
BD 1658 × *T. dicoccoides* S.Y. 20189	20.3	36.3
BD 1658 × *T. dicoccoides* S.Y. 20189	17.5	47.2

diseases and insect pests are the important factors limiting production. Current practices of land preparation, sowing, weed control and moisture conservation also need to be improved.

The objectives of the ICARDA/CIMMYT bread wheat improvement project is to produce genotypes for the region with resistance to the stresses and other factors described above, and to develop improved management practices in co-ordination with national programmes of the region. The bread wheat breeding programmes places special emphasis on multi-location testing of early segregating material. This is done by using a modified bulk method of selection in the F_2 generation and multi-location testing.

There are at present 2,584 accessions of bread wheats in the Genetic Resources Unit at ICARDA. These accessions consist mostly of breeding lines developed at ICARDA plus landraces collected during missions in the region. The landrace germplasm is undergoing evaluation and the data are being generated. Multi-location testing is an extension of the parallel selection initiated at ICARDA in 1978 on durum wheat evaluated under rainfed and supplementary irrigation within one location. The results of parallel selection lead to the conclusion that selection under low fertility, dry farming conditions is more successful in identifying drought tolerance genotypes than selection in the fertile high moisture environment (Srivastava *et al.*, 1983).

Whereas CIMMYT is mostly involved in breeding bread wheat and other wheat varieties for normal, irrigated or adequate rainfall areas, ICARDA concentrates on wheat research programmes aimed more specifically at helping farmers in the dry areas.

Bread wheat varieties developed in other wheat growing countries of North America and Europe, on the whole, did not have encouraging results in the region. These varieties, when tested at ICARDA, have a vegetative period similar to that of local cultivars up to heading time but their grain filling period is so extended that kernels shrivel because of temperature extremes and lack of rain during this important phase. In the high elevation areas locally adapted germplasm is being collected and used by the highlands cereal breeder in crosses which may transfer traits such as frost resistance, drought and low temperature tolerance to the high yielding varieties.

Constraints restricting greater use of collections

In general, the less-developed national programmes do not utilise landraces directly in their improvement activities for two reasons: a) they do not possess extensive facilities to collect and screen landraces on a large enough scale to identify suitable genetic material for breeding, and b) often landraces are adapted to very narrow environmental requirements and possess undesirable traits such as tallness and susceptibility to lodging and diseases which are difficult to eliminate.

Although breeders have used and continue to use landraces at ICARDA and in some national programmes in the region, there still exist certain constraints restricting greater use of germplasm collections. They are as follows:

1. Considerable germplasm has been assembled by genetic resources centres during the past decade but the available information on these accessions is limited. Without precise information on the minimum list of descriptors, breeders are unable to make use of the collected landraces. In order to detect genetic diversity for useful traits within landrace populations, multi-site evaluation is needed. Also landraces need to be collected from different ecological zones and characterised for different economically important traits (Damania & Porceddu, 1983).
2. Landraces and wild relatives used in crosses invariably result in a large number of variants often with undesirable characters. They manifest diseases and other traits which the breeder may need several seasons of careful selection to eliminate, due to linkages.
3. Some breeders feel that much more use of the germplasm could be made if the genetic resources centres were prepared to maintain single lines. However, the space needed and the cost involved for storing single lines separately from a population sample would be prohibitive. Therefore, breeders choose to maintain their own small collection of 'pure lines' obtained from landraces as well as other sources.
4. In some cases plant breeders are unaccustomed to using landraces in their breeding programmes because of their training where no emphasis was placed on the beneficial use of landraces.

Conclusions

Landraces, and for that matter also wild relatives of crop plants, will be increasingly used where specialised attributes are desired such as resistance or avoidance to drought, heat, salt and cold, which are not easily available in improved varieties particularly developed for high yields in favourable crop growing environments.

There is a positive trend towards utilisation of landraces and primitive forms in developing germplasm for stressful environments in the arid and semi-arid areas which has met with considerable success. With more exploration of such areas and proper evaluation, documentation as well as easy availability of the information for specific accessions of germplasm, an increasing number of plant breeders would find such genetic resources useful in their crop improvement programmes.

References

Ceccarelli, S. (1984). Utilization of landraces and *Hordeum spontaneum* in barley breeding for dry areas at ICARDA. *Rachis*, **3**, 8–11.

Ceccarelli, S., Nachit, M. M., Ferrara, G. O., Mekni, M. S., Tahir, M., van Leur, J. & Srivastava, J. P. (1985). Breeding strategies for improving cereal yield and stability under drought. In *Proceedings of Symposium on Improving Winter Cereals for Moisture Limiting Environments*, Capri, Italy.

Damania, A. B. & Porceddu, E. (1983). Variation in landraces of *turgidum* and bread wheats and sampling strategies for collecting wheat genetic resources. In *Proceedings of Sixth International Wheat Genetics Symposium*, pp. 123–36, Kyoto University, Japan.

Damania, A. B., Jackson, M. T. & Porceddu, E. (1985). Variation in wheat and barley landraces from Nepal and the Yemen Arab Republic. *Zeitschrift für Pflanzenzüchtung*, **94**, 13–24.

Duwayri, M., Tell, A. N. & Shqaidef, F. (1985). Breeding for improved yield in moisture limiting areas. The experience in Jordan. In *Proceedings of Symposium on Improving Winter Cereals for Moisture Limiting Environments*, Capri, Italy.

Fischbeck, G., Schwarzbach, E., Sobel Z. & Wahl J. (1975). Types of protection against barley powdery mildew in Germany and Israel selected from *Hordeum spontaneum*. In *Proceedings of Third International Barley Genetics Symposium*, pp. 412–17. Garching, Verlag Karl Thieming, Germany.

ICARDA (1983). *Annual Report 1983*, International Center for Agricultural Research in the Dry Areas, Syria.

ICARDA (1984). *Annual Report 1984*, International Center for Agricultural Research in the Dry Areas, Aleppo, Syria.

Jana, S. (1982). Canada–ICARDA collaboration for cereal germplasm conservation. *Plant Genetic Resources Newsletter*, **49**, 5–10.

Jana, S., Srivastava, J. P. & Gautam P. L. (1983). Evaluation of genetic resources of durum wheat for salt stress tolerance. In *Proceedings of Sixth International Wheat Genetics Symposium*, pp. 137–41, Kyoto University, Japan.

Jaradat, A. A., Jana, S. & Pietrzak L. N. (1986). Collection and evaluation of cereal genetic resources of Turkey and Jordan *Rachis* (in press).

Nachit, M. M. & Ketata, H. (1986). Breeding strategy for improving durum wheat in Mediterranean rainfed areas. In *Proceedings of the Fifth International Wheat Conference*, Rabat, Morocco.

Sharma, H. C., Waines, J. G. & Foster, K. W. (1981). Variability in primitive and wild wheats for useful genetic characters. *Crop Science*, **21**, 555–9.

Somaroo, B. H., Adham, Y. J. & Mekni, M. S. (1986). *Barley Germplasm Catalogue 1*. ICARDA, Aleppo, Syria.

Srivastava, J. P. (1977). Barley production, utilization and research in the Afro-Asian region. In *Proceedings of the Fourth Regional Winter Cereal Workshop – Barley*. Vol. 2, pp. 242–59, Amman, Jordan.

Srivastava, J. P. & Jana, S. (1983). Screening the wheat and barley germplasm for salt tolerance. In *Salt Tolerance in Crop Plants*. Staples, R., (ed.) John Wiley & Sons, USA.

Srivastava, J. P., Jana, S., Gautam, P. L. & Niks R. E. (1983). Parallel selection: An approach to increase grain yield and stability. In *Proceedings of Sixth International Wheat Genetics Symposium*, pp. 725–33, Kyoto University, Japan.

Tahir, M. (1986). Evaluation and utilization of *Triticum turgidum* L. var. *dicoccoides* for improvement of durum wheat. In *Proceedings of the Fifth International Wheat Conference*, Rabat, Morocco.

Weltzien, E. (1982*a*). Barley collection and evaluation in Syria and Jordan. *Plant Genetic Resources Newsletter*, **52**, 5–6.

Weltzien, E. (1982*b*). Observation on the growth habit of Syrian and Jordanian landraces of barley. *Rachis*, **1**, 6–7.

Weltzein, E. & Winslow, M. D. (1984). Resistance of durum wheat genotypes to saline-drought field conditions. *Rachis*, **3**, 34–6.

Witcombe, J. R. (1978). Two-rowed and six-rowed wild barley from the western Himalaya, *Euphytica*, **27**, 601–4.

Witcombe, J. R., Bourgois J.-J. & Rifie R. (1982). Germplasm collections from Syria and Jordan. *Plant Genetic Resources Newsletter*, **50**, 2–8.

7
Limitations to the use of germplasm collections

D. R. MARSHALL

Introduction

Over the past decade there has been a dramatic increase in the number of gene banks, especially in developing countries, with facilities for the long-term conservation of crop genetic resources (Plucknett *et al.*, 1983). There has also been a concomitant increase in the number of samples held in these gene banks (Lawrence, 1984; Holden, 1984). Indeed, it would appear that more germplasm is now available to breeders and there is less risk of its loss than at any previous time.

This does not mean that there are not important gaps in existing collections. These gaps are particularly evident in relation to collections of the wild relatives of the major crops and in some regionally important crops. As a result, the International Board for Plant Genetic Resources (IBPGR) has implemented a pragmatic and flexible programme aimed at filling these gaps and this is expected to be largely completed by the mid 1990s (Williams, 1984). However, the dramatic increase in material held in gene banks has not yet been matched by a concomitant increase in its use by plant breeders and experimental biologists. Consequently there is a growing perception that genetic resources collections are generally undervalued and underused (Holden, 1984; Peters & Williams, 1984). If this perception is valid and the situation remains unchanged and unchallenged, then the long-term survival of the global network of genetic resources centres could be jeopardised. The reason is simple. The recent rapid growth in germplasm collections has been based on the premise that they represent a valuable and useful resource for breeders and biologists. If this premise comes into question because of lack of use, financial and political support for their continuation could quickly diminish.

The above considerations raise several issues of concern:

1. Are collections generally underused? Or, alternatively, are the expectations of the critics unrealistically high?

2. If collections are generally underused does the fault lie with the collections and their curators, or the users, or both?
3. What are the major constraints, from the viewpoint of both curators and users, that are limiting the use of collections, and how can these be overcome?

Delineation and quantification of the problem

Claims that the use of germplasm collections by breeders is inadequate are not new. Indeed, more than 20 years ago Simmonds (1962) argued against gene banks, which he termed 'museum' collections, as a means of long-term conservation on two grounds. First, they were a wasting resource with often high rates of attrition. Secondly, they were poorly used by, and often inaccessible to their primary users, plant breeders. Simmonds' first argument is no longer valid because of advances in the technology of seed storage and the rapid worldwide growth in first-class facilities for such storage. However, his second argument is still as relevant today as it was in 1962 as evidenced by recent concerns about the size, accessibility and use of collections (Frankel & Brown, 1984; Holden, 1984; Peters & Williams, 1984).

However, because such claims are long standing does not necessarily mean that they are valid. Many breeding programmes around the world have been remarkably successful over the last two decades. These include programmes on rice at the International Rice Research Institute (IRRI), on wheat and barley at the International Center for Maize and Wheat Improvement (CIMMYT), several on winter wheat in Europe, on sunflower in the USSR, on soyabean and maize in the USA, on sorghum at the International Crops Research Institute for the Semi Arid Tropics (ICRISAT), on rapeseed in Canada and the Federal Republic of Germany, and that on lupin in Western Australia, to name a few. It may be argued that each of these programmes could have been far more successful if the potential existing in genebank collections had been better utilised. But in view of the results achieved this argument is hard to sustain.

What is adequate?

While there is a growing consensus that germplasm collections are often poorly used, there is less agreement on the crops or collections where the problem is greatest and on its causes. If programmes aimed at improving germplasm use are to be effective, the questions of where and why major problems occur must be answered. As an important first step a decision needs to be taken as to what constitutes a satisfactory or adequate level of use of germplasm collections. Critics who assert that current levels of use are inadequate, seldom specify what they perceive as an adequate level of usage. Yet presumably they have some figure or model collection in mind as a standard. What is it? What is a practical and achievable goal?

These are obviously important questions. They are also difficult to answer. Purists would argue that collections should be used to the maximum extent possible. While this is a worthy aim it is unachievable. No matter how well a collection is managed or used there will always be a case that utilisation is less than maximal. Clearly such a definition is of little value in deciding priorities when allocating limited resources to improve germplasm use. The first difficulty is the definition of an appropriate measure of use. One option is to use the total number of samples distributed. This statistic has the disadvantage that it can easily be biased upwards by one or a few large requests. A second option is to use the number of requests for accessions regardless of the total number of samples sought. Unfortunately, this statistic is less often reported in the literature. A third option is to try to define some measure of effective use, i.e. the number of accessions used in the development of new varieties. However, besides the obvious difficulties in assessing this statistic, it is likely to be of limited use because many of the gene banks in the IBPGR network have been operational for less than ten years and it is unrealistic to expect them to have yet had a significant impact in terms of released cultivars. A fourth option would be to assess use on the basis of the proportion of breeders who use collections.

This approach was adopted by Duvick (1984) in his analysis of germplasm use by US breeders. A deficiency in this approach is that the user community is far wider than just breeders, although they have traditionally been regarded as the primary users.

Of the options described above, the number of independent requests for accessions is probably the simplest and most appropriate measure of collection use (Peeters & Williams, 1984). This statistic is given for three collections in Table 7.1. For the years reported it can be seen that for both the collections of IRRI and the International Institute for Tropical Agriculture (IITA) collections, which are well-known international collections and presumably well used by any definition, about ten requests a year were received for every 1,000 accessions held. Obviously, these numbers vary from year to year. Even so, they are seldom less than five or more than 20 per 1,000 accessions held. On this basis it would seem reasonable to suggests that five requests a year per 1,000 entries held could be reasonably described as an adequate, or minimum desirable, level of use of a collection.

Table 7.1. *Number of requests, number of accessions distributed and total collection size for five germplasm collections*

Institute/Crop	Year	Number of requests	Number of samples	Total collection	Reference
IRRI/rice	1981	628	33,432	57,027	IRRI (1983)
IITA/all mandated crops	1982	235	9,812	20,362	IITA (1983)
CIMMYT/wheat	1983	—	3,998	20,183	CIMMYT (1983)
USDA/soyabeans	1985	—	20,000+	11,000	Palmer (pers. comm.)
Australian Cereals Collection/wheat	1984	277	6,772	21,000	Mackey (pers. comm.)

More information is available on the total number of samples distributed by active collections. Five examples are given in Table 7.1. It will be noted that the number of samples distributed varied from about 20 to 200 per cent of the total collection size. Wilkes (1983) presented comparable figures for nine major US collections for the year 1980. In eight of the nine cases the number of samples distributed represented between about 40 to 120 per cent of the total collections size. The sole exception was of the United States Department of Agriculture (USDA) rice collection which distributed only 765 samples, about 5 per cent, of the 15,000 samples held. From these data it can be argued that an adequately used collection would be expected to distribute on average approximately 20 per cent or more of the total accessions held each year. It must be emphasised that these figures or any alternatives that may be suggested, are not sacrosanct. They are put forward only as a first attempt to quantify by readily obtainable statistics what is meant by 'adequate' versus 'inadequate' use of collections.

Variation in collection use

The establishment of appropriate minimum acceptable levels of collection use leads to a second important question. How often are these minimum levels met in practice? How generally are collections underused?

The data in Table 7.1 and those given by Wilkes (1983), with few exceptions, refer to well-used collections. Overall, they form a highly biased sample and represent well-known collections of major crops in developed countries or international research centres. It is probable, therefore, that they give an overly optimistic picture and the level of use varies radically among collections. However, data to support this view are lacking, although Peeters & Williams (1984) have suggested that the newer small multicrop collections developed in recent years, often in the form of national gene banks, are a major area of concern. As a consequence, there would appear to be an urgent need to survey collections in the IBPGR network to establish existing use patterns. These data should be helpful in identifying problem areas (species or collections) and in determining priorities in the effort to improve germplasm use.

Spectrum of users

Another set of important questions relate to the spectrum of users, or the composition of the user population, of gene banks. It is generally assumed that plant breeders will be the primary and most prolific users of collections. Is this assumption valid? Or, are experimental biologists the primary users? Further, do well-used collections owe their success to the fact that they have attracted a greater number of experimental biologists, plant breeders, or both, in their clientele?

Again, appropriate data are rare or lacking in the literature. Thus, there appears to be a need to survey well and poorly used collections in the IBPGR network, once these are identified, to quantify the variation in the spectrum of users of different sorts of collections. This should assist curators of poorly used collections to develop policies which attract a larger and more diverse clientele.

Practical constraints on germplasm distribution and exchange

During the first decade of its existence from 1974–84 the IBPGR gave emphasis to the collection and conservation of genetic resources under serious threat of extinction. Less attention was given to the utilisation of this material. More recently, it has become apparent that the efficient management of working collections to ensure proper utilisation of germplasm will be far more difficult than anticipated (Peeters & Williams, 1984). Curators of active collections face additional problems to those encountered by managers of base collections, concerned solely with long-term conservation.

As a result, in recent years, IBPGR has sought (via consultants' reports and planning committees) to identify the major constraints limiting germplasm use and to develop practical strategies to overcome these constraints (Williams, 1984). From these studies it is clear that such constraints are many and varied. Further, as we might expect, their relative importance varies among germplasm stocks and among gene banks. These constraints can be divided into two broad categories – those that limit the availability of germplasm and those that limit its use, given that it is readily available. The first category is considered here.

Economic constraints

Seed multiplication

Economic constraints are a serious and common limitation to the availability of germplasm from collections, particularly from national gene banks in developing countries. Curators often have difficulty in obtaining sufficient funds to undertake seed increase of materials collected under IBPGR sponsorship for distribution to other gene banks and to users (Marshall, 1983). Although seed increase of collected samples is usually undertaken as rapidly as possible, if the number of samples is large and the funds available limited it can be several years before all entries in a collection are available to others.

It has been suggested that this problem could be significantly reduced in seed crops by collecting larger samples from the field so that sufficient material would be available for both long-term storage and distribution without the need for multiplication. This approach has obvious merit, particularly in cultivated plants, where it is often possible to collect large samples quickly. However, it also has obvious limitations. First, if this procedure is widely adopted, then collecting missions will have to be larger to handle the extra workload. As shown by Marshall & Brown (1975; 1983) the optimum sampling strategy, which maximises the amount of useful genetic variability collected, is to sample 50 to 100 individuals per population and to sample as many populations occupying as wide a range of environments as time and resources permit. Clearly, if much larger samples are collected at each site, then the number of sites and total variability collected will be reduced unless compensatory increases are made in the size of collecting teams. Secondly, this approach is clearly not applicable to wild and weedy species which shatter their seed.

The point is that genetic resources cannot be used unless adequate samples are available for distribution. This can be assured either by collecting larger samples, or by post collection multiplication, and a rational choice between these alternatives depends on many factors. However, both are costly and this cost must be allowed for if there is to be a smooth and rapid flow of new material from the field through the base and active collections to the user.

Sample preparation, packaging and distribution

Another economic constraint, again more particularly for national collections in developing countries, is sample preparation, packaging and mailing costs. This problem is particularly acute in some vegetatively propagated crops, for example, bananas, where the cost of preparing, packing, treating and sending a single clone may be as high as US$100. It is less serious in seed propagated crops. Nevertheless, curators are reluctant to meet blanket requests for the whole, or large parts, of their collections because they lack the resources, given other demands and priorities, to package hundreds of samples and mail them overseas.

It follows that large requests for material from national or regional collections are more likely to be met if the recipient is prepared to pay at least for the packaging and posting of the samples.

Plant quarantine

National quarantine regulations are a necessary precaution against the accidental spread of pests and diseases. However, they are also an impediment to the flow of germplasm among gene banks, and from gene banks to breeders. Numerous examples can be found of losses of material because of delays or errors in quarantine or of losses of variation in population samples because only a fraction of the original material was processed by quarantine authorities.

Such problems are not new and have been aired adequately at international conferences on genetic resources. New systems have been proposed to facilitate rapid entry of germplasm through quarantine (e.g. Chiarappa & Karpati, 1981; 1984) but have failed to gain sufficient support to be implemented.

A universally accepted scheme for the exchange of disease and pest free germplasm may eventually be developed. In the meantime, it is clear that plant quarantine will continue to be an impediment to germplasm exchange. To minimise this problem it is essential that curators of gene banks understand the quarantine requirements of their own countries and, where possible, establish personal contacts with quarantine officials. They should also be aware of the quarantine requirements of the countries with which they regularly exchange germplasm. Several countries hold national or international training courses for quarantine officers. Curators of collections should attend such a course and where possible, become licensed quarantine officers.

It is also important that collectors be aware of the quarantine requirements of the countries they visit as well as those to which they intend to send samples. Failure to process samples (i.e. by removing all weed seeds, soil, insect infested seeds, etc) in line with the quarantine laws of recipient countries prior to mailing can lead to their immediate destruction or substantial delays in their processing and hence, loss of viability or variability.

Legal constraints

Specific legislation restricting the export of propagating materials

Several countries have enacted legislation prohibiting the export of propagating material of specific crop which they consider to be of crucial importance to their national economies. Such legislation is clearly in conflict with the principle of free availability of germplasm and has been the basis of considerable controversy, and, on occasion, retaliatory refusal to supply germplasm to the countries concerned.

Fortunately, these restrictions generally apply only to industrial and plantation crops, e.g. cocoa, rubber, oil palm, manila hemp and jute. They seldom affect the major food crops. Further in most cases the ban is not complete and material may be obtained legally with permission of the appropriate Minister. Finally, such laws appear to have been often ineffective in preventing the loss of elite clones of industrial crops to competitor countries. Consequently, it is hoped that such restrictive legislation will diminish rather than increase in the foreseeable future.

Plant Variety Rights Legislation

Plant Variety Rights (PVR) legislation has also been claimed by its opponents to be a major constraint in germplasm exchange (e.g. Mooney, 1979). However, this view has been questioned by others (Frankel, 1981; IBPGR, 1983*a*). In theory, cultivars protected by PVR should be freely available for use for breeding and research purposes. In practice, however, breeders in countries with PVR legislation are often reluctant to send material to countries without such legislation because of the lack of legal restrictions preventing the large scale seed multiplication of a variety and its re-export to other countries. This is particularly true where material goes into a gene bank and is freely available to third parties. However, the only material likely to be affected are parents of current commercial hybrids and elite advanced breeding lines. Even with these, breeders in countries with PVR will consider supplying such material if this poses no threat to their commercial operations. Generally, they require that the recipient be an individual who is prepared to guarantee that the material will not be used commercially or provided to third parties without their specific consent.

Technical constraints

A variety of technical constraints are commonly regarded as significant limitations to the use of germplasm collections.

Curator imposed constraints

Since many collections are relatively new, curators vary markedly in their training and management experience. They also differ in their priorities and in their attitudes to supplying samples. In this way curators can be the cause of the non-availability of germplasm.

Emphasis also needs to be given in improving technical skills in such areas as seed testing, quarantine requirements and procedures, and computer information storage and retrieval systems.

Lack of specificity in requests for material from gene banks

A second technical problem of importance in limiting germplasm exchange is a lack of specificity in requests made by users to curators. Blanket requests for entire collections are one area of concern. These are often viewed with suspicion by curators who are understandably reluctant to commit a substantial fraction of their resources to meet a single request especially when the person is unknown to them and they are not confident the material will be used to good purpose. Vague requests for 'all the good lines' in a collection or requests where the characteristics specified are so basic that literally thousands of entries meet the specified criteria also pose problems even for experienced curators. As a result such requests are often either ignored, or met by sending out token samples, and this can lead to complaints of non-availability of germplasm.

Lack of documentation and description of collections

The lack of documentation and description of material held in germplasm banks is a third, and probably more important, technical constraint to the use of collections. All too often materials have been simply collected and stored in gene banks without taxonomic verification or characterisation. This procedure may have been acceptable in the past when much of the material collected by IBPGR-sponsored missions was under immediate threat of extinction. Such a procedure is less justifiable for material not under immediate threat and there is a need to provide greater resources for documentation and description, at least to the extent of providing full passport data on already collected samples as well as new collections made each year. IBPGR is fortunately well aware of this need and has designated documentation of collections as a key area for research funding and training (IBPGR, 1983*b*).

Another problem is the dissemination of this information once it is collected. Several excellent catalogues have been produced (e.g. for the Andean crop collections held in Peru) but these often have a limited distribution – to other genetic resources workers or centres rather than to plant breeders or other potential users. As a result, users who may be vitally interested in a collection may not be aware of its existence, its whereabouts or the availabilty of materials. This is likely to be less of a problems in the major food crops, such as wheat and rice, where the international agricultural research centres of the Consultative Croup on International Agricultural Research (CGIAR) maintain close contacts with the majority of breeders in both the developed and developing countries. It is, however, seen to be a major problem in the minor crops or those for which the international centres do not have a mandate. As emphasised by Smith (1981), this problem is exacerbated by the proliferation of national gene banks and by the increasing numbers of collections accorded the status of genetic resources. He further argued: 'Indeed, despite a decade or more of concentrated effort to rationalise the exploitation of plant genetic resources, the probability of users locating appropriate material with precision and ease may not have significantly increased.'

One approach to this problem would be to establish international registers of plant breeders for the 50 crops designated first priority by IBPGR and to ensure that documented information on collections held around the world goes to the breeders or each crop. While this may seem a daunting task, the Food and Agriculture Organization (FAO) has developed a World List of Plant Breeders in the past and this could be updated and revised.

Constraints on the utilisation of germplasm in breeding programmes

In the above section some important constraints limiting the full and free availability of germplasm, a basic tenet of the IBPGR programme, were outlined. Three factors which limit the use of germplasm, even when this ideal is met and germplasm is available for the asking, are considered below.

The numbers of plant breeders

By far the most important and serious factor restricting the use of available germplasm is a lack of plant breeders. This is particularly true of developing countries. For example, Dr. Sastrapradja at the 1981 International Conference on Crop Genetic Resources pointed out that at that time there was only the equivalent of $5\frac{1}{2}$ breeders in Indonesia to handle all crops (FAO, 1981, p. 54). As noted by Holden (1984) '...for the developing countries it is clear that the single most important prerequisite to the exploitation of genetic resources for the benefit of their agriculture is the training and organisation of a body of plant breeders and associated scientific specialists such as those who function as a team in the international centres of the CGIAR.' Indeed, unless this is done, no significant increase is likely in the use of the smaller, newer multicrop national collections often situated in developing countries and which Peeters & Williams (1984) indicated were underused. Further, unless more breeders are recruited in developing countries, then the benefits of the current moves to improve the efficiency of gene banks will principally flow to the developed countries which are well serviced by breeders and supporting experimental biologists.

Lack of evaluation of collections

This is widely seen as a major impediment to their utilisation (e.g. Frankel, 1977; Lyman, 1984; Frankel & Brown, 1984; Singh &

Chomchalow, 1984), particularly by breeders themselves (Marshall, 1983; Peeters & Williams, 1984). The philosophy underlying this viewpoint is that evaluation is a necessary prerequisite to utilisation and further, the more information that is available on a collection the greater its value to potential users (Frankel, 1970).

However, the assumption that lack of evaluation is a major constraint on germplasm use is difficult to validate. Indeed, the USDA small grains collection, one of the most widely used in the world distributing 99,000 samples from about 82,000 accessions in 1980 (Wilkes, 1983), is poorly evaluated and provides a striking counter example to this presumption. Further, there is little evidence to support the implicit assumption often made that the use of collections will increase in proportion to the quantity of information available. For example, considerable effort has gone into the systematic evaluation of the rice germplasm collection at IRRI (Chang *et al.*, 1982; Chang, 1984) and it is an outstanding example of a well-managed and -documented collection. Yet, the increased evaluation/documentation of the IRRI collection does not appear to have increased its use above that of other far less well-evaluated/documented collections.

Consequently, despite all that has been said and written concerning evaluation, the degree to which its absence impedes collection use (compared to lack of breeders, for example) is open to question. Indeed, it can be argued that if all collections were fully evaluated, this in itself would be unlikely to lead to a dramatic and immediate increase in their use. The reason is that breeders generally have sufficient variation for most traits in their working collections of elite adapted lines. If this argument is valid, then the contribution of lack of evaluation to collection underuse has generally been over-emphasised.

However, virtually all aspects of evaluation remain controversial and several of these points of controversy are discussed in detail by Marshall & Brown (1981) and Frankel (this volume).

Pre-breeding

A third factor limiting the utilisation of available germplasm by breeders is a lack of pre-breeding. This means the extraction of desirable traits, such as disease resistance or drought tolerance, from unadapted, unimproved or wild donors and their transfer into high yielding, adapted and improved backgrounds. The derived lines can then by used by breeders in crosses with other elite adapted lines, with the knowledge that they have a high probability of being able to select from such crosses commercially acceptable cultivars carrying the desired traits.

Pre-breeding is often a key factor in the utilsation of wild germplasm.

Because of the difficulties in using wild germplasm (Hawkes, 1977; Brown & Marshall, 1986), commercial plant breeders claim not to have the time, resources or expertise to go to the original sources for useful genes. Nevertheless, they are avaricious users of such genes if they are transferred to adapted genetic backgrounds.

However, while the problem is particularly acute for wild germplasm, it is also probably a major factor in the lack of use of landrace germplasm and unadapted exotic lines from collections. This may be especially true for breeders, public or private, in developing countries who face substantial pressures to produce improved varieties with limited resources and cannot afford to risk these resources in longer-term projects that have only a limited chance of success.

The international agricultural research centres have been a major success in terms of exploitation of diverse germplasm in several important crops. A key element in their success has been the provision of high yielding widely adapted germplasm to breeders carrying comprehensive resistance to major diseases and environmental stresses. This material can be used directly or in local breeding programmes. There would appear to be little doubt that the international agricultural research centres would have been far less successful had they merely provided computer lists of evaluated material highlighting desirable traits to local breeders.

This logic suggests, that, if the ultimate goal is utilisation of genetic resources, then this is unlikely to be achieved simply by evaluating collections for more and more characters. Rather the emphasis must be on *selective*, rather than systematic, evaluation and greater priority must be given to prepackaging the desirable genes identified into backgrounds easily incorporated into breeding programmes.

Conclusions

The use of the extensive collections held in the IBPGR global network of gene banks is now the focus of increasing attention. This has led to concern that the collections are generally undervalued and underutilised. However, analysis of the limited information in the literature suggests that the use of collections varies markedly – good use is made of some collections but not of others. Obviously efforts to improve use should be preferentially directed towards the areas of greatest need. To do this requires a rigorous definition of what is an adequate or acceptable level of collection use. It is suggested that adequate use could presently be defined in terms of the total number of samples distributed each year (with $\geqslant 20$ per cent of total accessions as a minimum acceptable level) or the number of independent requests a year (with a minimum acceptable level of five a year per 1,000 accessions). It also requires comprehensive surveys of existing collections to identify those that do not meet these minimum standards.

Once problem collections have been located, it should be possible to identify and remove the major limitations on their use. The constraints that may, or do limit utilisation of collections can be conveniently grouped in two broad categories:

- those that limit the availability of gene bank to users and,
- those that limit the use of readily available material by breeders and experimental biologists.

In the first category are the many factors which are directly or indirectly attributable to deficiencies in the management of the collections and they are rectifiable by providing better scientific training of curators and improved management procedures. The latter include:

- better information flow so potential users can identify appropriate collections and accessions.
- adequate financial support for seed multiplication and sample preparation, packaging and distribution.

The constraints in the second category are principally user related. Consequently, they can usually only be rectified by users, or users acting in concert with curators, and not by curators alone. The most controversial of these constraints is the lack of evaluation of collections. It has often been assumed that the use of collections will increase in direct proportion to the degree to which they are evaluated and, as a consequence, curators should have a direct and prominent role in evaluation. It is suggested that this assumption is largely invalid. As an alternative it is suggested that the use of variation in collections, particularly by developing countries, is dependent principally on the degree to which this variability is incorporated by pre-breeding into high yielding, widely adapted backgrounds. It is this pre-breeding, which of necessity must be preceded by evaluation but not evaluation alone, which has accounted for the outstanding success and impact of the international centres such as IRRI, CIMMYT, ICRISAT and IITA. Under this argument it is the extent of pre-breeding, preceded by selective evaluation, which must be seriously addressed if the use of collections by breeders in developing countries is to be significantly enhanced in crops where it is notably deficient.

References

Brown, A. H. D. & Marshall, D. R. (1986). Wild species as genetic resources of plant breeding. In *Proceedings International Symposium on Plant Breeding*, Lincoln, New Zealand.

Chang, T. T. (1984). Evaluation of germplasm-rice. In *Crop Genetic Resources: Conservation and Evaluation*, pp. 191–8, Holden, J. H. W. & Williams, J. T. (eds.). George Allen & Unwin, London.
Chang, T. T., Adair, C. R. & Johnston, T. H. (1982). The conservation and use of rice genetic resources. *Advances in Agronomy*, **35**, 37–91.
Chiarappa, L. & Karpati, J. (1981). Safe and rapid transfer of plant genetic resources: a proposal for a global system. *FAO/UNEP/IBPGR Technical Conference on Crop Genetic Resources*, p. 32. FAO, Rome.
Chiarappa, L. & Karpati, J. (1984). Plant quarantine and genetic resources. In *Crop Genetic Resources: Conservation and Evaluation*, pp. 158–62. Holden, J. H. W. & Williams J. T. (eds.). George Allen & Unwin, London.
CIMMYT (1985). *CIMMYT Report on Wheat Improvements 1983*. CIMMYT, El Batow, Mexico.
Duvick, D. N. (1984). Genetic diversity in major farm crops on the farm and in reserve. *Economic Botany*, **38**, 161–78.
FAO (1981). *FAO/UNEP/IBPGR International Conference on Crop Genetic Resources*. FAO, Rome.
Frankel, O. H. (1970). Evaluation and utilization – Introductory remarks. In *Genetic Resources in Plants – their Exploration and Conservation*, Frankel, O. H. & Bennett, E. (eds.), pp. 395–402. Blackwell, Oxford & Edinburgh.
Frankel, O. H. (1977). Genetic Resources. *Annals New York Academy of Science*, **287**, 332–44.
Frankel, O. H. (1981). Maintenance of gene pools – sense and nonsense. In *Evolution Today*. Proceedings of the Second International Congress of Systematic Evolutionary Biology, pp. 387–92, Scudders, G. G. D. & Reveal, J. L. (eds.). Hunt Institute of Botanical Documentation, Pittsburgh.
Frankel. O. H. & Brown, A. H. D. (1984). Plant genetic resources today: a critical appraisal. In *Crop Genetic Resources: Conservation and Evaluation*, pp. 249–57. Holden, J. H. W. & Williams, J. T. (eds.), George Allen & Unwin, London.
Hawkes, J. G. (1977). The importance of wild germplasm in plant breeding. *Euphytica*, **26**, 615–21.
Holden, J. H. W. (1984). The second ten years. In *Crop Genetic Resources: Conservation and Evaluation*, pp. 177–85. Holden, J. H. W. & Williams, J. T. (eds.). George Allen and Unwin, London.
IBPGR (1983*a*). *Plant Varieties Rights and Genetic Resources*. IBPGR, Rome.
IBPGR (1983*b*). *Annual Report* 1982. IBPGR, Rome.
IBPGR (1984). *Annual Report* 1983. IBPGR, Rome.
IITA (1983). *Annual Report* 1982. IITA, Ibadan.
IRRI (1983). *Annual Report* 1981. IRRI, Los Baños.
Lawrence, T. (1984). *Collection of Crop Germplasm: The first ten years 1974–84*. IBPGR, Rome.
Lyman, J. M. (1984). Progress and planning for germplasm conservation of major food crops. *Plant Genetic Resources Newsletter*, **60**, 3–21.
Marshall, D. R. (1983). Practical constraints limiting the full and free availability of genetic resources. Consultant report IBPGR (84/20). Rome.
Marshall, D. R. & Brown, A. H. D. (1975). Optimum sampling strategies in genetic conservation. In *Genetic Resources for Today and Tomorrow*, pp. 53–80. Frankel, O. H. & Hawkes, J. G. (eds.), Cambridge University Press, Cambridge.
Marshall, D. R. & Brown, A. H. D. (1981). Wheat genetic resources. In *Wheat Science – Today and Tomorrow*, pp. 21–40. Evans, L. T. & Peacock, J. W., (eds.), Cambridge University Press, Cambridge.

Marshall, D. R. & Brown, A. H. D. (1983). Theory of forage plant collection. In *Genetic Resources of Forage Plants*, pp. 135–48. McIvor, J. G. & Bray, R. A., (eds.), CSIRO, Melbourne.

Mooney, P. R. (1979). *Seeds of the Earth – a private or public resource.* Mutual Press, Ottawa.

Peeters, J. P. & Williams, J. T. (1984). Towards better use of genebanks with special reference to information. *Plant Genetic Resources Newsletter*, **60**, 22–31.

Plucknett, D. L., Smith, N. J. H., Williams, J. T. & Anishetty, N. M. (1983). Crop germplasm conservation and developing countries. *Science*, **220**, 163–9.

Simmonds, N. W. (1962). Variability in crop plants, its use and conservation. *Biology Review*, **37**, 442–65.

Singh, R. B. & Chomchalow, N. (1984). Evaluation and documentation of germplasm: Southeast Asian experience. In *Crop Genetic Resources: Conservation and Evaluation*, pp. 207–24. Holden, J. H. W. & Williams, J. T., (eds). George Allen & Unwin, London.

Smith, R. (1981). Principles and practice of germplasm distribution and exchange. *FAO/UNEP/IBPGR Technical Conference on Crop Genetic Resources*, p. 31. FAO, Rome.

Wilkes, G. (1983). Current status of crop plant germplasm. *CRC Critical Reviews in Plant Science*, **1**, 133–81.

Williams, J. T. (1984). A decade of crop genetic resources research. In *Crop Genetic Resources: Conservation and Evaluation*, pp. 1–17. Holden, J. H. W. & Williams, J. T., (eds.), George Allen & Unwin, London.

Part III

Size and Structure of Collections

8
The case for large collections

T.T. CHANG

Implications of large collections

Before discussing the size of germplasm collections, a number of terms needs to be mentioned to delineate the scope of the discussion in this paper. These are related to the mandate of the institution and the operational and managerial aspects. First, this paper will focus on the crop genetic resources centres (GRC) rather than seedbanks, since the former take on a comprehensive series of operations related to germplasm conservation, evaluation and use, while the latter are largely custodians of seeds deposited by a number of curators of smaller collections. Secondly, the size of crop collections refers to a single crop regardless of the crop-specific or multi-crop mandate of an institution, although the total holdings of the GRC will be affected by the number of crops involved and the complexity of operational matters in dealing with several crops. In a related manner, an international plant GRC differs from a national one in having to deal with a broader range of responsibilities and the required comprehensiveness of coverage.

Whether a crop collection is termed large or otherwise, should be based on the size of holding in relation to the total genetic diversity present in that crop and its wild relatives. For instance, the national rice collections of China and India, both numbering about 33,000, cannot be labelled large because their holdings have not been fully inventoried, the seeds were scattered over a number of storage sites, and the number of distinct accessions may be modest when compared to an estimated total of 100,000 rice cultivars in Asia (Chang, 1985). On the other hand, the sorghum collection 24,000-accessions of the International Crops Research Institute for the Semi-Arid Tropics may be considered large because it contains about 80 % of the distinct sorghum accessions in the world. Similarly, the 46,000 wheat accessions in the United States Department of Agriculture and the 63,000 wheats in the Vavilov Institute of the USSR are uniquely large. The same may be said of the 83,000 accessions of rice in the collection of the International Rice Research Institute (IRRI) because acquisition and field collection efforts have been implemented in the appropriate places and at opportune times before advanced genetic erosion occurred.

However, the above designation of existing collections as 'large' is empirical and lacks a strict scientific basis. The comprehensiveness of a crop collection should be determined by a group of scientists specialising in that crop after a thorough assessment of available information and supplemented by field survey, if needed. My ballpark estimate of 100,000 cvs. of Asian rice necessary in the collection was based on extensive discussions with rice workers of various countries as well as their reports made at two IRRI/IBPGR workshops (IRRI, 1978, 1983).

Biologically, the true size of a large base collection should be tallied after exhaustive efforts have been made to discount the obvious duplicates as well as the not-so-obvious duplicates. The general criteria used in rice for such an inventory are (1) variety name or code name, (2) country of origin/seed source, (3) other passport data and (4) key morpho-agronomic traits. At IRRI the first step was taken by planting those samples coming from the same country in adjacent plots and in alphabetic order by the variety names. Following comparisons in the field and laboratory, the obvious duplicates are bulked while the distinct samples are kept as individual accessions. We are now in the second stage of using statistical tests to further distinguish among those accessions of the same country of origin and having identical or similar variety names or similar morpho-agronomic traits. The third step is to use electrophoretic techniques. In short, numerical size should not be equated with comprehensiveness.

A question has been frequently raised: When would the IRRI collection cease to grow? The answer has to be based on the mandate given to the institute holding a base collection. Any curator of a base collection, in the fullest interests of genetic conservation, should not refuse to accept collected samples acquired by new expeditions or donations of national centres or lesser centres. The task is crucial because most GRCs in the humid tropics and subtropics lack dependable cold storage facilities and/or continuous human care. From 1974 to 1983, the International Rice Germplasm Center at IRRI has been receiving 4,000 to 7,000 new seed samples each year. During the first eight months of 1986, four thousand samples were deposited in the Center. Future plans include the receipt of as many as 25,000 from China and 20,000 from India. Moreover, only a small proportion of the wild species has been conserved. The means to check continuing growth of the base collection are by cutting down on the number of duplicates and by imposing stricter requisites in accepting breeding lines.

Several knowledgeable workers have pointed to the genetic erosion taking place inside genebanks or in specific crop collections (W. L. Brown in Anon., 1984; Goodman, 1984; Sun, 1986; Chang, 1986*a*). A large collection is likely to suffer losses as a result of one of the following factors: (1) shifts in policy, programme or funding, (2) turnover in personnel, (3) a lack, breakdown or ageing of refrigeration equipment, (4) unexpected setbacks in regeneration or processing operations, and (5) unexpected natural disasters. Such problems have led some critics (Anon., 1984; Mooney, 1985; Chang, 1986*a*) to question the validity of statistics on existing collection size gathered by surveys (Toll *et al.*, 1980; Plucknett *et al.*, 1983; Lyman, 1984). IBPGR has, for long, pointed out that numbers of accessions are only guides and are meaningless in estimating genetic diversity.

It is clear that the claimed size of most large collections considerably exceeds the sum total of distinct and viable accessions contained in them. Efforts in inventory, comparison and consolidation are sorely needed to reduce the redundancy in large (as well as smaller) collections so as to increase the effectiveness of conservation activities (see IRRI, 1978).

Roles and justification of large collection

The advantages of, and specific roles that, a large collection may play, are as follows:

1. Generally, a large collection is more diverse in genetic composition and more comprehensive in eco-geographic coverage, provided that indiscriminate redundancy has been removed to the extent possible. If well planned from the beginning and careful field collection implemented, a large collection can be expected to have a rich accumulation of useful alleles and a high frequency of such alleles in the collection (Marshall & Brown, 1975). Moreover, a large collection can supply accessions with several desired traits in different genetic backgrounds. Many rice researchers in newly developed rice areas have asked for accessions having the following combinations of traits: (a) extremely early maturity, cool temperate tolerance and blast resistance – for high latitude or hilly areas, (b) early maturity, blast resistance, stem borer resistance, drought resistance and moderate cool tolerance – for

upland areas, and (c) photoperiod sensitivity, drought tolerance at the juvenile stage, internode elongation ability and submergence tolerance – for deepwater areas. A striking example in evaluation is the following. The resistance gene (*Gsv*) to the strain 1 of grassy stunt virus was only found in one population of *O. nivara* originating in the Uttar Pradesh State of India, after IRRI virologists had tested under controlled inoculation 17,000 accessions of *O. sativa* and over one hundred wild taxa in the '*O. sativa* species complex' in a programme lasting four years. Only 3 out of 30 seedlings tested were resistant and these were immediately used in the breeding programme which has led to the control of this destructive disease by the use of resistant varieties. Subsequent searches among other wild taxa in the genus and in 10,000 more cultivars and breeding lines failed to yield another source of resistance (Chang *et al.*, 1975; IRRI, 1975, 1976, 1977). The returns from this rare allele have amounted to many millions of US dollars.

Unfortunately, a new strain of grassy stunt virus surfaced in 1982 (Hibino *et al.*, 1985). The *Gsv* allele was ineffective against the second strain. Similar to strain 1, plant resistant to strain 2 were found only in four wild species having the C genome, after thousands of cultivars and hundreds of wild taxa in the '*O. sativa* species complex' were screened (Cabauatan *et al.*, 1985). In the case of ragged stunt virus which emerged in 1976, only 3 cultivars out of 7,802 accessions tested showed resistance to the virus (IRRI, 1985*a*). Another finding from IRRI's evaluations was that the desired gene or genes vary markedly in their frequency and for some pests the sources of

Table 8.1. *Frequency of finding resistance to the whitebacked planthopper (*Sogatella furcifera *Horvath), in the IRRI collection, IRRI Entomology Department (IRRI, 1986)*

	Number		
Accessions	Tested	Selected	Per cent resistant
O. sativa	48,554	401	0.8
O. glaberrima	681	309	45.4
Wild rices	437	202	46.2

resistance/tolerance are few. The gene frequency is so low that only a large and diverse collection can yield a positive result from its evaluation. Table 8.1 illustrates this point in the case of the whitebacked planthopper (*Sogatella furcifera*). The insect is found in the cooler regions of South and Southeast Asia and is rapidly increasing in importance. A small number of *O. sativa* cultivars from South Asia are resistant to the insect, whereas some African cultivars introduced from Asia earlier show resistance, although the insect is not found in Africa.

One problem in screening large numbers of accessions for resistance to three major leaf-sucking insects has been observed at IRRI and elsewhere (Rezaul Karim & Pathak, 1982; Khush, 1984; IRRI, 1985*a*. Angeles *et al.*, 1986) For the most widely spread green leafhopper (*Nephotettix virescens*), 1,272 accessions (2.6 per cent) of the 49,000 accessions screened showed resistance. Seven genes have been found in 40 selected resistant varieties and no distinct change in the field composition of the insect has been noted. For the most destructive and variable brown planthopper (*Nilaparvata lugens*), the percentage of resistant accessions for each of the three biotypes ranged from 0.93 to 1.88 per cent of 76,000 accessions-samples tested. Six resistance genes have been identified from about 60 cultivars. For the white-backed planthopper, five resistance genes are known from allelism tests among 401 resistant parents. No relationship was apparent among the populational stability of the insect, the frequency of resistant accessions, and the number of resistance genes.

The results of large-scale mass screening of *O. sativa* cultivars to 11 additional insects by IRRI entomologists are summarised in Table 8.2. Results obtained from wild species are shown in Table 8.3. Although the numbers of wild species tested were rather small, they generally yielded a higher percentage of resistance sources than the cultivars. However, neither the cultivated nor the wild species can resist caseworm (IRRI, 1985*a*).

2. For an important staple food crop of the humid tropics such as rice, the area of cultivation is still expanding into new ecosystems as in Africa and South America and the intensity of multiple cropping is also increasing rapidly. The changing situation in pest damage, and edaphic and other ecological

stresses require a broad spectrum of genetic diversity to provide genetic protection. Specific examples are:

a) Rapid changes in the populational composition of a major pest associated with rapid turnover of major cultivars in the Asian tropics require a continuous supply of genes to cope with changes. The brown planthopper furnishes a striking example of shifting in prevalence from biotype 1 to 2 in Indonesia and Philippines. During 1975–6, the composition of field populations quickly changed when IR26 was continually planted in close succession or in staggered planting dates. This was soon followed by a shift from biotype 2 to 3 in the southern Philippines and to a different biotype in Indonesia after LR36 was extensively grown from 1978 to 1982. Fortunately, six resistance genes have been identified from *O. sativa* accessions (G. S. Khush, pers. comm.) But these are all of the vertical type. Only one Indonesian variety, Utri Rajapan,

Table 8.2. *Summary of screening* O. sativa *cultivars for resistance to insects, IRRI Entomology Department, 1962–84*

Insect	Number of accessions		Per cent resistant
	Tested	Resistant	
Striped stem borer (*Chilo suppressalis*)	15,000	23	0.15
Yellow stem borer (*Scirpophaga incertulas*)	22,920	26	0.11
Whorl maggot (*Hydrellia philippina*)	16,918	1	0.01
Green leafhopper (*Nephotettix nigropictus*)	527	66	12.53
Green leafhopper (*Nephotettix malayanus*)	158	129	81.65
Zigzag leafhopper (*Recilla dorsalis*)	2,383	36	1.51
Leaffolder (*Cnaphalcrocis medinalis*)	20,816	117	0.56
Caseworm (*Nymphula depunctalis*)	5,183	0	0.00
Thrips (*Stenchaetothrips biformis*)	237	78	32.91
Rice bug (*Leptocorisa oratorius*)	406	0	0.00
Black bug (*Scotinophara coarctata*)	300	2	0.01

appears to have the moderate resistance of the tolerance mechanism (Panda & Heinrichs, 1983).

b) Several diseases and insects have emerged from being obscure entities into being pests of major importance during the last 35 years. Paddock's 1950 book listed 24 diseases, but no virus disease was included. Two decades later, Ou (1972) described 59 diseases. Since 1972 ragged stunt virus, strain 2 of grassy stunt virus, and rice gall dwarf virus have appeared and caused considerable damage in Southeast Asia. The tungro virus now has the S and B forms. Three new virus diseases have been reported from West Africa. Other new diseases will certainly

Table 8.3. *Summary of screening wild rices for resistance to insects, IRRI, Entomology Department, up to October, 1984*

	Number of accessions		
Insect	Screened	Selected for resistance[a]	Per cent selected
Brown plant hopper (*Nilaparvata lugens*)			
Biotype 1	446	204	45.7
Biotype 2	445	168	37.8
Biotype 3	448	178	39.7
Whitebacked planthopper (*Sogatella furcifera*)	449	208	46.3
Green leafhopper (*Nephotettix* virescens	447	239	53.4
Green leafhopper (*N. nigropictus*)	91	54	59.3
Green leafhopper (*N. malayanus*)	30	26	86.7
Zigzag leafhopper (*Recilla dorsalis*)	422	218	51.7
Striped stem borer (*Chilo suppressalis*)	243	13	5.3
Yellow stem borer (*Scirpophaga incertulas*)	322	70	21.7
Leaffolder (*Cnaphalocrocis medinalis*)	338	8	2.4
Whorl maggot (*Hydrellia philippina*)	339	7	2.1
Caseworm (*Nymphula depunctalis*)	304	0	0.0
Thrips (*Stenchaetothrips biformis*)	85	12	14.0

[a] Damage ratings of 0 to 3.

be identified as rice workers became more alert to novel diseases (Chang, 1984).

Among the insects, the most serious pest in Asia, brown planthopper, was not known to most rice workers in the early 1960s. It was in 1964 and 1977 that the role of brown planthopper as a vector for two destructive virus diseases was established (Ling, 1972). Another widespread pest, whorl maggot, was identified as *Hydrellia philippina* as recently as 1968. A moderate level of resistance to whorl maggot is limited to one cultivar (IR40) after 16,000 cultivars were screened. A higher frequency of 15 per cent was found in 307 wild taxa, although the resistance sources are present in six species that are distantly related to *O. sativa* (Heinrichs *et al.*, 1985).

c) The problem of widespread zinc deficiency in rice soils was not known until the late 1960s. Earlier intensive studies in Japan associated the deficiency-syndrome of Akogare Type II soil disorder (red-withering of leaves) with ill-drained muck soils (Tanaka & Yoshida, 1970). Tolerant sources are present in several Indian and Thai varieties and a few tolerant IR varieties were bred as a result of crossing diverse parents. The genetic control of this trait is yet to be determined (Chang, 1986*b*).

d) When multiple and continuous cropping of rice or of rice and wheat in close succession continue to increase, we can expect more serious pest problems. The problem of the brown planthopper emerged when annual cycles of two or three rice crops were grown in the irrigated areas. The rice-wheat cropping sequence in central and east China and the spread of winter wheat cropping in a humid country such as Bangladesh causes concern. In Hunan Province of China, a sclerotial disease complex on rice can also attack wheat. This is a vulnerable situation which calls for hitherto unexplored genes and crop rotation.

e) A large collection is likely to include a greater level of genetic diversity than a smaller collection. As plant breeding and plant variety rights both progress at a rapid pace, the need to have access to diverse germplasm, even if only for a back-up purpose, is evident from the US experience with hybrid maize in the early 1970s.

We have identified about 150 short-statured accessions of unknown genetic make-up in the IRRI collection. Extensive tests of allelism with the sd_1 gene of Dee-geo-woo-gen (in nearly all IR varieties) showed that (1) about one-half of the semidwarf cultivars and induced mutants have the sd_1 gene, (2) seven geographically distinct cultivars share a complex locus with sd_1, and (3) about one-half of the short or shorter accessions carry other genes, but none of these is agronomically as desirable as the sd_1 type (Chang *et al.*, 1985). Nevertheless, there is a wide choice of semidwarfing sources to broaden the genetic base.

f) In the newly opened areas in the Amazon basin of Brazil, and in many parts of Africa, rice growers will certainly face emerging problems associated with new fauna and flora, or unusual soil problems. In other parts of the world, polluted water or air or both pose new problems. Only a large collection can be expected to furnish the desirable genes that can cope with the problems.

3. A large collection requires adequate physical facilities, a broader array of scientific staff, greater financial support, and a stronger supporting services than a small collection demand. Paradoxically, it is easier nowadays for a larger centre to obtain administrative and financial support for a large collection than for a smaller centre. This may be because the larger one is more visible to administrators and the public, and subsequently, will be able to get sustained support. Small centres located in remote areas are particularly handicapped in obtaining initial or sustained support.

   The IRRI germplasm bank may be an extreme case: its germplasm activities have drawn strength both from the collaborating disciplines in the Genetic Evaluation and Utilization (GEU) Program and a variety of essential services in database management, library and documentation, communication and publication, and from the supporting services in maintaining the physical facilities. For instance, it would be difficult for a small centre to have two-tier emergency generators, a team of refrigeration mechanics and electricians on 24-hour service call, and self-reliant repair services, as IRRI does.

If one traces the historical development of IRRI's collection, its rapid growth was concomitant with the growth of IRRI, particularly after the GEU Program was expanded in 1974. Meanwhile, the seed distribution services have grown proportionally (IRRI, 1985*b*). The high-yielding rice varieties (HYVs) of tropical Asia all have the sd_1 semi-dwarfing gene which was supplied by our Center. The HYVs now occupy 44 per cent of the planted rice area in South and South-east Asia.

The discussion about useful genes so far has been focussed on diseases and insect pests. The IRRI collection also contains a wealth of tolerances/resistances to deficiencies or excesses of water, adverse soil factors, and extreme temperatures. Such desirable sources have been enumerated in earlier papers (Chang, 1980, 1985; Chang *et al.*, 1982). Two outstanding sources of tolerance came from extensive field explorations of the early 1970s: the cool-tolerant Silewah from Indonesia and the floating rices of the Rayada group which can also resist the stem nematode (Ufra). The broad spectrum of genes thus obtained by field exploration and collection is essential in coping with climatic and edaphic stresses confronting subsistence farmers. The rice crop is now grown from 53 °N latitude to 40 °S latitude.

Efficacy of facilities and operation related to a large collection

In terms of inputs, it is more efficient in the long run to build, equip, and operate facilities that will accommodate a large collection than a small one. Greater initial investments in the building, insulation materials, refrigeration and dehumidification equipment, and seed packaging and storage equipment actually lead to lower operating costs over a period of time. Moreover, heavier equipment, if properly chosen and power-efficient, offers dependability, durability, and low running costs. Only a handful of moderately large-sized genebanks could afford to store seeds in hermetically sealed cans. Our Center is the only bank that uses non-rusting aluminium cans, solvent-based rubber gaskets for the can lids, sealing under partial vacuum, and compact storage of cans on mobile shelves. All these provisions mean added security, lowered power consumption, and increased storage capacity. It would be difficult for a small collection to acquire and maintain a facility that initially requires high-cost equipment. The cost of building the long-term seed storage room for a large collection is proportionally lower than that for a small collection with the same materials. When a prefabricated cabinet and associated standby generators are used, the cost for a room of 85 m^3 is nearly identical with that of a 113 m^3; when the volume is increased to 282 m^3, the estimated cost is only twice the latter (IBPGR, 1976).

The medium- and long-term storerooms at IRRI have a combined double storage volume of 684^3 m. The prefabricated cabinets, cooling units, five compressors, sea freight, and installation fees cost about $140,000. The refrigeration equipment included a dehumidification device.

In contrast, a walk-in refrigerated cabinet of 42 m^3 capacity and its dual compressors acquired by a neighbouring genebank cost about $70,000. The latter has no humidity control. A slightly larger walk-in two-compartment module recently acquired by the National Bureau of Plant Genetic Resources in New Delhi cost forty thousand pounds sterling.

In terms of operations, a large collection requires more germplasm-oriented workers and an efficient supporting staff. A large staff potentially makes more efficient use of human resources by job rotation, if and when needed. Upgrading of new staff members can be attained by implementing on-the-job training of the more competent and experienced workers. A larger collection also provides its workers with greater opportunities to carry out diversified research and thus increases the incentives for staff advancement.

Conclusions

1. Our experience with the large and diverse rice germplasm collection shows that a large collection is very effective in furnishing useful genes and certain rare alleles. But the size should refer to comprehensiveness in genetic diversity rather than numerical size.
2. Rice production under irrigated culture and continuous multiple cropping in the Asian tropics has led to rapid changes in varietal turn-over, pest dominance, and populational composition of pests. The expanding cultivation and the dynamic changes in rice ecosystems call for new genes and a greater diversity. A parallel situation is not found in other food crops of the humid tropics, some of which are gradually being replaced by rice.
3. Continuous growth in a large collection should be only allowed to conserve additions having distinctiveness. Meanwhile, serious efforts should be made to trim redundancy so that the conservation inputs would be more efficient.
4. Medium-sized national collections may double as regional collections to backup the base collection.
5. Wild relatives of rice (and also several other staple crops) should be further collected to add to the usefulness of the collections.
6. Conservation must be accompanied by multidisciplinary and systematic evaluation, full documentation, and effective communication among different disciplines, so as to enhance the usefulness of large collections and to justify long-term investments.

7. Much of the potential usefulness of genetic resources lies in the yet untested materials.

References

Angeles, E. R., Khush, G. S. & Heinrichs, E. A. (1986). Inheritance of resistance to planthoppers and leafhoppers in rice. In *Rice Genetics*, pp. 537–49, IRRI, Los Banos, Philippines.

Anonymous (1984). *Conservation and Utilization of Exotic Germplasm to Improve Varieties*. Report of 1983 Plant Breeding Research Forum. Pioneer Hi-Bred International, Des Moines, Iowa.

Cabauatan, P. Q., Hibino, H., Lapis, D. B., Omura, T. & Tsuchizaki, T. (1985). *Rice Grassy Stunt Virus 2: A New Strain of Rice Grassy Stunt in the Philippines*. IRRI Research Paper Series No. 106. IRRI, Los Banos, Philippines.

Chang, T. T. (1980). The rice genetic resources program of IRRI and its impact on rice improvement. In *Rice Improvement in China and Other Asian Countries*, pp. 85–106, IRRI, Los Banos, Philippines.

Chang, T. T. (1984). Conservation of rice genetic resources: luxury or necessity? *Science*, **224**, 251–6.

Change, T. T. (1985). Crop history and genetic conservation: rice – a case study. *Iowa State Journal of Research*, **59**, 425–55.

Chang, T. T. (1986*a*). The availability of germplasm. In *Bioscience in Crop Improvement International Symposium*, pp. 7–9, Royal Agricultural Society of England, U.K.

Chang, T. T. (1986*b*). Unique sources of genetic variability in rice germplasm. In *Proceedings of a Workshop on Biotechnology for Crop Improvement: Potentials and Limitations*. IRRI, Los Banos, Philippines.

Chang, T. T., Ou, S. H., Pathak, M. D., Ling, K. C. & Kauffman, H. E. (1975). The search for disease and insect resistance in rice germplasm. In *Crop Genetic Resources for Today and Tomorrow*, pp. 183–200, Frankel, O. H. & Hawkes, J. G. (eds.) Cambridge University Press, Cambridge.

Chang, T. T., Adair, C. R. & Johnston, T. H. 1982. The conservation and use of rice genetic resources. *Advances in Agronomy*, **35**, 37–91.

Chang, T. T., Zuno, C., Marciano-Romena, A. & Loresto, G. C. (1985). Semidwarfs in rice germplasm collection and their potentials in rice improvement. *Phytobreedon*, **1**, 1–9.

Goodman, M. M. (1984). An evaluation and critique of current germplasm programs. In *Report of the 1983 Plant Breeding Research Forum*, pp. 195–249, Pioneer Hi-Bred International, Des Moines, Iowa.

Heinrichs, E. A., Medrano, F. G. & Rapusas, H. R. (1985). *Genetic Evaluation for Insect Resistance in Rice*. IRRI, Los Banos, Philippines.

Hibino, H., Cabauatan, P. Q., Omura, T. & Tsuchizaki, T. (1985). Rice grassy stunt virus strain causing tugrolike symptoms in the Philippines. *Plant Diseases*, **69**, 538–41.

International Board for Plant Genetic Resources (1976). *Report of IBPGR Working Group on Engineering, Design and Cost Aspects of Long-term Seed Storage Facilities*. IBPGR, Rome.

IRRI (1975). *Annual Report for 1974*, IRRI, Los Banos, Philippines.

IRRI (1976). *Annual Report for 1975*, IRRI, Los Banos, Philippines.

IRRI (1977). *Annual Report for 1976*, IRRI, Los Banos, Philippines.

IRRI (1978). *Proceedings of the Workshop on the Genetic Conservation of Rice.* IRRI, Los Banos, Philippines.

IRRI (1983). *1983 Rice Germplasm Conservation Workshop.* IRRI, Los Banos, Philippines.

IRRI (1985*a*). *Annual Report for 1984*, p. 62, IRRI, Los Banos, Philippines.

IRRI (1985*b*). Genetic resources. In *International Rice Research: 25 Years of Partnership*, pp. 41–7, IRRI, Los Banos, Philippines.

Khush, G. S. (1984). Breeding rice for resistance to insects. *Protection Ecology*, **7**, 147–65.

Ling, K. C. (1972). *Rice Virus Diseases.* IRRI, Los Banos, Philippines.

Lyman, J. (1984). Progress and planning for germplasm conservation of major food crops. *Plant Genetic Resources Newsletter*, **60**, 3–21.

Marshall, D. R. & Brown, A. H. D. (1975). Optimum sampling strategies in genetic conservation. In *Crop Genetic Resources for Today and Tomorrow.* pp. 53–80. Frankel, O. H. & Hawkes, J. G. (eds.), Cambridge University Press, Cambridge.

Mooney, P. R. (1985). 'The Law of the Seed' revisited: seed wars at the Circo Massimo. *Development Dialogue*, **1**, 139–52.

Ou, S. H. (1972). *Rice Diseases.* Commonwealth Mycological Institute. Kew, Surrey, UK.

Panda, N. & Heinrichs, E. A. (1983). Levels of tolerance and antibiosis in rice varieties having moderate resistance to the brown planthopper, *Nilaparvata lugens* (Stal) (Hemiptera: Delphacidae). *Environmental Entomology*, **12**, 1204–14.

Plucknett, D. L., Smith, N. J. H., Williams, J. T. & Anishetty, N. M. (1983). Crop germplasm conservation and developing countries. *Science*, **220**, 163–9.

Rezaul Karim, A. N. M. & Pathak, M. D. (1982). New genes for resistance to green leafhopper, *Nephotettix virescens* (Distant) in rice, *Oryza sativa* L. *Crop Protection*, **1**, 483–90.

Sun, M. (1986). Fiscal neglect breeds problems for seed banks. *Science*, **231**, 329–30.

Tanaka, A. & Yoshida, S. (1970). *Nutritional Disorders of the Rice Plant in Asia.* IRRI, Los Banos, Philippines.

Toll, J., Anishetty, N. M. & Ayad, G. (1980). *Directory of Germplasm Collections. III. Cereals. 3. Rice.* IBPGR, Rome.

9
The case for core collections

A. H. D. BROWN

Introduction

If collections of plant germplasm are to be used more in the future than they are at present, then they will have to be better collections. This is the challenge of a new phase in the saga of genetic resources, and one which has led to several approaches. In other chapters, the importance of large collections, complete evaluation and expanding size, and the role of national collections, are considered. The thesis developed in this chapter is that a 'better' collection is one that is rationalised, refined and structured, around a small, well-defined and representative 'core'.

The numbers and scope of accessions in a collection, the information about them and the access to them are factors often argued as crucial to the use of collections in plant breeding. Paradoxically, however, the key to greater use may lie elsewhere. As recently pointed out (Frankel & Brown, 1984; Holden, 1984), germplasm collections have grown markedly in size, and may now be so large as to deter their extensive use for all but a few characters which are readily and rapidly discerned on single plants (e.g. some morphological traits, chemical differences detectable by spot tests, and major genes for disease resistance). Greater use of germplasm collections could be made, particularly for a wider range of characters, if a smaller number of accessions were to be given priority in evaluation and hybridisation.

Concept of the core collection

Given the need for economy of size, Frankel (1984) argued that a collection could be pruned to what he termed a 'core collection', which would represent, 'with a minimum of repetitiveness, the genetic diversity of a crop species and its relatives'. The accessions not included in the core would, apart from duplicates, not be jettisoned but retained as the 'reserve collection'.

The proposal is to set up a hierarchical structure to the collection. In simplest form, the hierarchy consists of the core and reserve fractions, but other levels may prove useful in future. The main purpose of the core fraction is to provide efficient access to the whole collection. The most important general criterion in setting it up is that, as a whole, its components represent and cover the major kinds of diversity present. Thus it would be the likely first source of genes to meet both present and future needs. This 'first look' may result in finding an accession in the core with the desired attribute. It may also guide the breeder to other better sources (more strongly expressing, or in more adapted genetic backgrounds) held in the reserve.

The word 'core' means the central or innermost part, the heart, and the most important part. It might be tempting to use the term 'core collection' in another context – that is to mean a subset of accessions for a single specific purpose. Thus a set of lines likely to confer tolerance of acid soils might be called 'the core for acid soil tolerance'. Similarly a selected set of lines for resistance to stem rust might be called 'the core for stem rust resistance'. I believe that use of the term for such sets of lines should be avoided. Such sets of accessions need no title other than the common phenotypic trait they are intended to express. Use of the term in this context leads to confusion over a fundamental issue. Consider two resistant lines A and B, and a susceptible line C. A 'disease resistant set' would include A and B and exclude C. A two-entry 'core' of these lines would include C with A or B and exclude the other resistant as more likely to be redundant (in the total absence of other evidence). Whether or not one can arrive at a true 'core' set for the collection by including all the entries that are a source for any useful character, is a moot point. This may or may not be the way to achieve the aim of representing the diversity.

Another very recent use of the term 'core collection' has entered official IBPGR parlance, regrettably. This usage is to denote that part of a gene bank collection which is freely available on request without restriction (Anon, 1985). The alternative part, or 'reserve' is that unavailable for general distribution because of restrictions by plant breeders rights, narcotic regulations, etc. This usage has neither precedence nor semantics in its favour, and is bound to confuse.

Reasons for setting up a core collection

Three different groups of arguments support the concept of a core collection. The first group includes statistical sampling considerations which essentially stem from the assumption that breeders, through crossing and selection, can recover desirable alleles when required. Hence in principle they need access to only one copy of such alleles. The second group covers those reasons relating to the genetic structure of plant populations in general, and germplasm collections in particular. The third group has to do with making easier the management of germplasm collections.

Statistical basis

A statistical rationale for the assembling of core collection follows from asking how adequate a representation of the whole collection does a restricted sample provide. Obviously some less of variation must attend sampling – the question is how much? If the allelic richness (i.e. the number of different alleles) in the reduced collection were but a pale image of that in the total, the case for core collections would be very dubious. Alternatively, if relatively little were lost, the case for core collections would be strengthened.

An exact answer to the question would require knowledge of the levels of genetic variation within and among the samples comprising the collection, as well as the method of selection to be employed. As such information is limited, general principles can first be based on the neutral allele model of Kimura & Crow (1964), and its sampling theory (Ewens, 1972; Nei, Maruyama & Chakraborty, 1975). Random sampling is assumed in order to establish baseline expectations. In practice the selection of core entries is systematic – deliberately aimed at covering the collection. Such systematic sampling should improve the effectiveness of recovery over that expected from random sampling.

Kinds of alleles

Let us assume that N is the total number of accessions in the collection, from which N_r accessions have to be sampled to form the core. Considering the collection as a whole, each allele could be classified as to whether its frequency within any one accession was ever greater than say 0.10, in which accession it would be regarded as *common*. Alternatively if the allele never reached significant frequencies in an accession it is classified as *rare*. Secondly, each common allele can be classified as to whether its common occurrences are *widespread*, or *localised* in very few accessions (at the extreme, in just one sample). A similar classification can be made for the rare alleles, based on their presence in few or many accessions. The result is that there are four conceptual classes of alleles (Marshall & Brown, 1975, 1983). The first class (*common, widespread*) are almost certainly included in a subsample of the collection, so need not concern us further. The fourth class, those which are *very rare and restricted* in occurrence, are sometimes put forward in debates about genetic resources as counter examples to proposed strategies (Bogyo *et al.*, 1980). In practice, the inclusion of such an allele in a collection is fortuitous, and remarkable. Obsession with the conservation of these alleles, or according them a priority above all other, does not lead to any guiding principles in handling resources – other than the impractical one of conserving everything.

The remaining two classes are relevant to formulating sampling strategies. Considering first the *rare, widespread* class, we assume that because these alleles are 'widespread', they behave as if the collection were one large bulk population, and each accession were a random subsample from it. What is the frequency distribution of alleles in the collection, or how many alleles are present which fall into various frequency classes? An answer to this question can be found for the neutral allele model. However, this model takes no account of the complex 'sampling' structure involved in the typical collection, so the answer is necessarily approximate.

Infinite neutral allele profiles

For the model, the expected number of alleles (n_a) with frequencies lying between p and q ($0 < p < q < 1$) in an equilibrium population of effective size N_e is

$$n_a = \theta \int_p^q (1-x)^{\theta-1} x^{-1} \, dx$$

The distribution depends on only one parameter θ ($= 4N_e u$, where u = mutation rate). The integration can be obtained for specific values of θ. For example if $\theta = 1$, there are 6.91 alleles per locus expected in the large collection with allele frequency (p) greater than 0.001. If this boundary is halved to 0.0005, then only 0.69 more alleles are added to this total (Table 9.1).

Turning now to the presumed random subsample as the core collection, what is the expected number of alleles retained (n_r)? From Ewens' (1972) sampling theory, and an integral approximation (Brown & Moran, 1980), the expected number of alleles is:

$$n_r \simeq \theta \log_e [\theta + 2N_r - 1)/\theta] + 0.5$$

with

$$\text{var}(n_r) \simeq \theta\{\log_e [(\theta + 2N_r - 1)/\theta] - 1\}$$

In Table 9.1, three specific values of θ are assumed and the expected number of alleles in the collection with frequencies greater than various values (p) shown. Table 9.1 also shows for the same θ, the expected number of alleles retained in the core in various samples of size N_r. As an example when $\theta = 1$, a sample of 1,000 individuals (presumed to be independent and unrelated with respect to the loci in question) would retain 8.1 alleles. These eight alleles would not usually be the eight most frequent at that locus. A two-fold increase in sampling size improves the retention only slightly to 10.4 alleles. The values in Table 9.1 show that very large increases in sample size have increasingly marginal effects on the number of different alleles kept.

The rare alleles at loci which, in this category, we are assuming to be little differentiated among accessions, thus clearly give a strong statistical argument for the concept of a core collection, provided their frequency distribution approximates that of the infinite allele model. This class and the above sampling theory for it, also give one rationale for the proportion to include in the core. An approximate (95 per cent) confidence interval for the number of alleles retained (n_r) can be computed, assuming approximation to the normal distribution. The lower confidence limit of

Table 9.1. *Expected number of neutral alleles (n_a) with frequency greater than (p) in the population: and the expected number of alleles (n_r) in a sample of size N_r*

	Level of polymorphism		
	$\theta = 0.5$	$\theta = 1.0$	$\theta = 2.0$
Frequency class (p)	**Number of alleles in the collection (n_a)**		
> 0.10	1.82	2.30	2.8
> 0.01	2.99	4.61	7.2
> 0.001	4.15	6.91	11.8
> 0.0005	4.50	7.60	13.2
> 0.0001	5.30	9.21	16.4
Sample size (N_r)	**Number of alleles in the sample (n_r)**		
50	3.15	5.1	8.3
100	3.50	5.8	9.7
300	4.05	6.9	11.7
1,000	4.65	8.1	14.3
3,000	5.20	9.2	16.5
10,000	5.80	10.4	18.9

n_r can then be expressed as a fraction of the actual number of alleles present in the original population, for a sample size (N_r), which is also expressed as a per cent of the original population. Fig. 9.1 relates these two values for a particular level of polymorphism ($\theta = 1$) and two sizes of collections, but the general conclusions hold for other values of θ.

In Fig. 9.1A and 9.1B the lowest curves apply for a single locus. They show that at any particular locus, we are 95 per cent certain of retaining at least about 20 per cent (Fig. 9.1A: 31 per cent) of alleles if 10 per cent of the original population is retained. But most loci will do much better than this. Thus, for the total number of alleles at 100 equally polymorphic loci, a sample size of 10 per cent retains at least 70 per cent (Fig. 9.1A: 77 per cent) of the alleles with 95 per cent certainty.

Fig. 9.1. The lower 95% confidence limit for the expected number of alleles retained (n_r) expressed as a fraction of the total number actually present (n_a) in the population, when a percentage of the total population (N_e) is kept. Figure 9.1A is for an original size of $N_e = 10^4$, and Figure 9.1B for $N_e = 10^6$

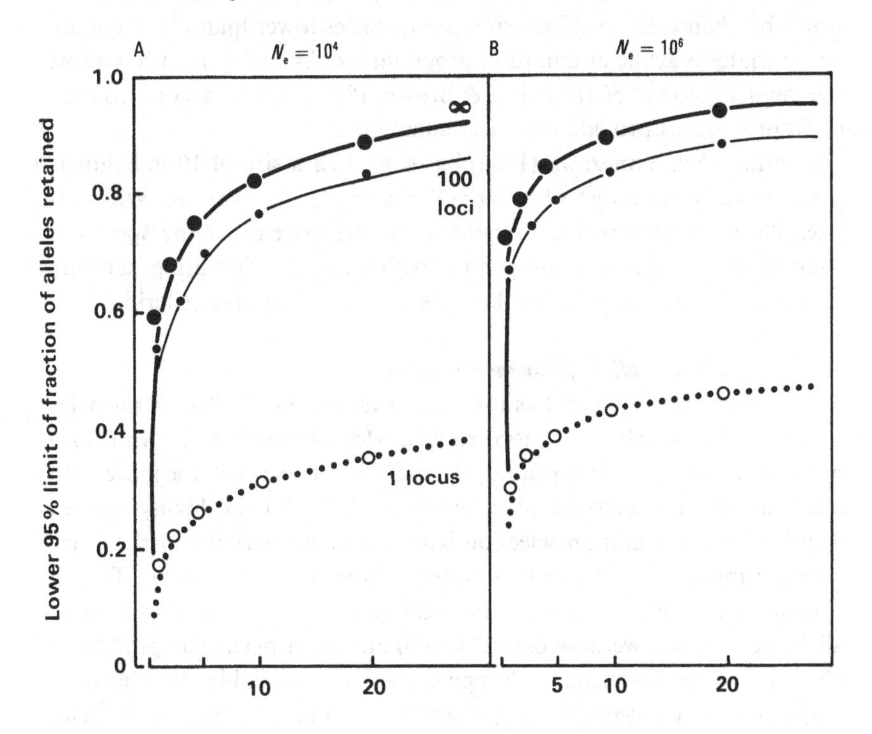

The curves in general have two distinct phases. As the percentage retention of initial population decreases, there is a slowly increasing loss of alleles. Somewhere below 10 per cent, however, there is a marked increase in rate of loss. The conclusion is that even quite drastic reductions in the size of the collections – which are assumed to be random with respect to the 'widespread rare' classes of alleles – will not seriously reduce the total number of such alleles. A convenient rule of thumb would be that the size of the core collection should be about 10 per cent of the entire collection.

The problem of including the widespread rare alleles also gives a rationale for a basic minimum size. Knowing how many alleles (n_r) are expected in a sample (of size N_r) the idea is to choose N_r to match this number (n_r) with the number present in the collection (n_a) above a given frequency. From Table 9.1, a sample of 3,000 is expected to contain 9.2 alleles (if $\theta = 1$). This number of alleles is as many distinct alleles at that locus as exist in the total collection with frequency greater than 10^{-4} (Table 10.1). As noted already, a few alleles in the sample would have population (collection) frequencies less than 10^{-4}. These alleles compensate for those with population frequency just greater than 10^{-4}, but which by chance escaped inclusion. Frequencies lower than this are more appropriate to artificial mutation programmes. Assuming a basic sample size per accession of 50 (Marshall & Brown, 1975), then 60 accessions each of 50 propagules provide the basic minimum.

Alternatively, Yonezawa (1985) has suggested a size of 10 individuals per site as a better target when strict limits apply to a total size. With this size, 300 accessions would be needed. At the extreme, if the species is completely inbred and all propagules within an accession are genetically identical, the minimum would be 3,000 accessions by this criterion.

Empirical allele frequency profiles

The above theory has used the distributions of allele frequencies generated by the infinite, neutral allele model. Marshall & Brown (1975) noted that such frequency profiles are intermediate between the more even distributions characteristic of heterotic models, and the highly skewed distributions of mutation-selection balance. Allelic variants in the latter type are much more difficult to preserve from loss in sampling. To test whether the above conclusions are very sensitive to the distribution of allele frequencies, we now consider sampling from particular profiles of observed allelic frequencies. Suppose data are available on the allele frequencies in a collection. It is possible to compute the number of alleles that a random sample of size N_r would be expected theoretically to retain.

At a single locus, assume k alleles are present with frequency profile $\{p_1, p_2, \dots, p_i, \dots, p_k\}$. Under a model of sampling with replacement (i.e. sampling from a very large population), the expected number of alleles is

$$n_a = k - \sum_{i=1}^{k} (1-p_i)^{N_r}$$

The analogous quantity for the model of sampling without replacement (which is appropriate when the collection is small), based on the hypergeometric distribution, is

$$n_r = k - \sum_{i=1}^{k} Q_i$$

where

$$Q_i = \prod_{j=1}^{N_r} [N(1-p_i)-j]/[N-j] \quad \text{if } p_i < 1 - N_r/N$$
$$= 0 \quad \text{otherwise.}$$

These formulae can be used with any set of allelic frequencies for many loci, and the expected retention for all the loci computed as the sum of the value of n_r for each locus. The proportion of alleles retained for a specific value of N_r can be computed.

The behaviour of these formulae is most affected by the smallest values of allele frequency. The above section used 10^{-4} as a lower boundary of interest. Then for a diallelic locus

$$n_r \simeq 2 - (0.9999)^{N_r}$$

and for a locus with allele frequency profile (0.9996, 0.0001, 0.0001, 0.0001, 0.0001)

$$n_r \simeq 5 - 4\,(0.9999)^{N_r}$$

A sample of 3,000 would be expected to retain ($n_r \simeq 1.26$, and $n_r \simeq 2.04$) 63 per cent and 41 per cent of the alleles in these two profiles. Hence, even for these severely skewed distributions, the retention properties of the suggested samples size is adequate.

Uneven distribution of diversity

The above sampling theory has suggested that a sensible target size for a core collection should be 5 to 10 per cent of the total collection, but at least greater than 3,000 individuals. However, this theory applies only to loci which are essentially undifferentiated among accessions in the collection (the widespread class). It further assumes that the level of polymorphism is evenly spread over the collection. This is generally unlikely to be the case, as collections would be expected to contain accessions which differ in their level of diversity.

To test the effect of uneven distribution of diversity on the retention properties in the core sampling, a collection consisting of two parts with markedly different values of θ is assumed. Suppose that, in the richer fraction, which comprises only one-ninth of the collection, $\theta = 5$ and in the bulk (8/9), $\theta = 0.5$. Using the above sampling theory, the base collection (if $N_e = 10^4$) is expected to contain about 29 alleles; or if $N_e = 10^6$, about 52 alleles. The number of alleles retained by various random fractions (one per cent, 2.5 per cent, ..., 50 per cent) of the collection and their standard errors were computed from the above formulae. The expected retentions as a fraction of the total alleles are shown in Figs 9.2A and 9.2B as well as the lower confidence limit for one locus. For comparison, the same retention statistics are shown for collections with an even distribution of diversity, giving similar values for the total number of alleles (for $N_e = 10^4$; $\theta = 3.5$ and $n_a = 28.7$; for $N_e = 10^6$, $\theta = 4.0$ and $n_a = 50.7$). As expected, the concentration of diversity in a small fraction of a collection will lead to lower levels of variation in the core (unless such rich samples are known to the curator and an increased fraction of them included). However, the effect is less marked than might be anticipated.

Fig. 9.2. Retention patterns of samples from uniform (θ constant) compared with uneven (θ variable) levels of diversity in the base population. The retention is expressed as in Fig. 9.1. Figure 9.2A is for $N_e = 10^4$, and Fig. 9.2B is for $N_e = 10^6$.

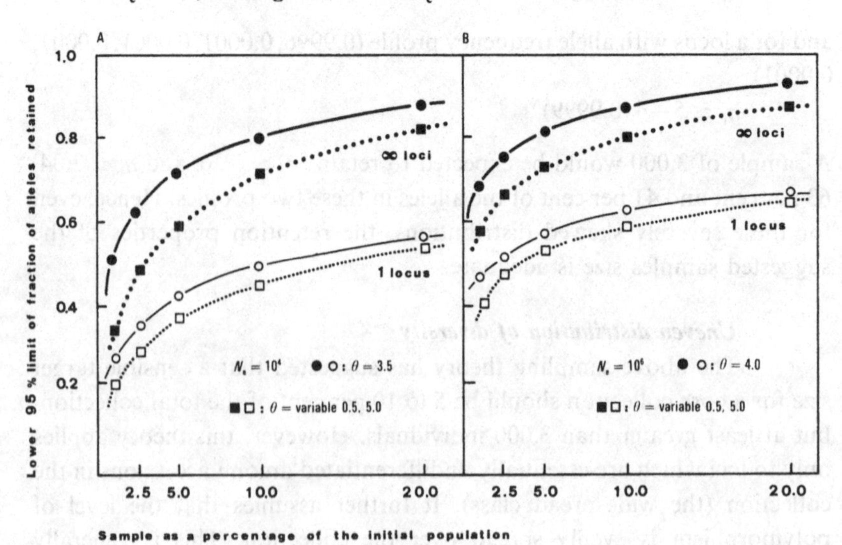

We now turn to the final class of alleles, those which are common yet highly localised.

The *common, localised* alleles are an important class to the user of collections. They include restricted sources of disease and pest resistance, and specialised ecological tolerances. Of the many examples, the occurrence of resistance to barley yellow dwarf virus in Ethiopian barleys (Qualset, 1975) or the genes for dwarf stature in Japanese wheats (Marshall & Brown, 1981) illustrate these cases. For such alleles, the issue is one of scale. Highly localised alleles are less likely to be included in the core collection. A sample of such genes will be included and the problem in core designation resolves itself into maximising this component while keeping accession numbers within the accepted limit. This is done by ensuring coverage of the major geographic, ecological zones and morphological discontinuities in the collection. As an aside, it can be urged that the remaining, highly localised alleles which escape inclusion, are less likely to confer broad adaptability on their descendants in breeding programmes, than are their more widespread counterparts. If there are none of the latter type (e.g. the only resistance source is to be found in one accession), and the need is great, then one must resort to the whole collection as a reserve. However, such an extreme example cannot be uppermost in mind when framing strategies for increased germplasm use in plant breeding.

The genetic basis of the core concept

The principal thesis to which all the evidence of experimental plant population genetics points, is that the genetic diversity of a species is not randomly dispersed within and among populations, but is organised to varying degrees. The extent to which this organisation is caused by systematic selection pressures, as opposed to random processes (such as isolation by distance) and the extent to which this organisation is itself adaptive, are topics of disagreement and the focus of research. However, the fact of organisation, whatever its causes, adds strongly to the case of the core collection, because it means that carefully chosen subsamples of a species will contain more of the species genetic diversity than would the random subsamples discussed already. Thus it is this basic structure of species variation which makes small samples relatively efficient at including a disproportionately large fraction of the variation of an entire species.

Conversely, this pervasive structure means that carelessly chosen subsamples of a species might well reduce the level of diversity in the core, below that expected in a random fraction. This can happen when choice is governed unduly by any character state (e.g. rust resistance) or by one geographic region (e.g. a centre of genetic diversity).

Genetic structure can occur on two basic axes, the first of which is geographic. Populations of a species may diverge genetically to varying extents. Thus two individuals coming from different populations may be more likely to differ at a locus than were they to come from the same population. However, when populations differ among themselves in their level of diversity, two individuals drawn from a particular, highly polymorphic population may be more likely to differ at a locus, than two individuals from different populations. The problem in this case is knowing which populations are the more highly polymorphic.

The second axis of genetic structure is genomic. When two gametes (from one population) are known to differ at one locus, they may more often differ at a second locus than would be the case if they were identical at the first locus. Clearly these two kinds of association of genes can vary in degree, from being barely perceptible to being extremely strong, according to a range of factors (Table 9.2). One further complication is that associations along both axes interact so that two gametes drawn from two different populations may be more likely to differ at locus B if they are known to differ at locus A, than are two gametes either from the same populations but differing at A, or from different populations but identical at locus A. It is this reinforcement which is likely to be apparent in germplasm collections.

These are the conceivable kinds of organisation of genetic diversity in populations. What is the extent of such organisation in plants?

The extent of population differentiation in plants was recently reviewed by Loveless & Hamrick (1984). Their survey of the electrophoretic evidence from natural populations of plant species related the degree of differentiation with several biological attributes (breeding system, mode of pollination and seed dispersal, life history, geographic range, successional stage, etc). All these factors individually appear to have a significant effect on the extent to which populations differ from one

Table 9.2. *Two axes of population genetic structure*

Geographic patterns (population differentiation)	Genomic associations (linkage disequilibrium)
founder effect	linkage
migration	mutator systems
breeding system	breeding system
diversifying selection	epistatic selection
population size	population size

another. However, the primary variable appears to be the breeding system, especially outbreeding as opposed to uniparental breeding systems (self-fertilisation or obligate apomixis). As measured by diversity indices, outbreeding populations are much less differentiated, whereas selfers or apomicts are strongly differentiated. However, populations of outbreeders can contain different alleles, especially rare alleles, which contribute little to diversity indices. For example, 163 distinct isozyme alleles were recorded in a survey of 94 accessions of maize representing 34 Mexican races. Of these alleles, about half occurred in only five or fewer collections (Doebley *et al.*, 1985). Thus plant populations are distinctive from one another to varying extents.

While inbreeders show more intense population differentiation than do outbreeders, they are also more prone to show an uneven distribution of their genetic diversity among populations (Brown, 1978). In principle, this makes their optimal sampling more difficult, and any evidence on the geographic pattern of their diversity levels would greatly improve sampling efficiency.

Concerning the second axis of structure (genomic), far fewer studies are available because of the complexity of studying the associations among several genes at once, and because sampling problems are more severe than for single-locus studies (Brown, 1984). The few studies which can be collated for an overview have shown that the breeding system again is the critical variable. Autogamous species generally show marked multi-locus associations, whereas allogamous species show associations only between closely linked genes.

Only a few studies have been made of the genetic structure of germplasm collections as such (Jain, 1979). Kahler & Allard (1981) found both geographic patterns of allelic variation and the prevalence of some allelic combinations at four esterase loci in the barley collection of the United States Department of Agriculture (USDA). Multivariate analysis of variation at 40 isozyme loci, clustered rice accessions into two main groups with some intermediate types (Second, 1982). In the USDA's collection of *Triticum turgidum*, diversity for six morphological polymorphisms was patchily distributed among countries (Jain *et al.*, 1975). Detailed studies of the isozyme variation in the maize races from the Americas (Goodman & Stuber, 1983; Doebley *et al.*, 1985) have related allelic frequencies with racial classification, and partially with altitude of origin. It is already clear from these data that the above two dimensions of genetic structure; among groups of accessions of similar origin, and between loci, are apparent in germplasm collections, particularly in self-fertilising crops.

Pragmatic reasons for a core collection

Most activities in the running of germplasm collections require the curator to make choices, or set priorities among accessions. When the centre receives new samples, the curator has to decide whether the new receipts merit adding to the collection. This decision is based on whether the samples represent new variation, or whether they would be redundant. Choice is also involved in selecting which accessions are to be evaluated, tested for viability, or regenerated. When responding to a request, the curator may have to decide the number and scope of accessions to dispatch.

The designation of a core set of accessions can be a formal aid to this decision making. In the first example above, a core would provide a ready reference set of accessions for assessing redundancy. (Does the new sample resemble any current core item? If so, how many of this type are in the reserve collection already? If not, should the sample be a new core item?) Evaluation for new characters could be made first on this set, and more extensive secondary surveys of the reserve collection made more efficient. Indeed, the development and testing of new procedures for screening could be done on the core. Priorities for conservation would be clear. (For example, the core would form the set of accessions which should first go to a secondary centre as a duplicate holding, or the set of accessions which could be held under several storage conditions if such are available.) The viability of the core accessions could be monitored to indicate when groups of related accessions might need regeneration. The designation of a core is needed for testing general combining ability with locally adapted germplasm in the search for yield enhancement (Frankel & Brown, 1984). Such evaluation has always been an option for outbreeding species, but is now likely to be available to inbreeding species too, with the aid of male gametocides or chemical hybridising agents. Finally documentation, and communication of results informally and through the scientific literature, would be facilitated if the number of accessions were cut substantially.

The main idea behind all these possibilities is that resources are saved by reducing the number of accessions to be handled in any one operation. These resources can then be devoted to a greater diversity of activities, in greater depth and thoroughness.

Selection of the core entries

Having developed the aims of, and justification for, the core collection, we now consider the steps to set one up. Of course, the actual details will differ markedly between species and even between collections of the same crop. Yet several issues are typically met, so some general guideliness can be given.

The first major issue is size. Two approaches suggest themselves. The first is to decide upon an almost arbitrary proportion. In the light of the above theory, five to 10 per cent would be appropriate. The second approach is to nominate an upper limit for a category of accessions (e.g. 3,000 per species). Such a limit could be justified as natural, or arise from restricted resources, but the size is not chosen as a specific proportion of the total collection. The decision between these two ways to set the total size is not obvious. The first approach seems preferable when the collection essentially covers all the known morphological and geographic range of a species.

Classification

Once the size is decided, the problem is to analyse the collection to identify degrees of genetic similarity or communality among the accessions. This operation is essentially a hierarchical cluster analysis, where every accession is sorted into a related subgroup from which a representative sample will be drawn. To identify these groups, three types of data can be used, namely

1. The origin of accessions, the collection site or breeding station.
2. Characterisation data including taxonomic data and genetic markers, both morphological and biochemical.
3. Evaluation of agronomic traits.

Ideally, there should be sufficient descriptors to define about as many groups as there are to be chosen for the core collection. The hierarchy of descriptors used for selecting a core of the Australian Winter Cereals Collection was taxonomy (species, cytotype); category of material (wild, landrace, old or recent cultivar, breeder's line); country; region; morphological traits.

In several situations, multivariate methods such as numerical cluster analysis, principle component analysis and/or network analysis, will be valuable tools. This is particularly so when many characters have been evaluated on most accessions. Such an approach has been favoured for pasture plant introduction (Burt, 1983). The methods can also be applied to passport information provided the minimum data of latitude and longitude are known and that these variables can index a large body of climatic, geological or ecological data so that hierarchical classification of sites of origin might indicate communality among accessions. It is really the ability of multivariate techniques to cope with large numbers of accessions, or several data points for each accession, that is relevant here.

When groups of similar accessions are obtained by numerical techniques, it is important that a biological basis can be seen to these groupings if they are to be important in designating the core collection. For example, Doyle & Brown (1985) used a numerical analysis of isozyme patterns to cluster accessions of *Glycine tomentella* into related groups. Subsequent study of the members of the groups revealed cytogenetic as well as geographic affinities (Doyle *et al.*, 1986), thus confirming the grouping as a biological basis for deciding on a core collection for this wild perennial relative of soybean. Lower levels of grouping, even if they did reveal subtle patterns of similarity were judged far less significant.

Problems of data

Typically, several problems with the database of the whole collection will make the task of selecting the core more difficult and less automatic. These problems are that 1) data are uneven, varying greatly in extent among accessions; 2) data vary in reliability; 3) most traits do not measure genetic similarity directly (Brown & Clegg, 1983); and 4) the numbers to include from each group defined by the analysis must still be decided arbitrarily (e.g. landraces as opposed to cultivars). I do not believe that the answer to these problems is to ignore the assembled information and choose accessions at random. This amounts to assuming that all accessions in a collection are of equal value, and there is no reason to assume so. Nor should a group necessarily be represented in the core in the proportion which it occurs in the collection. Partial data, at least, give evidence of distinctiveness, even when such data cannot be trusted to provide a reliable measure of similarity.

Another general principle when choosing between accessions with uneven data, is that preference could be given to accessions with more extensive or reliable data provided the overall representation is not infringed. A typical example is the present holding of *Triticum dicoccoides* in the Australian Winter Cereals Collection. There are 272 well-documented samples from Israel and 16 others (from USSR or unknown). A set of 30 accessions of this taxon, 28 from Israel, was chosen for the core collection. Although the origin of the Israeli material is better known, two samples of the other lines were included to retain coverage. At some future date, authentic material from elsewhere in the species range should replace these samples and the Israeli representation be reduced.

Bulking

Bulking of distinct accessions to form an entry in the core is not recommended. Bulking has a role to play in amalgamating clearly redundant accessions and possibly in coping with numbers in the reserve collection if these are excessive. This has been its role in the several collections of maize races in Latin America. Timothy & Goodman (1979) commented how important it was for rational use of this germplasm to retain a few strains per country chosen as 'types' and to maintain these separately. Equal amounts of the remaining samples of each race were bulked into racial composites. 'Only when individual samples cannot be maintained should composites be initiated. Compositing should be done only on a biologically systematic basis with as many categorical units as possible.'

Collections of asexual clones

Collections of species which must be vegetatively propagated (e.g. banana, edible aroids, yams, sisal) presents a dilemma for the core concept. On the one hand, the cost of maintaining such collections places a severe restriction on their size. This amounts to a strong incentive to construct a core. On the other hand, their asexuality means they will be used directly as genotypes and not as the source of desirable genes. One cannot use the rationale of access only to a single copy of each allele. Therefore in designating a core collection for an asexual species, phenotypic characters in the relevant environments will be more important, and measures of genetic differences, or information on origin will be relatively less important than in designating a core for a sexual species. The same point holds when a core is to be chosen to assist evaluation of a large collection for direct use, as in the case of okra described by Hamon & van Sloten (this volume).

Example – perennial wild relatives of soybean

Over the past decade, a large collection of perennial species of the genus *Glycine*, many of which are indigenous to Australia, has been built up. As of June 1986, this collection comprises 1,400 accessions of at least 12 species. To facilitate duplication at overseas institutions, research on hybridisation and evaluation, and dispatch to users, a core set of accessions was designated. The criteria in choosing these accessions were to include: 1) at least a few accessions for each species to provide some 'replication' at the species level for generalisations; 2) geographic coverage of each Australian State, and as broad a scatter and range of habitats as possible; 3) known morphological, cytological and isozyme groups where intraspecific variation has been studied; and 4) when choices were still to be made, preference went either to accessions which had been used for research in the past, or were authentic, first-hand collections.

The numbers in the core collection and in the total are shown in Table 9.3. they show, in achieving the overall target of about 10 per cent, that the proportion of each category (species × state) varies dramatically. Hence the core, constructed in this way, is very different from a random set.

Evolution of the core

The core set of accessions should be a dynamic rather than a static set of accessions. In the course of time, changes in content and size of the core are to be anticipated, indeed sought, by the curator. The factors which will cause such changes are:

1. Receipt of new accessions into the collection from distinct new areas or of new taxa.
2. Replacement of accessions of questionable authenticity with new samples from a presumably comparable source.

Table 9.3. *Core and total collections of perennial* Glycine *species, by state of origin (June 1986)*[a]

Glycine species	Q[b]	NSW	Vic	Tas	SA	WA	NT	Other
arenaria	—	—	—	—	—	3 4		—[c]
argyrea	3 15	—	—	—	—	—	—	—
canescens	8	6 65	1 1	—	2 5	3 8	3 16	—
clandestina	3 20	11 158	[d]	1 1	2 3		—	
curvata	4 7	—	—	—	—	—	—	—
cyrtoloba	5 27	1 4	—	—	—	—	—	—
falcata	5 23	—	—	—	—			—
latifolia	6 14	5 10	—	—	—	—	—	—
latrobeana	—	—	3 9	2 5	1 1	—	—	—
microphylla	2 14	5 85	1 2	1 1		—	—	
tabacina	5 42	8 151			1 2			2 38
tomentella (2×)	4 52		—	—	—	1 21		1 9
tomentella (4×)	5 139	3 40	—	—	—	2 5	1 8	3 15

[a] I wish to acknowledge the assistance of Mr J.P. Grace in selection of the core accessions.

[b] Abbreviations: Q = Queensland, NSW = New South Wales, Vic = Victoria, Tas = Tasmania, SA = South Australia, WA = Western Australia, NT = Northern Territory, Other = Southeast Asia, Papua New Guinea, South Pacific Islands.

[c] A hyphen indicates that the species is not known to occur in that State.

[d] A blank space indicates that the species occurs but is still to be collected in that state.

3. Revision of categories or affinities in the light of new data about accessions from evaluation, research, or indeed 'passport' information obtained from later queries to the original sources.
4. Review of breeder's priorities and needs.

However, alteration of the core should be a relatively rare process. One aim of the core is to build up a body of information on a restricted 'reference' set of lines, and too rapid a flux of accessions through the core would defeat this aim.

Thus the task of maintaining a core is ongoing for the curator of a collection. It resembles that of a systematic botanist, in continually updating and revising groupings and affinities among accessions. The analogue of the taxonomist's characters are the indicators of genetic distinctiveness and coverage. As summarised above, these indicators are origin, morphology and, ultimately, evaluation data.

Role of the reserve collection

If a core is designated for a collection, will not the remaining accessions, despite being retained as the 'reserve', wither and die of neglect? Will not this major fraction of the collection be rendered vulnerable to the action of unwanted economies?

To prevent these undesirable outcomes, the reserve must be seen as an integral part of the genetic resources of the collection. It should serve at least three functions. First, it is a back-up resource which can be screened for a needed variant, if such is not present in the core entires. Second, it may provide alternative sources of the same character, which may be in somewhat different genetic backgrounds. These supplementary sources may be known as such from previous evaluation, or they may be potential alternatives because they share origin or affinity with the core item. Third, it may provide a reserve source of seed for an item in the core collection if this is temporarily unavailable, and urgently needed.

The general rationale for core designation, as developed here, has been as an efficient way into the total collection for its greater use. In this scheme, the reserve is an essential part. This conception separates as two distinct issues whether a collection is too large to use easily, as opposed to being too large to maintain. The former problem points to the need for a core. The latter problem points the need to reduce redundancies by bulking, if necessary, to lower numbers in the reserve (as first discussed in Frankel & Soulé, 1981).

The arguments and roles for a core collection are not to be seen as reasons for scrapping 90 per cent of the germplasm in collections. Rather they are pointers as to how we can now use the treasury of genetic resources, acquired at great expense. This use should increase, rather than lessen, the will to preserve such material for future generations.

Conclusions

1. To encourage greater use of a germplasm collection by breeders, a core collection needs to be designated by the curator. This core would represent the genetic diversity in the collection.
2. Statistical considerations suggest that the core should consist of about 10 per cent of the collection, up to about 3,000 entries.
3. Because genetic diversity is organized in plants, such a restricted subsample of a species can, if chosen carefully be very efficient in retaining variation.
4. The choice of core entries should be on data (i) about the origin of the accessions, (ii) from characterisation; and finally (iii) from evaluation. Multivariate analysis of the set of data can be a useful tool.
5. Inevitably data will be uneven in extent and reliability. Yet choice based on evidence of genetic divergence will be better than purely random choice.
6. The composition of the core itself will change with time, as new material, data or requirements come along.
7. The remaining accessions form the reserve collection. They should be conserved as secondary sources.

References

Anon (1985). *Report of the Third Meeting of IBPGR Advisory Committee on Seed Storage*. IBPGR, Rome.

Bogyo, T. P., Porceddu, E., & Perrino, P. (1980). Analysis of sampling strategies for collecting genetic material. *Economic Botany*, **34**, 160–74.

Brown, A. H. D. (1978). Isozymes, plant population genetic structure and genetic conservation. *Theoretical and Applied Genetics*, **52**, 145–57.

Brown, A. H. D. (1984). Multilocus organization of plant populations. In *Population Biology and Evolution*, pp. 159–69, Wöhrmann, K. & Loeschcke, V. (eds.). Springer-Verlag, Berlin.

Brown, A. H. D. & Clegg, M. T. (1983). Isozyme assessment of plant genetic resources. In *Isozymes: Current Topics in Biological and Medical Research*, vol. 11, pp. 285–95, Rattazzi, M. C., Scandalios, J. G. & Whitt, G. S. (eds.). Alan R. Liss, New York.

Brown, A. H. D. & Moran, G. F. (1980). Isozymes and the genetic resources of forest trees. In *Isozymes of North American Forest Trees and Forest Insects*, pp. 1–10, Conkle, M. T. (ed.). USDA.

Burt, R. L. (1983). Observation, classification and description. In *Genetic Resources of Forage Plants*, pp. 169–81, McIvor, J. G. & Bray, R. A. (eds). CSIRO, Melbourne.

Doebley, J. F., Goodman, M. M. & Stuber, C. W. (1985). Isozyme variation in the races of maize from Mexico. *American Journal of Botany*, **72**, 629–39.

Doyle, M. J. & Brown, A. H. D. (1985). Numerical analysis of isozyme variation in *Glycine tomentella. Biochemical Systematics and Ecology*, **13**, 413–19.

Doyle, M. J., Grant, J. E. & Brown, A. H. D. (1986). Reproductive isolation between isozyme groups of *Glycine tomentella* (Leguminosae) and spontaneous doubling in their hybrids. *Australian Journal of Botany* **34**, 523–35.

Ewens, W. J. (1972). The sampling theory of selectively neutral alleles. *Theoretical Population Biology*, **3**, 87–112.

Frankel, O. H. (1984). Genetic perspectives of germplasm conservation. In *Genetic Manipulation: Impact on Man and Society*, pp. 161–70, Arber, W., Llimensee, K., Peacock, W. J. & Starlinger, P. (eds). Cambridge University Press, Cambridge.

Frankel, O. H. & Brown, A. H. D. (1984). Current plant genetic resources – a critical appraisal. In *Genetics: New Frontiers*, vol. IV, pp. 1–11, IBH Publ. Co. New Dehli and Oxford.

Frankel, O. H. & Soulé, M. (1981). *Conservation and Evolution*. Cambridge University Press, Cambridge.

Goodman, M. M. & Stuber, C. W. (1983). Races of maize VI. Isozyme variation among races of maize in Bolivia. *Maydica*, **28**, 169–87.

Holden, J. H. W. (1984). The second ten years. In *Crop Genetic Resources: Conservation & Evaluation*, pp. 277–85, Holden, J. H. W. & Williams, J. T. (eds.). George Allen & Unwin, London.

Jain, S. K. (1979). Biosystematic studies of populations in germplasm collections. *Proceedings Conference on Broadening the Genetic Base of Crops*, pp. 103–10. Pudoc, Wageningen.

Jain, S. K., Qualset, C. O., Bhatt, G. M. & Wu, K. K. (1975). Geographical patterns of phenotypic diversity in a world collection of durum wheats. *Crops Science*, **15**, 700–4.

Kahler, A. L. & Allard, R. W. (1981). Worldwide patterns of genetic variation among four esterase loci in barley (*Hordeum vulgare* L.) *Theoretical and Applied Genetics*, **59**, 101–11.

Kimura, M. & Crow, J. F. (1964). The number of alleles that can be maintained in a finite population. *Genetics*, **49**, 725–38.

Loveless, M. D. & Hamrick, J. L. (1984). Ecological determinants of genetic structure in plant populations. *Annual Review of Ecology and Systematics*, **15**, 65–95.

Marshall, D. R. & Brown, A. H. D. (1975). Optimum sampling strategies in genetic conservation. In *Crop Genetic Resources for Today and Tomorrow*, pp. 53–80, Frankel, O. H. & Hawkes, J. G. (eds.). Cambridge University Press, Cambridge.

Marshall, D. R. & Brown, A. H. D. (1981). Wheat genetic resources. In *Wheat Science – Today and Tomorrow*, pp. 21–40, Evans, L. T. & Peacock, W. J. (eds.). Cambridge University Press, Cambridge.

Marshall, D. R. & Brown, A. H. D. (1983). Theory of forage plant collection. In *Genetic Resources in Forage Plants*, pp. 135–48, McIver, J. C. & Bray, R. A. (eds.). CSIRO, Melbourne.

Nei, M., Maruyama, T. & Chakraborty, R. (1975). The bottleneck effect and genetic variability in populations. *Evolution*, **29**, 1–10.
Qualset, C. O. (1975). Sampling germplasm in a center of diversity: an example of disease resistance in Ethiopian barley. In *Crop Genetic Resources for Today and Tomorrow*, pp. 81–96, Frankel, O. H. & J.G. Hawkes (eds). Cambridge University Press, Cambridge.
Second, G. (1982). Origin of the genetic diversity of cultivated rice (*Oryza* spp.): study of the polymorphism scored at 40 isozymic loci. *Japanese Journal of Genetics*, **57**, 25–57.
Timothy, D. H. & Goodman, M. M. (1979). Germplasm preservation: The basis of future feast or famine. Genetic resources of maize – an example. In *The Plant Seed: Development, Preservation an Germination*, pp. 171–200, Rubenstein, I. *et al.* (eds.). Academic Press, New York.
Yonezawa, K. (1985). A definition of the optimal allocation of effort in conservation of plant genetic resources – with application to sample size determination for field collection. *Euphytica*, **34**, 345–54.

10

The role of networks of dispersed collections

P. M. PERRET

Introduction

One way to improve the conservation and use of plant genetic resources is to set up a regional network which, for a particular crop, includes its dispersed collections and the breeders they serve. This chapter shows how the implementation of such crop-based networks allows the extent of redundancy in collections to be determined. This is an essential preliminary to better conservation, and to the reduction of unnecessary duplication in evaluation studies. It discusses benefits which can be derived from crop-based networks by the users of genetic resources. Concrete examples are drawn from the European Cooperative Programme for the Conservation and Exchange of Crop Genetic Resources which IBPGR has been co-ordinating for the past four years.

The importance of crop working groups

A common general interest in crop genetic resources is insufficient to ensure productive international co-operation. It is now widely recognised that the natural unit of collaboration in international programmes is the single crop rather than a plurality of crops.

The organisational basis for such international collaboration is an expert Crop Working Group. In setting up such a group it is essential to ensure:

(i) that the chosen crop is a major one for the particular countries thereby attracting support

(ii) that the members of the Working Group are recognised specialists, and

(iii) that collaboration exists with other organisations dealing with the crop in the home region, as well as collaboration with similar Working Groups on different continents.

Each Crop Working Group sets its priorities and objectives depending on the nature of the crop and the history of collections within the countries (previous collecting, conservation, characterisation, evaluation and documentation). Information on the nature, origin and status of the accessions held in the dispersed collections, is needed to formulate sound workplans (whether the priorities of the Crop Working Group are centred on further collecting and maintenance of materials, or on better characterisation and evaluation). This information is available through passport data which accompany the accessions. A computerised database is often needed when the amount of data to be dealt with is large. The Crop Working Group with both potential users from different regions and curators of collections is best able to draw a realistic workplan which, may have to be a compromise between constraints of curators and expectations of breeders. The latter may not be informed on the limitations of gene banks and on the possible benefits to users of an international database. The breeders can take part in evaluation projects which may not necessarily bring them any immediate benefits.

The European cooperative programme for the conservation and exchange of crop genetic resources (ECP/GR)

The network of the ECP/GR includes both formal national genetic resources programmes as well as a range of smaller collections and institutes. The ECP/GR originated during meetings between the United Nations Development Programme (UNDP), Food and Agriculture Organization of the United Nations (FAO) and the EUCARPIA Gene Bank Committee from 1977 to 1979. It has developed in three phases – preparation, consolidation and operation. Following the recommendations of a review conducted in March 1982, Phase II (1983–86) was focussed on more intensive work for six crops to achieve more meaningful results. Six Crop Working Groups were established for *Allium*. *Avena*, forage grasses and legumes, barley, *Prunus* and sunflower. The reports of the several Working Groups can be obtained from the IBPGR Headquarters. The programme was, at the request of the participating Governments, put under the aegis of the IBPGR, as a special project. At present, 26 European countries participate and fund the programme in an amount equal to the contribution of the UNDP. A third phase will start in 1987 entirely funded by the participating countries to continue the activities started on the six crops. This international programme will be used below to exemplify the operation and the benefits of co-operative networks.

Efficient and economic conservation of accessions

The rationalisation of collections

Table 10.1 gives the total number of accessions for which passport data have been registered in the European databases, the number of named and unnamed accessions, and for some crops the number of presumed duplicates. The search for duplicates covers only named accessions, including cultivars and genetic stocks, mutants and a few important breeders lines. The rationalisation of collections is easiest for this first category of accessions, as each represents a specific genotype with a well-known set of characters. The accession chosen from among a group of duplicates should be the one which best represents the known set of characters. As a guide (with few exceptions) the accessions kept in the country where the cultivar was bred will be the most similar to the original genotype.

The first counts showed that 48 per cent of barley cultivars (8,900 accessions) and 36 per cent of *Avena* entries (2,800 accessions) are duplicates, but these figures are underestimates. Once all named duplicates are confirmed, curators need not regenerate or multiply the redundant accessions, and the distribution to users should proceed only from gene banks holding the preferred source.

Table 10.1. *Number and type of accessions registered in European databases* (June 1986)

	Named accessions	Unnamed accessions	Total	Number of[1] named duplicates
Allium	1,553	1,413	2,966	
Avena	7,915	1,591	9,506	2,831
Barley	22,018	28,336	50,354	8,900
Prunus spp.	9,036	309	9,345	1,739
Sunflower (wild)		1,463	1,463	
(cultivated)	972	366	1,338	
Forages[2]			> 20,000	
Festuca spp.	415	1,806	2,221	101
Lolium perenne	491	2,279	2,770	
Trifolium pratense	996	117	1,173	233

1 Duplicates identified only by similar names

2 The following species are registered in different European databases: *Bromus* spp., *Dactylis* spp., *Festuca* spp., *Hedysarum* spp., *Lathyrus* spp., *Lolium* spp., *Phalaris* spp., *Phleum* spp., *Poa* spp., *Medicago* spp., *Trifolium* spp. and *Vicia* spp.

Table 10.2. *Percentage of available information in European databases for a few passport descriptors and a few crops* (June 1986)

Passport descriptor	Barley Collected accessions	Barley Named accessions	Barley Wild species	*Allium*	*Avena*	*Festuca* Collected accessions	*Festuca* Named accessions
Donor institute	84	23	84	69	39	29	73
Donor number	50	23	33	41	12	12	2
Other number associated with accession	4	4	15	26	15		
Collector's number	25		14	34		27	
Collecting institute	29		23	48	3	61	
Country of origin	73	82	84	84	74	71	58
Province state and/or Location of collection site	39		35	35	8	57	
Altitude	2		1		6.0		
Growth habit	68	87	83		84		

The forage and *Allium* Working Groups considered the rationalisation of collections as not urgent because the total European holdings are small. In contrast to those, sunflower collections need rationalisation because the crop is wind-pollinated. However, duplicates apparent from passport data still need checking with evaluation data.

The unnamed accessions consist mainly of landrace populations and wild species. In this case, presumed duplicates can only be identified by checking detailed data (collecting number, year of collection, etc) which are unfortunately lacking in the majority of collections (Table 10.2). This group of accessions is rationalised by pooling potential genotype duplicates. However, all ECP/GR Working Groups consider this a lesser priority.

For the present, all accessions of wild species are to be maintained because of their scarcity. In the past, the samples of wild species distributed to other collections were so small that only a limited part of the genetic variation of the original population was sent to any one curator.

The saving in funds and efforts due to the rationalisation of collections, are indisputable, but what matters still more is that the reduced material can be regenerated to higher standards in the future.

The publication of comprehensive lists of accessions held in the active collections gives the curator of the base collection the means to check which material is not yet duplicated for safety into the base collection. Further identification of unique accessions will avoid unnecessary storage and germination tests in the base collections. The ideal situation is when the crop database accompanies the base collection. In the ECP/GR programme, this is the case for the *Allium* collection of the National Vegetable Research Station, Wellesbourne, UK.

The identification of gaps for further collecting

The Working Group can identify the material which is of most potential value and that which is currently lacking in collections. Priorities for collecting missions can then be clear for participating countries to act upon, possibly supported by international organisations such as IBPGR. Such lists have been established in the ECP/GR Working Groups for *Prunus*, forages, *Allium* and *Avena* (wild species). For barley and cultivated oats, the groups decided that no urgent action was called for. Further collecting should only be launched when the comprehensive holdings of European collections (together with proper data on locations) are registered in the databases.

Clearly, optimum decisions on further collecting require a careful analysis of the available passport data. Further, certain evaluation data linked with the geographical origin of these accessions may indicate the source of material of value to users, but poorly represented in collections.

Gene bank management

Apart from the benefits of having to cope with fewer accessions, the management of gene banks is not automatically improved when a co-operative network is set up. This is particularly true when the network has no funds to support activities of the participating collections and when the network is based on goodwill, as is the case for the ECP/GR networks. Curators have real economic constraints and, due to the many accessions in their collections which need regeneration, may not be able to meet desirable scientific standards. Nevertheless, the meetings of the Working Groups and their reports can spell out these problems for the attention of decision makers who provide the means to achieve, for instance, proper regeneration. Also, any lack in scientific knowledge for proper regeneration is revealed. The Crop Working Group may have no material or political means to resolve the problem as such but it can speak to the international community and governments. This applies especially to *in vitro* conservation of *Prunus* and *Allium*.

The benefits to users

Documentation in international databases

Most of the passport descriptors which are necessary for rationalisation of collections and selected by the Crop Working Groups are of no direct value for breeders. They give little indication of valuable specific traits or adaptations. A few, such as location of the original collection or the pedigree are useful. However, these data are far from complete (Table 10.2). It is hoped that more research may reveal correlations between passport data and some properties of the material, and that this will further enhance the value to users of the European databases.

In addition, an international database can collate all characterisation and evaluation data which are available in the co-operative network. This would allow statistical analyses to classify the total diversity into groups by defining the association of characters, to investigate the geographical patterns of characters such as disease resistances, and to study the stability of these characters in different test environments (Valentes Soares & Vanderborght, 1986).

A more realistic and pragmatic approach was adopted by the ECP/GR.

The first step was to survey European *Allium* and barley breeders, and to ask them to rate the value of the individual passport characterisation and evaluation descriptors. For both crops, breeders rated descriptors for resistance to diseases and stress as being most important, together with a few agronomic descriptors – this was despite the admitted problems of genotype × environment interactions for these descriptors.

Table 10.3 shows the selected set of characters recommended for registration in the central database. In indicates a compromise in developing strategies for co-operation along the following guidelines: (1) material should be identified properly. (A large number of descriptors have been selected for proper identification rather than breeding interest in the case of outbreeders.); (2) characters should be readily observable, or easy and cheap to test; and (3) only a few characters can be recorded for a large number of accessions.

In the case of barley, the Working Group left each national gene bank to decide, with its users, a list of five evaluation descriptors for registration in the European Barley Database.

The rationalisation of collections eliminates or reduces the unnecessary duplicates in evaluation studies. In the case of *Avena*, the eight evaluation characters for the central database apply only to the unique original accessions. The registration of data in a central database implies the standardisation and the clarity of the information. Thus the user has access to a wider scale of genetic diversity and knows what the true characteristics of the material are.

Standardised evaluation and better taxonomic identification

For polygenic characters such as plant height, yield, flowering time, a well-known, standard cultivar is essential as a reference. This allows some control over interactions due to the different growing conditions in each year, and interactions due to environment. If each crop collection uses its own standards, comparison between different material and locations is severely limited. The Sunflower Working Group recommended the inbred line HA89 as a common standard for days to flowering and plant height. Each sunflower collection in Europe is to receive seed originating from the same source, so that the same genotype is used everywhere.

For cultivated plants, an infraspecific taxonomic distinction often gives a relevant description of a precise set of morphological characters which are highly heritable. Additionally these characters may be linked with useful traits or agronomic properties. With wild species, an accurate and standardised taxonomy is important for the user.

Table 10.3. *Characterisation/evaluation descriptors selected for further registration in the European databases*

			Allium		
	Onion, shallot and close wild relatives		garlic, pearl onion/ great headed garlic		leek, kurrat and *A. fistolusum*
4.1.3*	Leaf erectness	4.1.3*	Leaf erectness	4.1.3	Leaf erectness
4.1.10	Shape of full-grown bulbs	4.1.7	Presence of bulbils (topsets)	4.1.1	Foliage colour
4.1.11	Uniformity of bulb shape				
4.1.12	Bulb skin colour	4.1.12	Bulb (clove) skin colour		
4.2.2	General fertility	4.1.14	Number of cloves/compound bulb		
4.2.6	Mode of reproduction	4.2.1	Ability to flower		
6.1.1	Bulb skin thickness			6.1.4	Length of leaf sheath
6.1.2	Bulb flesh colour	6.1.2	Bulb flesh colour		
6.1.7	Dry matter content of storage organ			6.1.5	Median diameter of the leaf-base pseudostem
6.1.8	Storage life of storage organs	6.1.8	Storage life of storage organs		
6.2.3	Time of flowering relative to a standard variety				
6.2.4	Day length requirement	6.2.4	Day length requirement		
6.2.6	Cold requirement for bolting			6.2.6	Cold requirement for bolting
		7.1	Susceptibility to low temperature	7.1	Susceptibility to low temperature
				7.2	Susceptibility to high temperature

Table 10.3. (*cont.*)

Avena			
4.1.5	Plant height	6.3.6**	Percentage oil content of caryopses (%)
6.1.6	Lodging at mature stage	7.1	Low temperature damage
6.2.2	Days to harvest	7.5	Winter kill
6.3.2	1000 grain weight (g)	8.2.1	*Erysiphe graminis avenae*
6.3.4	Percentage of husk (%)	8.2.3	*Puccinia coronata avenae*
6.3.5**	Percentage protein content of caryoposes (%)	8.4.1	Barley yellow dwarf virus (BYDV)

* The numbering follows the IBPGR descriptor list for the respective crop, which provides detailed information

** The percentage protein and oil may be measured on the kernel; it should be clearly specified if such analyses have been performed on the kernel or on the caryopsis

Table 10.3. (*cont.*)

Barley

List of descriptors and descriptor status used in each collection.
Five evaluation descriptors selected by each national gene bank in consultation with its users.

Prunus

	Almond		*Apricot*
4.2.1*	Season of flowering	4.2.1	Season of flowering
4.2.2	Harvest maturity	4.2.2	Harvest maturity
4.3.2	Kernel shape[2]	4.2.3	Flesh colour
6.2.13	Nut shape	4.3.1	Kernel taste
6.2.15	Marking of outer shell[2]	4.3.2	Separation of stone
6.2.18	Softness of shell		
6.3.4	Kernel taste		
	Cherry		*Peach*
4.2.2	Harvest maturity	4.2.2	Harvest maturity
4.2.3	Fruit skin colour	4.2.3	Flower type (shape)
4.2.4	Cropping efficiency	4.2.5	Flesh colour
6.2.3	Fruit size	6.1.5	Petiole gland shape (nectaries)
		6.3.3	Stone adherence to flesh of fully ripe fruit
	Plum		*Rootstocks*** (all crops)
4.2.2	Harvest maturity	4.1.1	Propagation method
6.2.5	Fruit size	4.1.4	Dwarfing
6.2.8	Ground colour	4.1.5	Induction of precocious bearing scions
6.3.2	Stone shape (lateral view)	6.1.4	Scion/rootstock compatibility
UPOV G.L. 22[1]	Shape of blade leaf[2]		
UPOV G.L. 37[1]	Length of flower peduncle[2]		
UPOV G.L. 64[1]	Hairiness of ovary of a very young fruit[2]		

* The numbering follows the IBPGR descriptor list for the respective crop, which provides detailed information
** The descriptor numbers refer to the IBPGR cherry descriptor list
[1] UPOV Guidelines for the Conduct of Tests for Distinctness, Homogeneity and Stability (Plums)
[2] Strongly recommended if data on this descriptor are available

Table 10.3. (*cont.*)

Sunflower

a)	*Cultivated sunflower*	b)	*Wild species*
5.4*	Sowing date /Month/Year	5.4	Sowing date /Month/Year
4.1.3	Plant height (at full flowering)	4.1.3	Plant height (at full flowering) Average from 10 main heads
4.2.1	Days to flowering	4.2.1	Days to flowering
4.2.8	Head shape (grain side)	4.2.10	Type of branching
4.3.4	Weight of 1000 seeds	4.3.8	Oil percentage
4.3.8	Oil percentage	6.1.1	Root type
6.3.7	Seed yield	6.2.4	Bract shape
8.1.1**	*Homeosema nubullela*	6.2.7	Number of ray flowers
8.2.1**	*Plasmopara helianthi*	6.2.8	Shape of ray flowers
		6.2.9	Colour of ray flowers
		8.2.1**	*Plasmopara helianthi*
		8.2.13**	*Erysiphe cichoracearum*
		10.3	Chromosome number
		10.4	Restorer fertility genes

* The numbering follows the IBPGR descriptor list, which provides detailed information for recording of the data

** Scored as 0 for absence of attack, 1 for susceptible, 2 for segregation of the character into the population

The European database for wild sunflower adopted the taxonomic system recommended by Heiser *et al.* (1969) and converted any other botanical names for accessions into this classification. Similarly the perennial *Medicago* European database has adopted a unique taxonomic system (Lesins & Lesins, 1979) and provided a list of common synonyms. The Barley Working Group agreed on the use of Mansfeld's system (Mansfeld, 1950) for the European barley database; additionally the barley database agreed to classify any material sent by the collections participating in the ECP/GR network.

Better links between curators and breeders

Links between breeders and curators must develop voluntarily. Despite some lack of co-operation in the past, there is now increasing recognition of the need to interact. Co-operation varies greatly in degree among crops and countries. One example of strong co-operation for forage plants is at the Welsh Plant Breeding Station in the UK where an autonomous genetic resources unit works closely with the breeding programmes of the station. Many of the sunflower collections in Europe are held by breeders who nevertheless ensure the conservation of their genetic resources. In The Netherlands, the recently established Center for Genetic Resources (CGN) has established a network of collaborators. Private breeders participate in the evaluation of the material of the CGN, and they handle about 50 per cent of the regeneration of the crops held by the CGN (Hardon, 1986). Finally, the staff members of the gene bank in the Central Institute for Genetics and Cultivated Plant Research of the Academy of Sciences of the German Democratic Republic are integrated with the specialists from all over the country involved in the evaluation of their material into crop specific breeder collectives (Lehmann, 1984). In these, current and planned breeding programmes are extensively discussed.

Two factors give a strong impetus in this direction in the ECP/GR. First, many Crop Working Groups have required for their databases, evaluation data which need specialist help (Table 10.3). Thus gene banks must implement more active collaboration with breeders. Second, the Crop Working Groups agreed that European databases should distribute the contents of their data bank (preferably in computerised form) only to one institution in each country (the national gene bank or the most significant collection in the country for each respective crop). Thereafter, these institutions deal with queries and distribute the information to their national breeders. This increases the dialogue between curators and users.

European breeders are interested in the material coming from other gene banks more than in the material held in their own national gene bank. Indeed they know and have already exploited their own landraces, and they often know what is available in their national gene bank. The fact that numerous small dispersed efforts are integrated, first by the drawing of a large co-ordinated programme and then by collation of results, is indeed a key role for a network of small dispersed collections. A breeder may not wish to evaluate to a set of descriptors for a few accessions in which, with fair probability, no outstanding characters will be found. However, if he knows that more than 20 colleagues are simultaneously screening a different set of material for the same characters, or observing the same material as his but for a different set of characters, his attitude towards this workload will be substantially modified.

The identification of duplicates allows evaluation to concentrate on unique original material. In this matter, co-operative programmes will have to adopt realistic approaches. One approach is to characterise or evaluate the unique original material for a few characteristics, in accordance with the descriptor states recommended by the Working Groups. The main problem will consist in choosing characters valuable for the largest array of users. Later on, the collation of passport data and a few meaningful characterisation/evaluation data will allow selection of the more interesting material for detailed evaluation.

Conclusions

The experience from the ECP/GR network has already shown that positive results can follow from networks of co-operation between dispersed collections, based on Crop Working Groups. Each Group provides a major stimulus to co-ordinated rationalisation and evaluation of accessions of that crop.

As these activities depend on the passport data associated with each accession, the data should be centralised and standardised in a database. The identification of duplicates, the choice of characters for evaluation and the collation of data are fostered by the Crop Working Group.

The primary duty of a programme on small dispersed collections is to 'clean its house' and create a sound base and favourable conditions for better use of genetic resources. Concerted efforts by gene banks and breeders can focus on major problems which limit progress in breeding. For example, an ECP/GR Barley Workshop agreed that the limiting factor in barley breeding for higher yield was that short straw is apparently always linked with short roots. Only an 'umbrella' programme may first collate all available information relevant to such a problem and thereafter screen the material to determine conclusively whether or not the combination short straw/long roots is available in the collections.

Finally, the key to better utilisation of genetic resources lies in the willingness of the personnel of gene banks and breeders to collaborate more closely. The implementation of networks creates favourable conditions for such collaboration and additionally offers greater possibilities to the user.

References

Hardon, J. J. (1986). Handling and utilization of genetic resources in the Netherlands. In *Proceedings of the 15th Anniversary of the Activities of the Braunschweig Institute in Genetic Resources*, pp. 174–80, Braunschweig, FRG.

Heiser, C. B., Smith, D. M., Clevenger, S. B. & Martin, W. C. (1969). The North American sunflowers (*Helianthus*). *Memoirs of the Torrey Botanical Club*, **22**, 1–218.

Lehmann, C. (1984). Germplasm evaluation at Gatersleben: the relationship between gene bank and breeder. In *Crop Genetic Resources: Conservation and Evaluation*, pp. 202–6. Holden, J. H. W. & Williams, J. T. (eds.). Allen and Unwin, London.

Lesins, K. A. & Lesins, I. (1979). *Genus Medicago (Leguminosae). A taxogenetic study*. Dr. W. Junk Publisher, The Hague, Boston, London.

Mansfeld, R. (1950). Das morphologishe System der Saatgerste, *Hordeum Vulgare* L. sl. *Züchter*, **20**, 8–24.

Valentes Soares, E. & Vanderborght, T. (1986). Analysis of plant genetic resources data for *Phaseolus vulgaris* L. Final report to IBPGR, Rome (mimeographed).

Part IV

Evaluation

11
Characterisation and evaluation of okra

S. HAMON AND D. H. VAN SLOTEN

Introduction

Okra (*Abelmoschus esculentus*) is an important vegetable crop throughout the tropics and subtropics. Its genetic resources, including related cultivated and wild *Abelmoschus* species, were the subject of joint IBPGR/ORSTOM projects,which resulted in a status report largely based on available literature (Charrier, 1983, 1984). For details on taxonomy, geography, cytology, inter- and intra-specific crossability, etc, reference should also be made to this report. The species nomenclature used in this paper is summarised in Table 11.1.

The world collection of okra

The composition of the joint ORSTOM/IBPGR okra collection based in Côte d'Ivoire, is shown in Table 11.2. There are 2,283 accessions. Clearly, the African continent (with 2,029 accessions) and West Africa in particular (with 1,769) is far more heavily represented than other continents and countries. Wild and cultivated species, other than *A. esculentus*, from Asia are absent in the collection and require collecting in the next few years.

Worldwide, cultivated okra is largely species *A. esculentus*, but *A. manihot* and *A. moschatus* may be grown as well (see Table 11.1). A major discovery was an undescribed cultivated species, collected mainly in Côte d'Ivoire (Siemonsma, 1982*a*, 1982*b*). Hamon & Yapo (1986) detail further the distribution of this latter species which we refer to here as 'West African taxon' (WAT), after its present known area of distribution.

Figs. 11.1 and 2 map the sampling sites for *A. esculentus* and WAT. These figures, covering West Africa and part of Central Africa show each species separately to emphasise the differences in cultivation areas throughout the four major climatic zones. From north to south these are:

desert (village or oasis cultivation), Sahelian (north of latitude 12°N), savannah (between latitudes 8 °N and 12 °N) and rain forest climatic zones.

A. esculentus (Fig. 11.1) is primarily distributed throughout the intermediate savannah zone between the rain forest and the arid Sahel. The species is less frequently found in the rain forest zone but is, on the other hand fairly well represented in the Sahel zone. With one exception, the WAT (Fig. 11.2) does not occur in the Sahelian zone since it has a long life-cycle and usually requires abundant, continuous rainfall. The eastern boundary of its distribution is difficult to determine due to lack of samples from Central Africa. At present, the most distant sampling sites are in Cameroon. Since a natural interspecific hybrid of the two cultivated species occurs in the central part of Sudan, WAT is possibly more widely distributed than currently known.

Information on the collection

Table 11.2 summarises the coverage of passport data in the ORSTOM/IBPGR okra collection. It is comparatively well documented (cf. Peeters & Williams, 1984). Elevation data are missing because they are not very important in West Africa. Local names are listed frequently but not systematically translated and therefore often unusable. Missing passport data largely relates to material obtained from other genebanks before 1981.

Samples acquired from multi-crop collecting missions fail to list the number of fruits, fruit characteristics, and comments on local traditions indicating crop associations. This produces a major data gap. Further information can be obtained from local names when they are systematically collected and translated. Fig. 11.3 has been prepared on the basis of a translation of local names from the Togo/Benin collecting mission (Hamon & Charrier, 1983). The relative frequencies are shown in decreasing order of importance from top to bottom. Asterisks represent the most commonly used characteristic for a given category. Interest focuses primarily on the harvest period. Contrast between early varieties (*A. esculentus*) and late varieties (WAT) is the most usual distinction. The date of planting and the length of plant cycle are secondary.

It is not unusual to identify a variety by colour or shape of the fruit, comparing the fruit with a part of some familiar animal (e.g. antelope horn, agouti cheek, rat tail) or of a human being.

Names describing plant characteristics (height, leaf type, etc) are less frequent, and tend to be used by ethnic groups who already know a great

Table 11.1. *Nomenclature of* Abelmoschus *species used in this paper*[1]

Species	Chromosome number	Cultivated/wild
1. *A. moschatus*		
1a) *A. moschatus* subsp. *moschatus* var. *moschatus*	72	±Cultivated
1b) *A. moschatus* subsp. *moschatus* var. *betulifolius*	?	Wild
1c) *A. moschatus* subsp. *biakensis*	?	Wild
1d) *A. moschatus* subsp. *tuberosus*	38	Wild
2. *A. manihot*		
2a) *A. manihot* subsp. *manihot*	60–68	Cultivated
2b) *A. manihot* subsp. *tetraphyllus* var. *tetraphyllus*	130–8	Wild
2c) *A. manihot* subsp. *tetraphyllus* var. *pungens*	138	Wild
3. *A. esculentus*	66–144	Cultivated
4. *A. ficulneus*	72–8	Wild
5. *A. crinitus*	?	Wild
6. *A. angulosus*	38	Wild
7. West African Taxon (WAT)	185–98	Cultivated
8. *A. tubercalatus*	58	Wild

[1] Species 1 to 6 following van Borssum-Waalkes (1966)
West African Taxon described by Chevalier (1940), Siemonsma (1982*a*, 1982*b*) and Hamon & Yapo (1986)
A. tuberculatus described by Pal *et al.* (1952) in India only

Table 11.2. *Current status of okra 'world collection' at ORSTOM, Centre d'Adiopodoume, Côte d'Ivoire*[1]

Region/Country	Number of samples[2]						
	Abelmoschus esculentus (1)	West African Taxon (2)	Hybrids (1) × (2)	Mixed samples (1) + (2)	*Abelmoschus moschatus*	Total	Samples received, but no germination
West Africa							
– Benin	213	64	6	2	12	297	26
– Burkina Faso	144	30	2			176	4
– Cameroon		23				23	
– Congo		1				1	
– Côte d'Ivoire	88	244			1	333	5
– Ghana	24	23				47	21
– Guinée Conakry	97	94	1	4		196	8
– Liberia		5				5	1
– Mali	19					19	4
– Niger	31					31	
– Nigeria	49	24		1		74	25
– Togo	206	165	8		6	385	86
– Zaire	2					2	
North Africa							
– Algeria	1					1	
– Egypt	35					35	

East Africa							
– Sudan	128		1	1		130	
Southern Africa							
– Zambia	24					24	
– Zimbabwe	70					70	10
America							
– Cuba	3					3	
– Guatemala	2					2	
– Mexico	1					1	
– Peru	2					2	
Mediterranean							
– Turkey	116					116	
– Yugoslavia	13					13	
Middle-East							
– Afghanistan	8					8	
– Iran	16					16	
– Pakistan	7					7	
– Saudi Arabia	1					1	
– Syria	4					4	
Asia							
– China (Taiwan)	4					4	
– India	61					61	
– Philippines	6					6	
Totals	1375	673	18	8	19	2093	190

[1] The majority of the accessions in this collection have been obtained from IBPGR and/or ORSTOM germplasm collecting missions (largely multi-crop) carried out during the period 1980–5.

[2] No information on country of origin available for *A. manihot* (3 accessions) and *A. moschatus* (1 accession); the collection also includes 10 standard international cultivars

Fig. 11.1 Geographical distribution of *A. esculentus* in West Africa

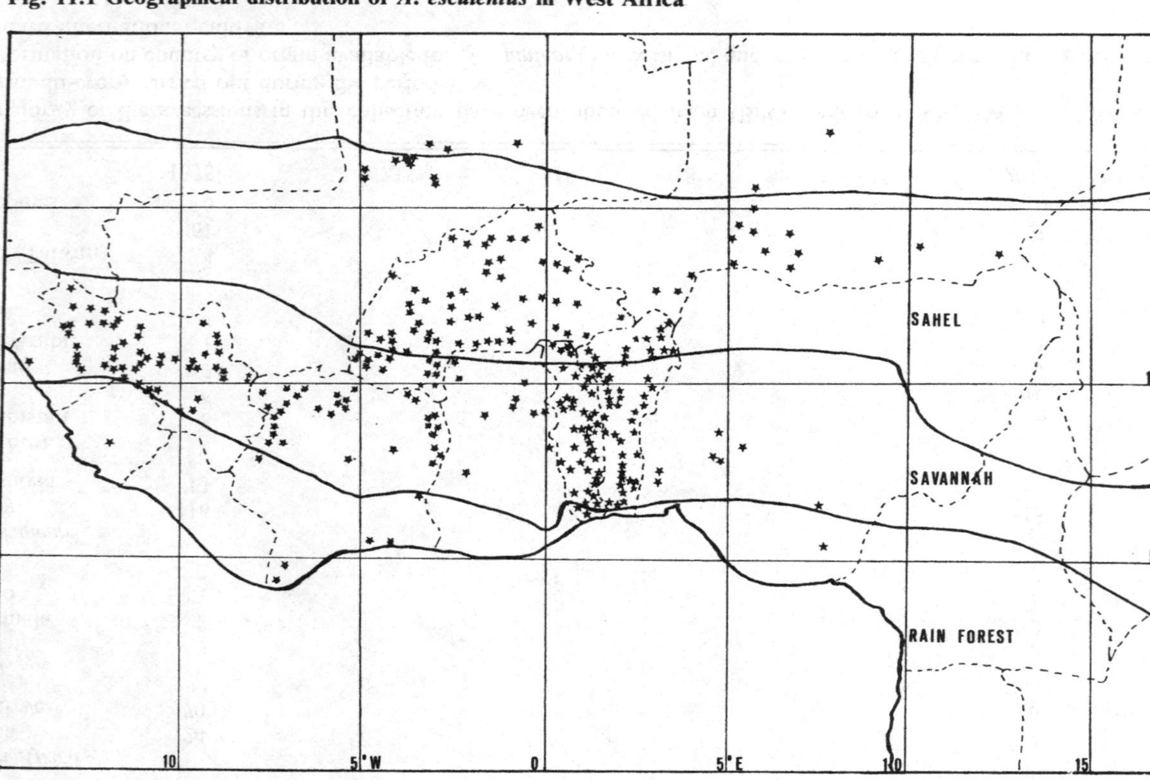

Fig. 11.2. Geographical distribution of the West African Taxon (WAT)

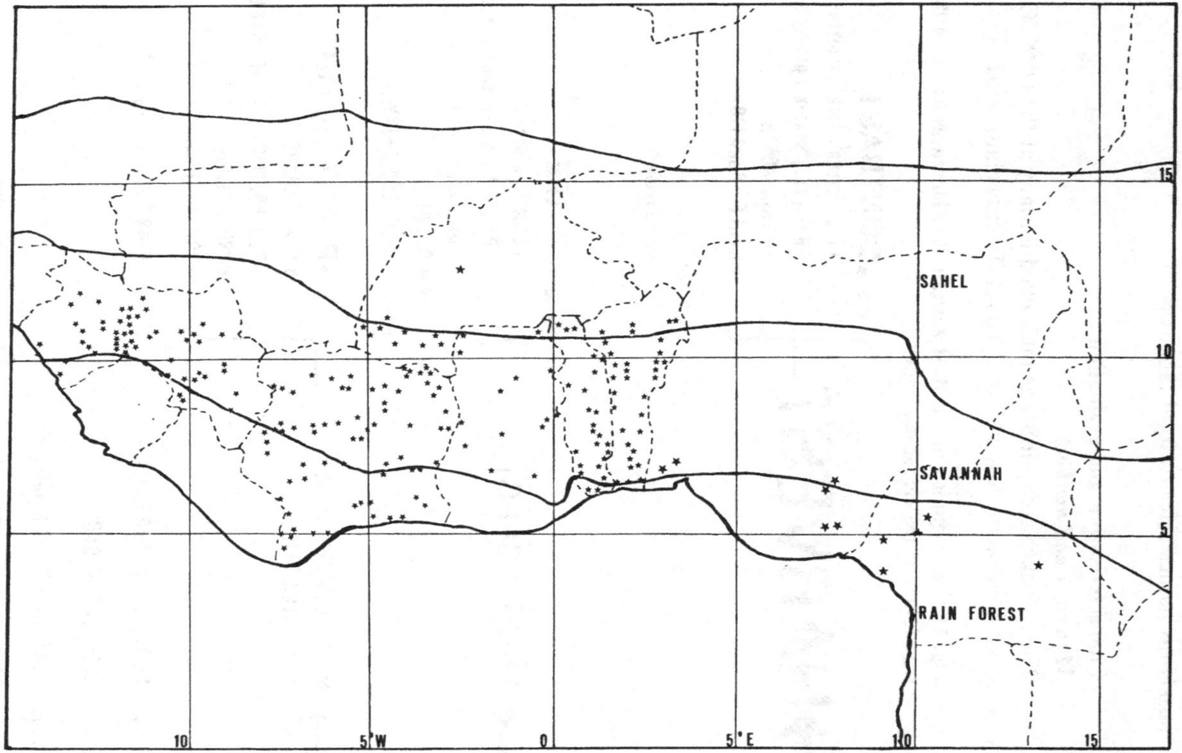

★ Sampling sites of WAT collected in West Africa

deal about the plant. Where a collecting mission finds such names it should ask very detailed questions.

Very occasionally, a local name may refer to some aspect of the plant's provenance, culinary properties or some associated crop. Only *A. moschatus* names refer to food taboos.

Characterisation and evaluation

Materials and methods

The full okra collection was multiplied initially at the ORSTOM Station in Adiopodoumé (lower Côte d'Ivoire). Twenty plants were planted

Fig. 11.3. Recognition of the West African okra landraces based on the translation of the vernacular names

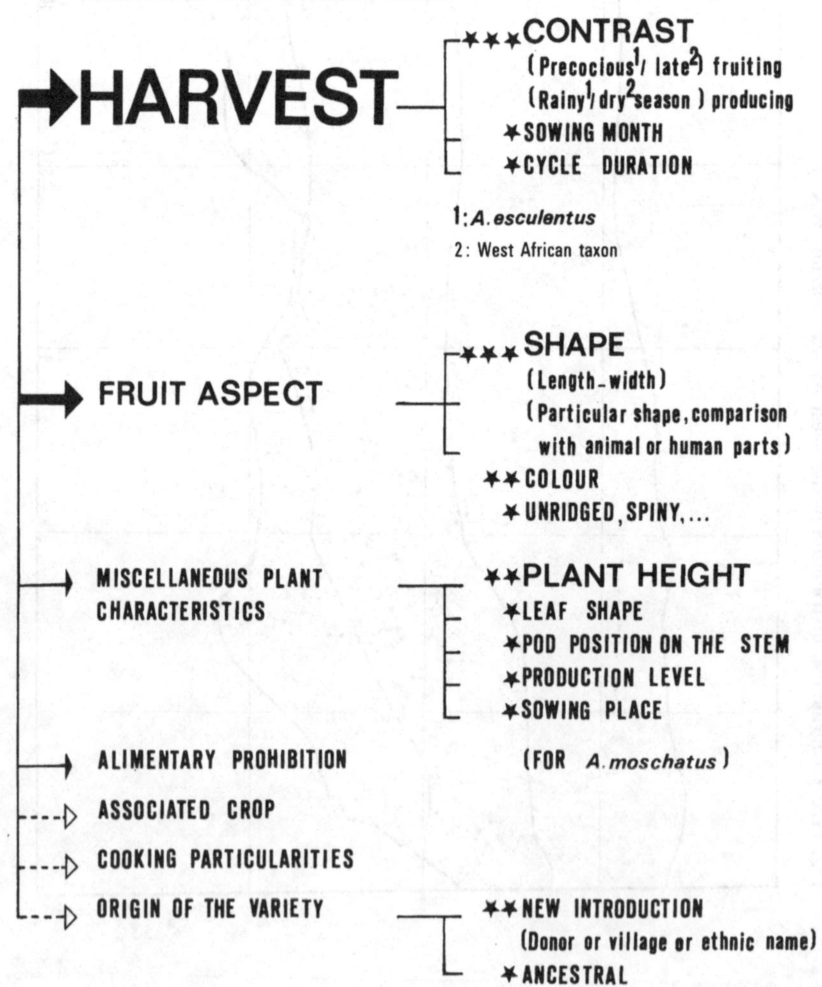

in rows, with the control varieties Clemson Spineless (international *A. esculentus* cv.) and WAT (primitive cv. ORS 520) were planted every twenty rows. Fungal pathogens, insect pests and nematodes were controlled chemically. However, leaf curl virus, transmitted by a whitefly, *Bemissia tabacci*, is very prevalent during the first six months, peaking between February and June. It is impossible to obtain an *A. esculentus* plant at that time which will top 50 cm by the end of its cycle. All parts of the plant are deformed, inflorescences poor and seeds malformed.

Descriptors used in the characterisation and evaluation fall into three categories: (i) quantitative, (ii) qualitative and (iii) enzymatic. In general the published IBPGR/ORSTOM descriptor list (Charrier, 1983, 1984) was used, but for the qualitative descriptors more descriptor states were used.

Quantitative descriptors: Plant morphology and development were characterized by plant height, number of internodes, stem diameter and branching. Measurements were taken systematically between 80 and 100 days after planting, coinciding with the end of the growing cycle of the Clemson Spineless control cultivar. For very long-cycle plants, mainly WAT accessions, a second series of measurements was made two or three months later.

The day of first flower opening, the height at first flowering and first fruiting, and the number of internodes were also noted for each plant. Fruit-setting parameters (average total number of fruits per plant and distribution along the main stem and branches) and fruit characteristics (length, width, number of ridges) and seeds (weight of one thousand seeds) were also noted.

Table 11.3. *Proportion of missing passport data in ORSTOM/IBPGR okra collection*

Passport descriptors	Percentage unknown
Collecting organisation	11
Collector	20
Collector's number	20
Country of origin	0.2
Town/Province	30
Latitude	33
Longitude	33
Altitude	99
Vernacular names	60

Qualitative descriptors: There are three main types of qualitative descriptors: colour, shape and other features. The specific descriptors were the colour of the main stem, the petal base, the leaf petiole, the veins, the lamina and the fruit (unripe); the shape of the leaves; and the position of the fruit on the main stem. These descriptors tend to be highly subjective.

Enzymatic descriptors: Isoenzymatic electrophoresis can provide a description virtually unaffected by the environment and fairly easy to determine.

Electrophoresis has been carried out with starch gel according to Second & Trouslot (1980). For okra, eleven systems could be used, in decreasing order of resolution:

Excellent: Alcohol dehydrogenase (Adh), phospho-glucose-isomerase (Pgi), phosphoglucomutase (Pgm), inositol

Fig. 11.4. Electrophoretic discriminating patterns of the two main okra cultivated species

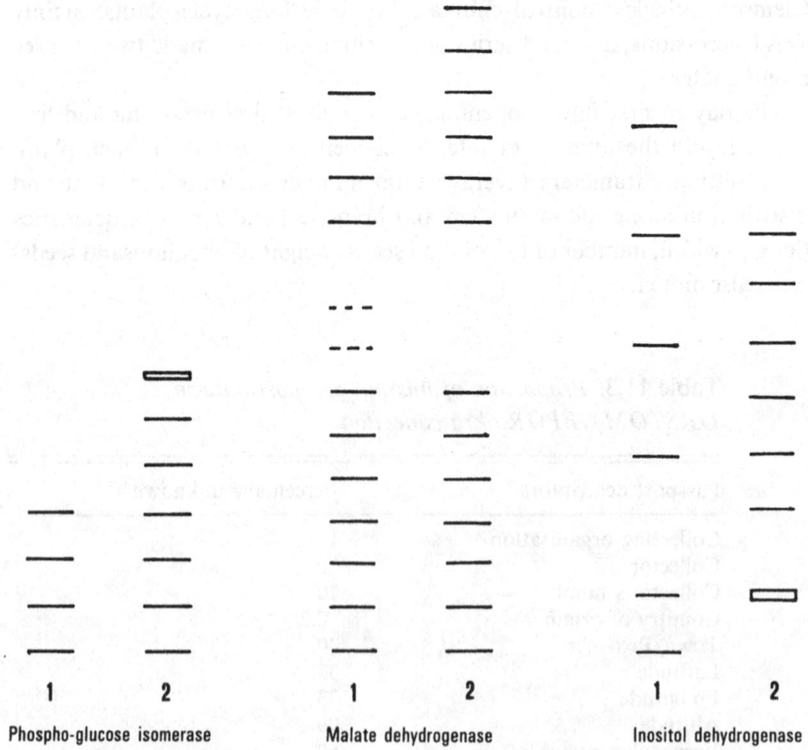

1: *A. esculentus* – 2: West African taxon

dehydrogenase (Idh), 6. phosphogluconic acid dehydrogenase (6. Pgd).

Good: Shikimic dehydrogenase (SKdh), Glutamate oxalo-acetate transaminase (GOT).

Fair to poor: Esterase, acid phosphatase, peroxidase, catalase.

Within the cultivated species (*A. esculentus* and WAT) enzymatic variability hardly exists. The electrophoretic patterns of both species are distinct (Fig. 11.4), and hence the method can be used for species identification (Hamon & Yapo, 1986).

Problems encountered during characterisation and evaluation

(*i*) *Species identification*

Species identification prior to planting is essential to proper organisation of the work, and passport data on this descriptor are often incorrect. Species identification can be made on fruits at the time of collection (whole fruits are the most common method of okra preservation).

Fig. 11.5 shows the main types of fruit found: fruit types 1 to 5 WAT; fruit types 6 to 10 *A. esculentus*; fruit type 11 the control cultivar Clemson Spineless.

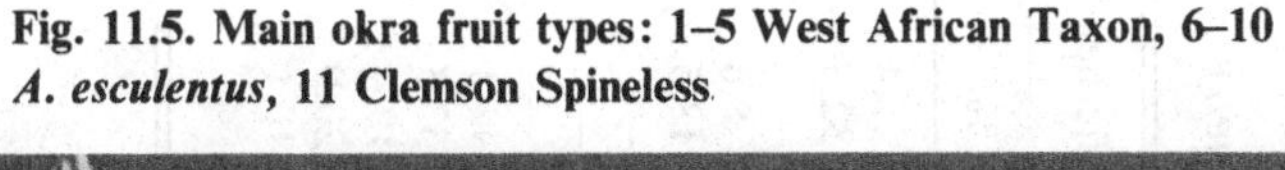

Fig. 11.5. Main okra fruit types: 1–5 West African Taxon, 6–10 *A. esculentus*, 11 Clemson Spineless

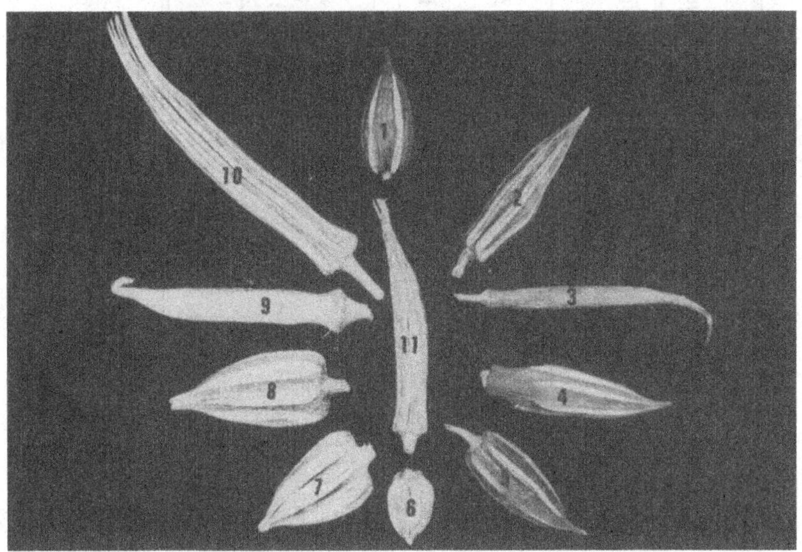

Table 11.4. *Okra accessions' level of heterogeneity*

Country of origin	Number of cases	Intraspecific heterogeneity				Interspecific heterogeneity				Total
		ESC	%	WAT	%	M1	%	M2	%	
Benin	255	181	15.4	65	7.7	8	3.1	1	0.4	16.4
Burkina Faso	131	102	32.3	25	8.0	4	3.0	—	—	29.7
Cameroon	23	—	—	23	0.0	—	—	—	—	—
Ghana	45	23	8.6	21	14.2	1	2.2	—	—	13.3
Guinea	181	81	18.5	95	9.4	4	2.0	1	0.5	16.0
Ivory Coast	21	—	—	21	0.0	—	—	—	—	—
Liberia	4	—	—	3	—	—	—	—	—	—
Mali	17	17	—	3	—	—	—	—	—	—
Nigeria	22	18	11.1	4	0.0	—	—	—	—	9.1
Sudan	35	34	32.3	—	—	1	2.8	—	—	34.3
Togo	352	172	19.4	163	27.6	15	4.2	2	0.6	25.5
Zambia	6	6	20.0	—	—	—	—	—	—	—
Zimbabwe	59	59	10.1	—	—	—	—	—	—	—
Total	1130									20.1

M1: The two species are found in the same sample
M2: As M1 but with the hybrids between them
ESC: *A. esculentus*
WAT: West African Taxon

Direct identification by seed sampling is impossible for both species, but electrophoresis of single seeds can quickly distinguish between the two species. The zymograms in Fig. 11.4 are invariant within each of the two species and different between them. Some samples were found to be mixtures of the two species (Table 11.4). The species can also be identified from the seedling, but this requires expertise and is not conclusive. Species identification by morphological criteria is definitive only on the basis of flower characteristics (number and shape of epicalyx segments). This is effective, but it does have the drawback that it occurs late in the sequence of the trial.

(ii) Seed germination

The problem of poor germination was the most disturbing factor in establishing trials. Limited numbers of seeds, low rate of germination, pest attacks and other damage produced trial imbalance for a fairly large number of accessions.

(iii) Heterogeneity

Okra, owing to its floral structure and the absence of self-incompatibility, produces much of its progeny through selfing. However, cross-pollination is frequently mentioned in the literature. The extent of outcrossing varies according to the variety, the cropping season, and the location (Chandra and Bhatnagar, 1975; Martin, 1983; Tanda, 1985), ranging from 0 to 60 per cent. There is a close correlation between cross-pollination and the presence or absence of insects. With strict pest control, contamination is severely restricted. But this is not the case in the traditional agricultural setting where pesticide treatments are non-existent. The continuous flowering and the special constraints of the evaluation procedure make both systematic bagging and isolation impossible. There is, therefore, a certain risk factor.

Table 11.4 lists heterogeneity rates for species and provenances studied in 1984–5. There are three major types of heterogeneity:

1. Interspecific heterogeneity. This is due to mixing of seeds of two distinct species during collecting;
2. Intraspecific heterogeneity refers to accessions which are mixed or segregating. This can result from crossing or from mixing of fruits by the donor or collector. One possibility might be to identify a whole fruit as an accession, but this would cause an enormous increase in the number of accessions. Table 11.4 shows that 18 per cent of WAT and 22

per cent of *A. esculentus* accessions fall into this category;

(3) Partial intraspecific heterogeneity refers to heterogeneity for only one or two traits.

An accession which is homogeneous is a rare occurrence in a plant which is reproduced by seeds and is not strictly autogamous.

Results

(i) Uni-variate analysis

Quantitative descriptors. Table 11.5 lists the statistics of quantitative descriptors for the two species. There are marked inter-species similarities and dissimilarities for particular descriptors. Discriminant analysis (See page 189), clearly brings out the differences between the two species.

Table 11.5. *Statistical parameters observed on data recorded as quantitative descriptors*

Descriptors	Min.	Max.	Mean	SD	CV	
Plant height	16	137	64.2	24.1	37.5	ESC
	24	144	67.7	22.6	33.4	WAT
Number of internodes	5.0	24.0	9.6	1.3	13.5	ESC
	8.2	28.8	19.0	4.2	22.1	WAT
Stem	6.0	32.0	15.3	4.6	30.4	ESC
Diameter at base	10.0	36.0	20.3	4.5	22.4	WAT
Number of branches	0.0	14.0	1.8	0.36	20.0	ESC
per plant	1.0	22.4	8.6	3.9	45.3	WAT
First flowering day	34.0	89.0	47.3	5.7	12.1	ESC
	48.0	101.0	67.4	10.3	15.2	WAT
First fruit producing	4.2	19.5	7.1	2.0	28.4	ESC
node	5.6	27.0	12.7	4.5	35.1	WAT
First flowering node	3.0	19.4	6.9	0.8	11.5	ESC
	4.8	36.0	11.8	2.7	22.9	WAT
Flowering amplitude	13.0	59.0	17.5	9.7	55.6	ESC
	8.0	210.0	34.5	25.0	74.7	WAT
Fruit length at maturity	5.0	30.0	14.4	5.6	39.1	ESC
	5.0	17.0	10.2	2.3	22.3	WAT
Fruit diameter at	0.7	4.8	2.1	0.7	35.1	ESC
maturity	1.2	—	2.5	0.5	21.7	WAT

Min. = Minimum value Max. = Maximum value Mean = Mean value
SD = Standard deviation CV = Coefficient of variation
ESC = *A. esculentus* WAT = West African Taxon

Qualitative descriptors: The main markers are:

Species specific traits:

– Specific to *A. esculentus*

Bronze (7) stem colour, fruit colour (14), sea green being characteristic of Sudan accessions; darker colour of fruit ridges.

– Specific to WAT

Colour of floral spot always internal; blackish green fruit colour, fruits slightly or very pendulous; fruits may be prickly; seeds may have a reddish fuzz;

Difference of frequencies:

Differences of frequencies basically concern three plant aspects. The following are most common in *A. esculentus*:
green stems, petioles, fruits
entire leaves, no clearly marked lobes
shorter branches.

(ii) Bi-variate analysis

Correlations among quantitative variables. Correlations between quantitative variables were calculated. The strongest and most persistent correlations are between early flowering and first flowering and fruiting nodes; next, between the first fruiting node and plant structure (stem diameter, number of internodes and number of branches). An unexpected relationship involves the weight of 1,000 seeds which is lower for fruits set at a node high on the stem.

In *A. esculentus*, plant height is closely correlated with flowering and fruiting parameters and structure. This is much less or not at all true of the WAT; height is postively correlated with total seed and fruit production at 80 days. This last descriptor is itself correlated with total fruit production. It can also be seen that the close correlation between fruiting on the stems and on the branches of *A. esculentus* disappears in the WAT.

It is evident that the distribution of quantitative and qualitative variables and their correlations vary between species. Some correlations are not constant between accessions, even within the same species.

Relations between early flowering and latitude of sampling site. Fig. 11.6 depicts early flowering as a series of cumulative frequencies whilst also showing the environmental zone of the sampling site for both species. We have grouped the data in accordance with climatic zones as schematised in Figs. 11.1 and 2. Fig. 11.6 shows that precocity increases as one moves from humid to more arid zones. It also shows that *A. esculentus* flowers earlier that WAT at the same latitude.

Under the evaluation conditions in lower Côte d'Ivoire, the WAT accessions from Cameroon exhibited major flowering problems; some accessions had grown two metres in one year after planting without putting forth a single flower. This was also true to a lesser extent for some accessions at less than 7° from the equator. Generally speaking, early flowering in both species means less development – smaller size, shortened cycle and, for the WAT, fewer branches.

Fig. 11.6. Flowering behaviour of West African okras

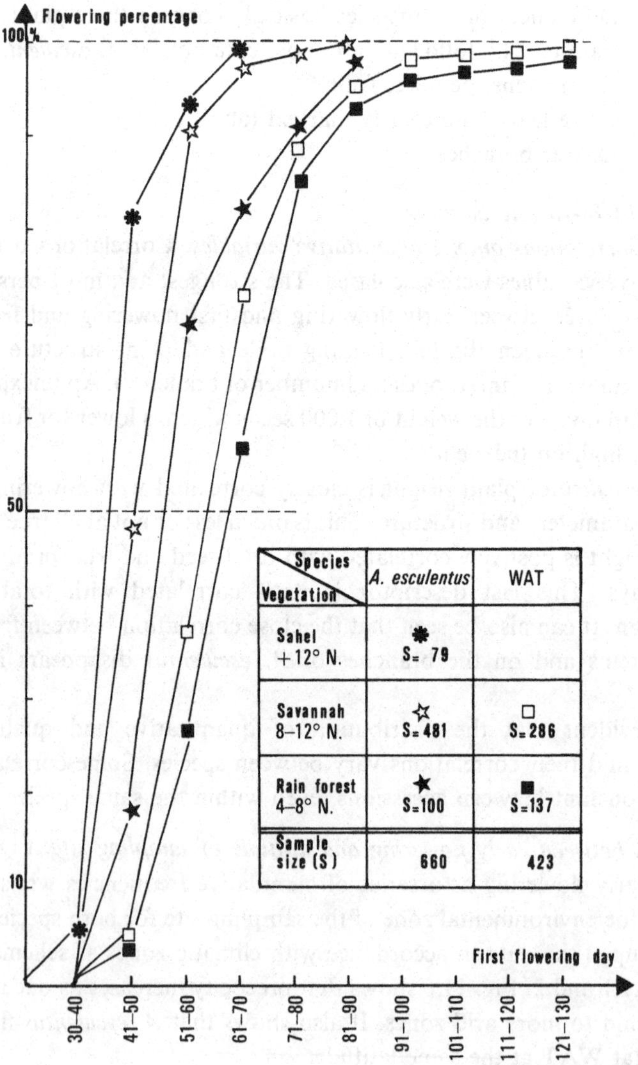
Flowering percentage
100 %
50
10
Species
Vegetation
A. esculentus
WAT
Sahel
+12° N.
S=79
Savannah
8–12° N.
S=481
S=286
Rain forest
4–8° N.
S=100
S=137
Sample size (S)
660
423
First flowering day
30–40
41–50
51–60
61–70
71–80
81–90
91–100
101–110
111–120
121–130

(iii) Multivariate analysis

Comparison between the USDA and Côte d'Ivoire collections. Factor analysis was carried out on the collection available in 1982. In Fig. 11.7 the scatter diagram represents 45 per cent of the total variability. All variables were involved in the analysis, but the only ones represented in the figure are those which actually contributed to the axes.

A clear contrast between the two cultivated species is apparent along horizontal axis 1. Vertical axis 2 shows intraspecific variability, particularly in colouration, seed production, fruit width and branching. It shows clearly the contrast between the USDA *A. esculentus* accessions and those from Côte d'Ivoire. The Côte d'Ivoire accessions are much more polymorphic.

Fig. 11.3 shows the varietal recognition methods as deduced from translating the local names of the samples collected. There is a strong similarity between the local names and the factorial variables. This shows that careful attention should be given to peasant systems of variety classification.

Geographic distribution of A. esculentus *variability.* The Côte d'Ivoire *A. esculentus* collection is much more variable than the USDA collection. To understand such differences, a prior comparative analysis of several countries is needed. Principal component analysis of the quantitative variables was undertaken, and Fig. 11.8 shows the scatter diagram containing 50 per cent of the total variability. The countries included are Benin, Burkina Faso, Guinea, Mali, Sudan, Togo, Zambia and Zimbabwe. Only the limits of variability encountered for each country are shown.

A big difference in balloon size is immediately apparent between Togo and Benin with peak variability, and Mali with the smallest balloon. Axis 1 contrasts two types of plants; one early, small and unbranched with heavy seeds, and the other much hardier. It is interesting to note that Axis 2, fruit production, is independent of the okra type and that the most productive plants come from Sudan or Burkina Faso.

Comparison of A. esculentus *and WAT.* The discriminant analysis using quantitative variables, fulfils two major objectives. The first is a test classification of individuals into pre-defined groups according to a selected criterion. The second is to establish a classification of variables in descending order of discrimination, in order to select a minimum number which will suffice.

A comparison was made between the two cultivated species. First, all quantitative variables were compared and, next, only the three most discriminant variables (number of internodes, plant height and total plant

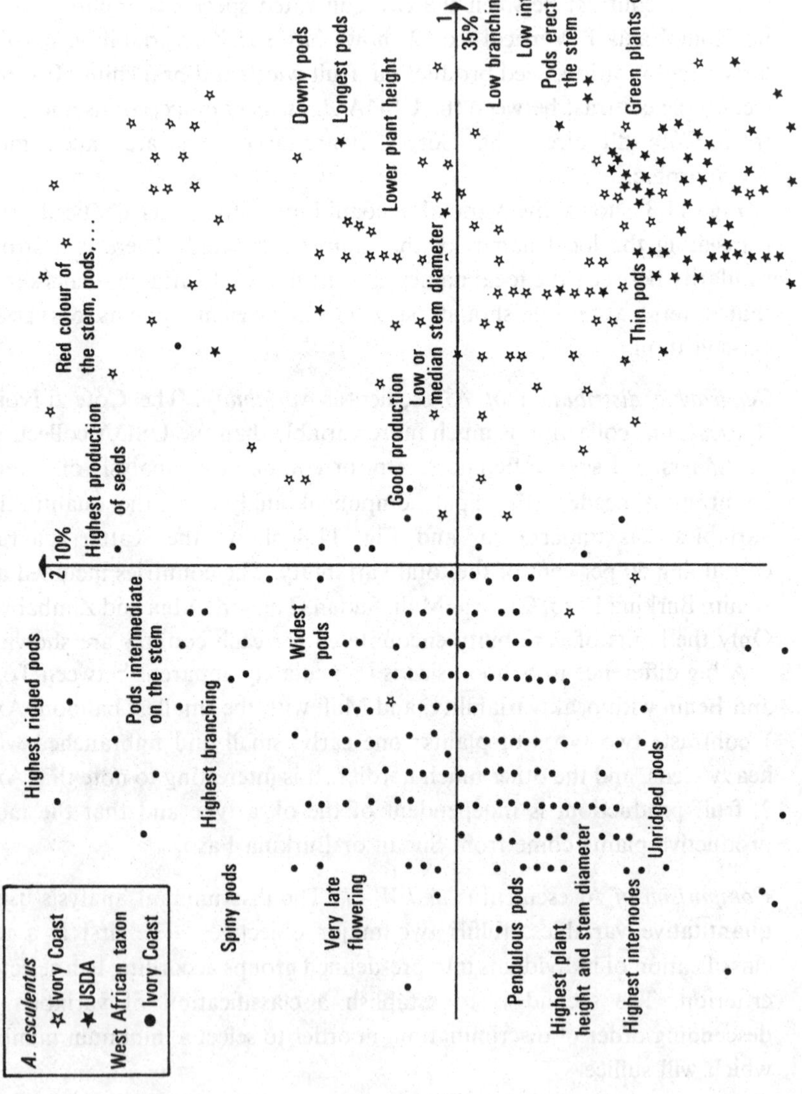

Fig. 11.7. Factor analysis on the ivorian and USDA okra collections

Fig. 11.8. Principal component analysis on *A. esculentus* introductions

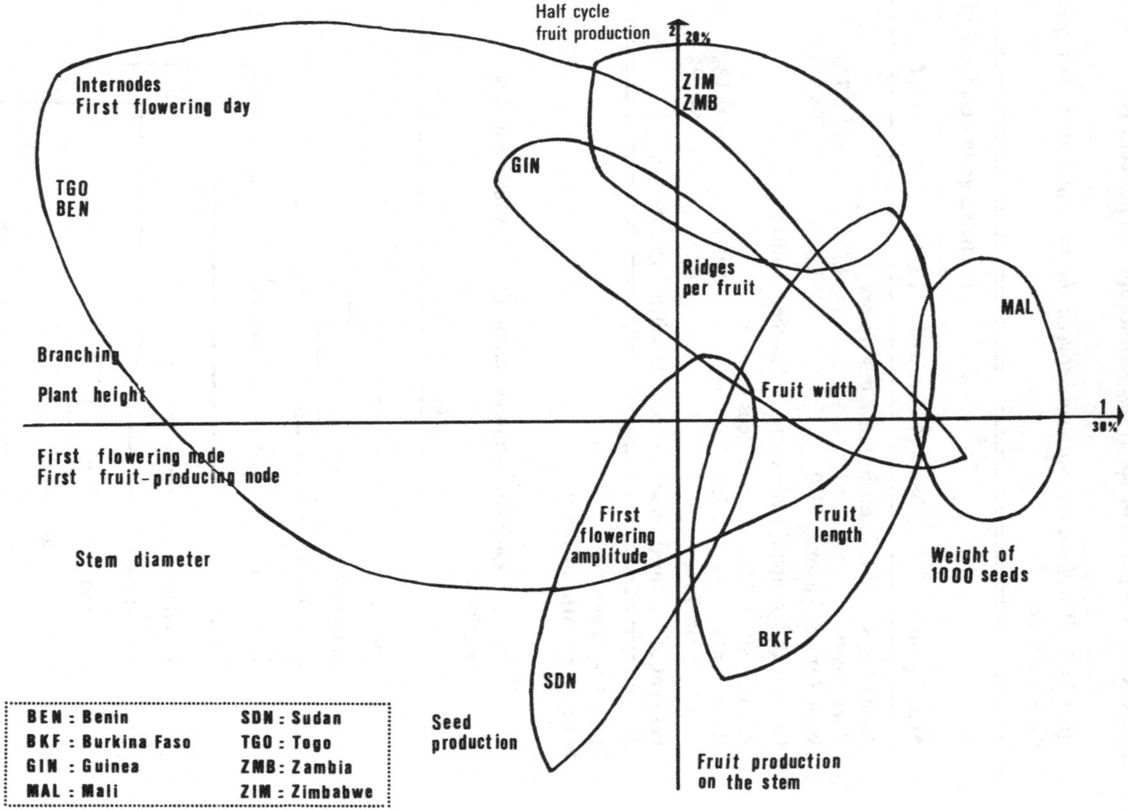

fruit production) were selected. The percentage of properly classified variables was determined for each. The results are found in Table 11.6.

As can be seen, the margin of error is usually small and the estimation with the three variables produces quite comparable results. There is a difference of 0.4 per cent for *A. esculentus* and 5.4 per cent for the WAT.

Table 11.6. *Discriminant analysis between the two cultivated okra species in West Africa*

Actual group	Number of cases	Predicted group membership ESC	WAT
ANALYSIS 1: Whole quantitative descriptors			
A. esculentus	591	99.3%	0.7%
West African Taxon	310	5.5%	94.5%
Per cent of 'grouped' cases correcly classified: 97.67%			
ANALYSIS 2: Maximum of steps equal 3			
A. esculentus	702	98.9%	1.1%
West African Taxon	422	10.9%	89.1%
Per cent of 'grouped' cases correctly classified: 95.20%			

ESC: *A. esculentus*
WAT: West African Taxon

Fig. 11.9. Euclidian distances (variance criteria) between okra descriptors

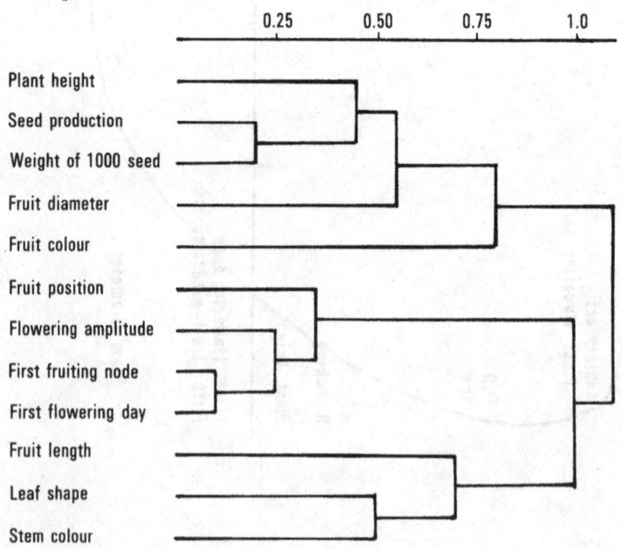

The slight differences can be explained by specific adaptations or perhaps by introgression from the other species. This is an extremely useful method which can be applied to different kinds of groups such as: plant type, geographic origin, etc.

Regrouping individuals and/or variables. Hierarchical clustering methods allow data to be classified by descriptors or by individuals. In Fig. 11.9 a classification scheme is presented for 12 descriptors, on the basis of factor analysis. Descriptors which are very strictly correlated, fruiting or colour of the petiole and stem, have not been taken into consideration.

This classification shows that the choice of descriptor depends on the required level of precision in describing variability. At the Euclidian distance level of 0.80 three distinct groups can be distinguished (Fig. 11.9). Already at this level one can see a grouping of descriptors translating the variability encountered. A choice at random of one variable in each group would already provide a strong indication of available variability. A choice at level 0.30, which appears acceptable in our situation, would allow the elimination of four descriptors.

Establishment of core collection

The principal ideas behind the concept of the 'core' collection are described in Frankel & Brown (1984); see also chapter by Brown in this volume. The need to reduce collections does not appear directly at the level of the base collection which is multiplied at least once and conserved in its entirety. However, considering the lack of data on many genetic resources collections, and the extremely costly and time-consuming characterisation and evaluation of large collections, there is an obvious need for reduction (Peeters & Williams, 1984; van Sloten, 1987).

With regard to the okra collection, it was decided in 1983 to establish such a core collection (from 200 to 300 accessions). This number was not selected in terms of percentage of the entire collection, but rather in accordance with the following objectives:

1. to have a manageable collection scaled down to the needs of the breeder and/or other user; and
2. to include the widest possible range of variability.

Towards the end of 1985, an okra core collection of 189 accessions was established on the basis of representative variability as described by passport, characterisation and evaluation data, but also including rare types. This core collection has already been distributed to several countries for further evaluation.

In this connection it should be noted that there are relatively few okra breeding programmes in the world and therefore limited opportunities for gene introductions, i.e. the transfer of specific genes in breeding programmes. There is, however, enormous scope for using the core collection in adaptation trials in a large number of different environments (direct plant introduction).

Conclusions

On the basis of the experience gained in the evaluation of the ORSTOM/IBPGR okra collection, we are attempting an overview of the problems, of the methods of characterisation and evaluation, and possible solutions which may have application in other crop plants.

The quality of the information obtained at the collecting site is an extremely important factor. It goes without saying that information on geographic co-ordinates is an absolute necessity. It is possible to improve the level of information during collecting by observing the following:

1. restrict the mission to one or a very limited number of species;
2. take sufficient time to become familiar with the local conditions and customs;
3. request a systematic translation of local names;
4. ensure the involvement of women farmers.

The cultivation system used is an important source of information, which becomes more important when the crop has a long tradition in the particular country. One can therefore not expect to obtain the same quality of information in all areas, but one should know how to profit most from information available. We have seen that the graphical representation of the variability by means of factorial analysis corresponds closely to the farming systems used and the latter therefore provide a certain orientation in the choice of morphological descriptors.

The choice of descriptors is the second critical stage. Isozyme descriptors are now being used more and more. Their use eliminates the environmental influences, the necessity for large areas for cultivation, and the method is fairly simple. The authors agree with Crawford (1985) who underlines the limitations of the use of such markers and considers the electrophoretic information as complementary to standard characterisation. Morphological descriptors are important, since they are of most interest to the agronomist and breeder. Morphological polymorphism is not necessarily associated nor correlated with enzymatic polymorphism. Davis & Gilmartin (1985) emphasize that substantial morphological variation could be associated with only minor enzymatic changes. At the same time, adaptation plays an important role in the differentiation of ecotypes. We have observed the disappearance of the WAT from arid zones and a decrease in variability in areas where cultural practices have become restrictive.

The reduction in the number of descriptors can be obtained through multivariate analysis. A step-wise approach consists of running a principal component factor analysis which projects the variability in a limited dimension. Then a clustering analysis is done which will be later tested by a discriminant analysis. In this approach the choice made by the researcher leads to a deliberate, but controlled loss of information.

An effective reduction of a collection, on the understanding that the original collection still needs to be conserved, is a crucial problem. In okra, a collection reduced to 200–400 well-described accessions is likely to be of a size which can be properly used.

References

Borssum-Waalkes, J.van (1966). Malesian Malvaceae revised. *Blumea*, **14**, 1–251.

Chandra, S. & Bhatnagar, S. P. (1975). Reproductive biology of *A. esculentus*. 1. Reproductive behaviour – floral morphology – anthesis and pollination mechanism. *Acta Botanica India*, **3**, 104–13.

Charrier, A. (1983). *Les Ressources Génétiques du Genre* Abelmoschus *Med. (Gombo)*. International Board for Plant Genetic Resources, Rome.

Charrier, A. (1984). *Genetic Resources of the Genus* Abelmoschus *Med. (Okra)*. International Board for Plant Genetic Resources, Rome (English translation of Charrier, 1983).

Chevalier, A. (1940). L'origine, la culture et les usages de cinq *Hibiscus* de la section *Abelmoschus*. *Revue de Botanique Appliquée et d'Agriculture Tropical*, **20**, 319–28.

Crawford, J. L. (1985). Electrophoretic data and plant speciation. *Systematic Botany*, **10**, 405–15.

Davis, J. I. & Gilmartin, A. J. (1985). Morphological variation and speciation. *Systematic Botany*, **10(4)**, 416–25.

Frankel, O. H. & Brown, A. H. D. (1984). Current plant genetic resources – A critical appraisal. In *Genetics*: *New Frontiers*. Proceedings of the XV International Congress of Genetics. Volume IV, pp. 3–13, Chopra, V. L., Joshi, B. C., Sharma, R. P. and Bansal, H. C. (eds.), Oxford & IBH Publishing Co., India. Reprinted In *Crop Genetic Resources*: *Conservation and Evaluation*, pp. 249–57, Holden, J. H. W. & Williams, J. T. (eds.), Allen and Unwin, London.

Hamon, S. & Charrier, A. (1983). Large variation of okra collected in Benin and Togo. *Plant Genetic Resources Newsletter*, **56**, 52–8.

Hamon, S. & Yapo, A. (1986). Perturbation induced within the genus *Abelmoschus* by the discovery of a second edible okra species in West Africa. *Acta Horticulturae*, **182**, 133–44.

Martin, F. W. (1983). Natural outcrossing of okra in Puerto Rico. *Journal of Agriculture of the University of Puerto Rico*, **67**, 50–2.

Pal, B. P., Singh, H. B. & Swarup, V. (1952). Taxonomic relationships and breeding possibilities of species of *Abelmoschus* related to okra (*A. esculentus*). *Botanic Gazette*, **113**, 455–64.

Peeters, J. P. & Williams, J. T. (1984). Towards better use of genebanks with special reference to information. *Plant Genetic Resources Newsletter*, **60**, 22–32.

Second, G. & Trouslot, P. (1980). Electrophorèse d'enzymes du riz (*Oryza* sp.). Traveaux et documents de l'ORSTOM, ORSTOM, Paris.

Siemonsma, J. S. (1982*a*). West African okra – morphological and cytogenetical indications for the existence of a natural amphidiploid of *Abelmoschus esculentus* (L.) Moench and *A. manihot* (L.) Medikus. *Euphytica*, **81**, 241–52.

Seimonsma, J. S. (1982*b*). *La culture du Gombo* (*Abelmoschus spp.*), *Légume – fruit tropical – avec référence spécial à la Côte d'Ivoire*. Thesis Agricultural University, Wageningen, The Netherlands.

Sloten, D. H. van. (1987). The role of curators, breeders and other users of germplasm in characterization and evaluation of crop genetic resources. Vth SABRAO Congress.

Tanda, A. S. (1985). Floral biology, pollen dispersal, and foraging behaviour of honeybees in okra. *Journal of Agricultural Research*, **24**, 225–7.

12
Evaluation of cereals in Europe

G. FISCHBECK

Introduction

Breeding of the major cereals grown in Europe traces back to intensified efforts to produce better seeds by means of careful selection within landraces about 120 years ago. It took four to five decades for genetic recombinations based upon artificial crosses to become the major sources for further progress, except for the outbreeding crop rye. Nevertheless, if a ten-year time scale for a conventional cross breeding programme is assumed, wheat, barley and oat cross breeding in Europe has now reached its fifth to seventh cycle.

Landrace base of important progenitors in European wheat breeding

Reliable and relevant evaluation of potential parents forms the corner-stone of each breeding programme. From the early steps in cross breeding of cereals in Europe, it is obvious that a large proportion of the more important parents for later breeding cycles descended from rather diverse parentage (see Fig. 12.1 and Table 12.1). Carsten VIII, a very important progenitor for wheat breeding in Germany, traces back to eight landrace selections derived from four different countries. The same holds true for Capelle, the most prominent progenitor included in the parentage of recent cultivars in most west European countries. Heine VII, the first cultivar well adapted to combine harvesting in Germany, combines six landrace sources out of five different countries, but does not carry a German landrace in its pedigree. Derenburger Silber, another important parent in winter wheat breeding in Germany, came from a winter × spring wheat cross including three non-German landrace sources.

Two other important aspects are included in Table 12.1. All of the four key breeding sources have at least one landrace source in common – the

Fig. 12.1. Pedigree of winter wheat cultivar *Caribo* and landrace parentage of cultivar *Carstacht*

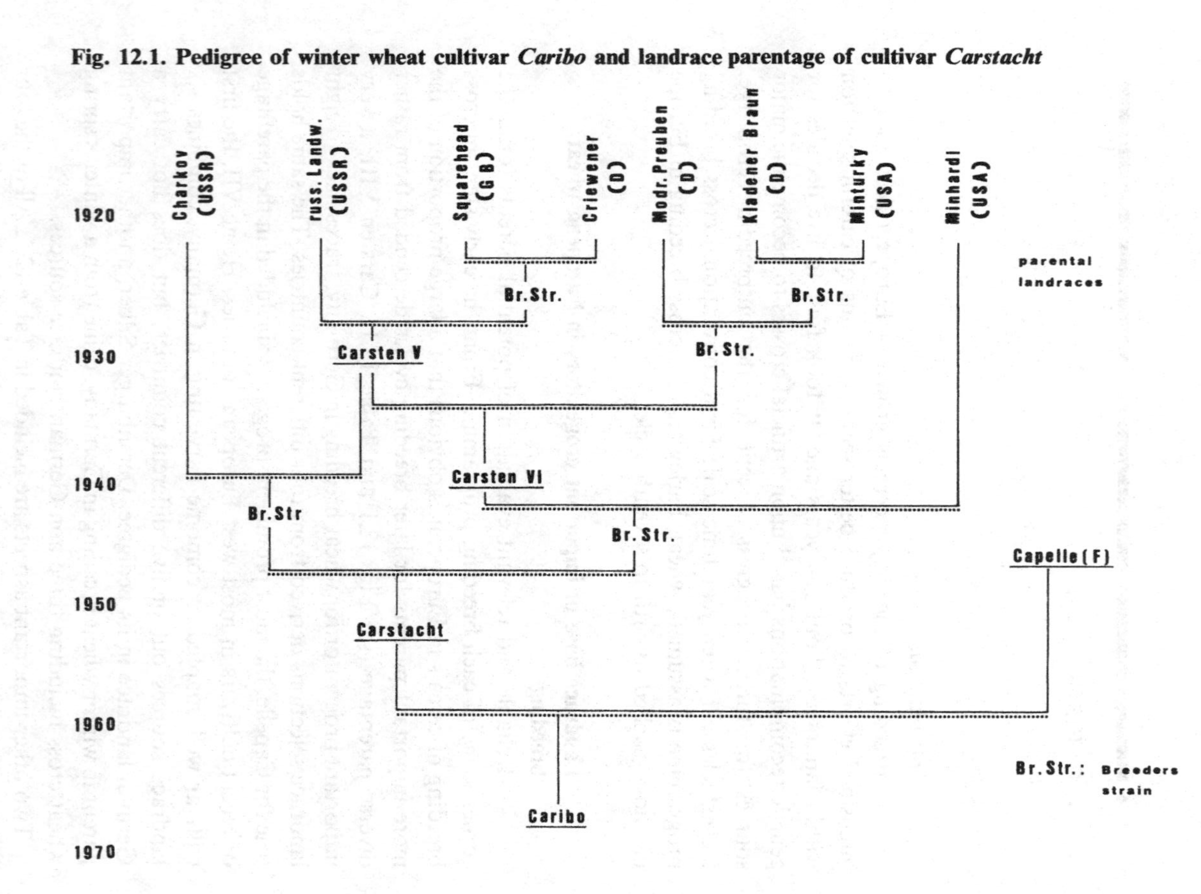

Table 12.1. *Origin of landraces of wheat contributing to the pedigree of selected cultivars of winter wheat**

	England	USSR	USA	France	Germany	Sweden	Italy	Total
Capelle (released 1946) (Vilmorin 27 × Hybr. Jonquois)	4	1	—	2	—	—	1	8
Carsten VIII (released 1952) (Breeders strain × Breeders strain)	1	2	2	—	3	—	—	8
Heine VII (released 1943) (Hybr. a courte paille × Sv. Kronen)	1	2	1	1	—	1	—	6
Derenburger Silber (released 1941) (Panzer III × Peragis)	1	—	2	—	—	—	—	3
Total	7	5	5	3	3	1	1	25 (19)

	Identical landraces in pedigrees of			Identical landraces other than Squarehead		
	Carsten VIII	Heine VIII	Derenburger Silber	Carsten VIII	Heine VII	Derenburger Silber
Capelle	1	3	1	0	2	0
Carsten VIII	—	1	1	—	0	0
Heine VII	—	—	2	—	—	1

* Extracted from pedigrees given by Wienhues & Giessen (1957) for major genitors of later breeding cycles.

famous English Squarehead mutation. But, apart from Squarehead, only a few common ancestors remain between Capelle and Heine VII, as well as between Heine VII and Derenburger Silber. Therefore, of the 25 landrace sources included in the parentage of the cultivars listed in Table 12.1, 19 are unrelated. Capelle and Carsten VIII share only one landrace source (Squarehead) but each of them traces back to seven more unrelated sources. As indicated by these examples, the breeding of winter wheat in Europe is based on a large pool of landrace diversity which had already been assembled in the early phases of cross breeding.

Furthermore, it will be noted that the preceding cross from which the four cultivars listed in Table 12.1 were derived included well known cultivars or breeders' strains developed and evaluated by the breeder himself. This system applies to European breeding programmes with other self-pollinating cereals as well and has maintained its dominance ever since, except for special characters (e.g. dwarfing genes and genes for disease resistance) which have been added step by step to the common gene pools mainly by backcross procedures.

In summary, an early phase of increased genetic diversification and recombination can be recognised. Progress made in later breeding cycles appears to be based principally upon pyramiding the potential of genetic recombinations created by the range of genetic diversity assembled in the early phases of cross-breeding.

Working collections of cereal breeders

With the advent of hybridisation, breeders could no longer rely solely on their own breeding material but needed to collect and develop working collections of potential parents. Some representative data for the present-day situation in European barley breeding are available from a survey (Table 12.2) which was conducted as part of a multinational co-

Table 12.2. *Size of breeders' collections* (*number of accessions*)

	Number of breeders
0— 100	2
100— 500	14
500—1,000	6
1,000—5,000	12
5,000—10,000	4
> 10,000	—
n	38

Source: UNDP/IBPGR, 1986

operation between countries in Europe which joined the UNDP/IBPGR European Programme for the Conservation and Exchange of Crop Genetic Resources (UNDP/IBPGR, 1986; see also Perret, this volume). This survey indicates that the preferred size of working collections is 100–500 accessions which are carefully evaluated for future use in a given breeding programme. It is likely that the figures for wheat and oat breeding would be similar.

Obviously breeders carry a substantial work load to evaluate genetic resources most suitable for their own breeding programme. The database assembled within the UNDP/IBPGR European Programme reveals some more detailed information about this process (Table 12.3). The major sources for yield improvement are extracted from the breeders' own working collections. This is not very different from choosing sources for stress tolerance, but, for disease resistance, gene bank material receives much more attention. About 30 per cent of the major sources used in present-day barley breeding programmes in each of the three sectors comes from 'other sources', which are mainly based upon co-operative and mutual exchange of advanced material between breeding stations. The material derived from gene banks concentrates upon new sources for disease resistance; it is less frequently used in efforts to improve stress tolerance and makes up only 15 per cent of the major sources used for yield improvement.

If the status of material in present-day breeding programmes is reviewed, it becomes very clear that wild and weedy sources are rarely

Table 12.3. *Major sources (%) used for individual breeding needs*

	Disease resistance	Stress tolerance	Yield increase
Own breeding collection	28	41	55
Other sources	31	30	30
Gene bank materials	41	28	15
replies (*n*)	54	47	44
	Status of the material (%) used		
Wild	6	6	—
Weedy	4	3	—
Primitive cultivar or landrace	22	28	4
Breeders lines	38	37	44
Advanced cultivar	30	26	52
replies (*n*)	81	68	54

Source: UNDP/IBPGR, 1986

included in breeding programmes even in relation to disease resistance and stress tolerance. Primitive cultivars and landraces receive much more attention, almost equal to breeders' lines and advanced cultivars in breeding for disease resistance and stress tolerance, while material for yield improvement is largely confined to breeders' lines and advanced cultivars.

This survey provides a useful basis for discussing the needs and experiences with evaluation of cereals collected and maintained by European gene banks, which have reached very sizeable numbers.

Size of cereal collections within European gene banks

One of the co-operative projects within the UNDP/IBPGR European Programme aims to establish a European Barley Data Base (EBDB). Passport data on 49,800 accessions from 25 collections maintained in 21 countries had been recorded by July 1985 (UNDP/IBPGR, 1986). These included 26,800 unnamed accessions originating from 70 countries (12,400 from Ethiopia, 2,300 from Turkey, 2,000 from Nepal, and 10,000 accessions from 67 other countries) and 23,000 'cultivars' (named accessions) which quite often carry equal or similar names and are apparently duplications. This leaves 14,000 differently 'named' accessions which at a first glance may represent unique genotypes. Summing up, with the unnamed accessions about 41,000 different genotypes of barley are maintained within the group of European genebanks. However, only a small fraction has been thoroughly evaluated for potential use in barley breeding programmes.

The collection of plant genetic resources assembled at the Institut für Pflanzenbau und Pflanzenzüchtung der Forschungsanstalt für Landwirtschaft (FAL) in Braunschweig-Völkenrode (Federal Republic of Germany) may serve as an example for the size of individual gene banks in Europe. In early 1986 it held a total of 46,300 accessions, 57 per cent (26,300) of which belonged to the major cereal species, comprising 15,600 wheat samples and 9,000 barley samples, which leaves a total of 1,800 accessions for oat, rye and other species (Seidewitz, 1986).

Status of evaluation of cereal collections in European gene banks

Evaluation of gene bank accessions in Europe has largely been organised at national level and therefore differs according to the institutional framework for plant breeding and agricultural research in each country.

Much information is collected during the regeneration process (Lehmann, 1984) sometimes in co-operation with private breeders

(Dambroth, 1986). However, specific evaluation trials have also been organised in some countries (Bares *et al.*, 1985; Bartos *et al.*, 1985). In addition, at least parts of the collections have been subjected to specific tests for disease reaction at institutes with the appropriate expertise. For example, Lehmann (1984) reported that between 2,000 and 6,000 accessions of the Gatersleben (German Democratic Republic) barley collection have been tested against mildew, leaf rust, yellow rust and loose smut respectively, sometimes with different races of each disease. A number of accessions, mainly of Ethiopian descent, showed combined resistances against mildew, leaf and stripe rust and loose smut (Nover & Lehmann, 1973; Walther & Lehmann, 1980).

The results of another attempt to accelerate evaluation of gene bank material, and at the same time speed up the interest of breeders in the immediate use of it, are shown in Table 12.4. About 3,100 non-exotic accessions of the collection of winter wheats held at Braunschweig-Völkenrode were included in a trial series where individual sets of 50 to 200 accessions were distributed to 10 to 15 private breeders and public research institutions in the Federal Republic of Germany, over three consecutive years. They were sown in small observation plots, evaluated for a number of characters in comparison to a set of standards made up from recent winter wheat cultivars. Based upon the notes taken at the individual sites, about 25 per cent more promising accessions have been

Table 12.4. *Results of co-operative evaluation trials of 'non-exotic' accessions of the winter wheat collection at Institut für Pflanzenbau und Pflanzenzüchtung, FAL, Braunschweig-Völkenrode*

Total number of accessions (*Triticum* and *Aegilops*)	15,552
T. aestivum accessions	11,472
'Non-exotic' accessions	3,102
Accessions evaluated in repeated tests	256

	Number	
Mildew	256	11.3% superior to best standard cultivar (M)
Leaf rust	229	16.1% superior to best standard cultivar (L)
Glume blotch	227	10.6% superior to best standard cultivar (GB)
Yellow rust	93	24.7% equal to best standard cultivar (Y)

Combined resistances

	Number			Number			Number	
M+L	227	0.3%	Y+M	93	7.5%	Y+M+L	93	1.1%
M+GB	227	0.04%	Y+L	93	2.1%	Y+M+GB	81	2.5%
L+GB	227	0.04%	Y+GB	81	2.5%	Y+L+GB	81	0.0%
M+L+GB	227	0.0%						

selected from each set to be included in replicated tests carried out at several stations in the following year. From 256 accessions tested in more than one year and at different locations some useful evaluation data are summarized in Table 12.4. Between 10 and 16 per cent appeared to be superior to the best standard cultivar as far as mildew, leaf rust and glume blotch resistance is concerned, and 25 per cent did not show more yellow rust than the best standard cultivar which is rated highly resistant. Much less favourable results are obtained if one looks for combined resistances. Only one accession appears to combine improved resistance against mildew and leaf rust at the high level of yellow rust resistance which is already available in commercial cultivars of winter wheat. Since all observations are based upon spontaneous disease attack occurring in field trials in comparison with a set of standard cultivars, no information is available on the spectra of resistance against individual pathotypes. It remains to be seen whether this trial series will stimulate the immediate use of gene bank material by the participating German plant breeders. Since so-called 'exotic' accessions were purposely excluded from this trial series, the chances are not very high of detecting many more samples of interest to commercial breeders if the trials were extended to the remaining 8,400 *Triticum aestivum* accessions held in the Braunschweig-Völkenrode wheat collection. The exotics will probably be much less adapted to the growing conditions in Germany than the set which has been tested so far.

Evaluation needs of European cereal breeders

Criticism about inadequacy of gene bank databases has come from many breeders and it probably will be impossible to satisfy all their wishes and needs. Within the UNDP/IBPGR European Barley Data Base project some broader-based information about this issue has been assembled (UNDP/IBPGR, 1986). As seen from Table 12.5 about 60 per

Table 12.5. *Breeders' appraisal on quantity and quality of information available on samples in genebanks*

	Quantity (%)		Qualtiy (%)
More than sufficient	5	Excellent	0
Sufficient	10	Good	11
Just sufficient	25	Adequate	44
Insufficient	55	Inadequate	39
Quite in sufficient	5	Quite inadequate	6
Number of replies	38		36

Source: UNDP/IBPGR, 1986

cent of the breeders rated the quantity of information about gene bank samples insufficient. In addition, 45 per cent felt also that the quality of information is not adequate for breeding purposes.

A rating of the relative importance of descriptors for gene bank accessions of barley by European breeders from a survey within the European Programme is given in Table 12.6. It clearly demonstrates how the relative importance of a particular descriptor varies with the location of the evaluation trials and the origin of the accessions. For more adapted accessions, information collected at growing conditions similar to the breeding station (local importance), growth habit (spring/winter type), resistance to lodging, powdery mildew reaction, grain yield and 1,000 kernel weight received the highest ratings. In contrast, for less adapted accessions, information collected at different growing conditions (international importance), growth habit, powdery mildew reaction, winter-hardiness, dwarf leaf rust reaction and low temperature resistance were placed in the top ranks. Of 17 descriptors related to disease resistance 16 received higher rankings in the column for international importance, only the highest ranking for mildew reaction remained unchanged. On the other hand, for 17 descriptors related to plant type and yield, or yield

Table 12.6. *Mean ratings of individual barley descriptors (1–9; with 9 = most important)*

	Local importance	Importance in an international database	Number of replies
Plant type, yield and yield components, quality			
Growth habit	8.6	8.6	35
Plant height	6.9	6.5	35
Days to flower	6.9	6.4	34
Row number/spike type	7.1	7.5	34
Spike density	5.0	5.0	35
Number of spikelets/spike	4.8	4.8	33
Awns/Hoodedness	5.8	6.0	34
Awn roughness	4.2	4.4	33
Length of rachilla hairs	3.5	3.8	33
Kernel covering	6.7	6.7	32
Lemma colour	4.0	4.3	34
Pericarp colour	4.8	4.9	34
1,000 grain weight	7.8	6.9	33
Grain yield	7.9	6.7	33
Protein content	6.4	6.0	33
Lysine/protein ratio	6.1	6.1	32
Content in dry malt	5.8	5.8	31

components, the ranking in local importance and international importance remained unchanged within rather narrow limits, except for yield and 1,000 kernel weight. For the latter two characters, local importance rated markedly higher than international importance. The interpretation of these results becomes clear if reference is made to the data given in Table 12.3. They show that European barley breeders are in need of new and often 'exotic' sources for disease resistance which they expect to come from gene bank collections. But, apart from the fact that the status of evaluation of European cereal collections for disease reactions is far from comprehensive, new sources of disease resistance which are present in existing collections will only enter the breeding process if sufficient efforts are invested in a relevant pre-breeding phase.

In view of the advanced state of cereal breeding in Europe, there is only a small chance for gene bank accessions to prove useful for yield improvement. This will be different only if gene banks provide samples of commercial varieties from neighbouring countries as is the rule in East European countries. The national gene banks in these countries have extensive programmes to collect the latest cultivars of neighbouring countries and evaluate them under local conditions (Bares *et al.*, 1985; Lekes *et al.*, 1985). In some West European countries a commercial cultivar of cereals will be transferred to the national gene bank only when the plant breeders' rights connected with it are either withdrawn or expire. But seeds of all commercial varieties are sold on the market and exchanged between breeders. Furthermore, commercial cultivars receive extensive evaluation in the official variety testing systems, the results of which are published and therefore available to everyone interested.

From the pedigrees of commercial cultivars of the major cereal species in Europe, it is evident that the characteristic system of later breeding cycles demonstrated in Table 12.1, still prevails. In most cases, recent cultivars originate from crosses between breeders' lines and the most successful cultivars from other breeders. Again, certain cultivars appear frequently in the pedigree of successful crosses based upon outstanding levels of general combining ability. For example, 19 of 55 winter wheat cultivars named in the 1985 German list descend from crosses with cv. Caribo, which was released in 1968 from the cross Capelle × Carsten VIII (see Fig. 12.1). Even of the seven cultivars released in Germany in 1986, five are based upon Caribo crosses. Obviously, the base of genetic diversity laid down in the early phases of cross breeding of wheat in Europe still provides sufficient chance for further progress. At the same time the high level of adaptation and yield potential reached during several breeding cycles makes it more difficult to introduce new sources of genetic diversity into commercial breeding programmes. But, if the indication is correct that cereal breeding in Europe still profits from genetic diversity introduced in the early days of cross breeding, a new phase of accelerated increase in genetic diversity eventually will be needed, which certainly will also affect the needs for evaluation of germplasm collections.

Deficiencies in the availability of suitable germplasm became obvious in the lack or the insufficient effect of disease resistance in cereal breeding. Increased disease pressure connected with intensified cereal culture favoured disease attack, as did insufficient genetic diversity in disease resistance from gene bank accessions from other countries, often unadapted to the local growing conditions as well as to the biotypes of the parasites. For improving resistance against powdery mildew, European barley breeders exploited only three major sources of resistance between 1930 and 1960 and about eight additional sources since 1960. Only the first one was derived from a German landrace of spring barley, the others were screened mainly from unadapted foreign accessions held in different collections, with the exception of the *mlo* type of mildew reaction which has repeatedly arisen from mutations that have occurred at the *mlo* locus on barley chromosome 4 (Schwarzbach & Fischbeck, 1981). The situation with other diseases and cereal crops resembles the experience with mildew resistance of barley.

'In depth' evaluation

In a EUCARPIA symposium on 'Evaluation for the Better Use of Genetic Resources Material', held in March 1985, at Prague, European breeders expressed their needs more explicitly as shown with the following citations from Eucarpia Bulletin No. 15.

> Evaluation for each crop (should) be roughly divided into two categories:
>
> a) that which is largely or wholly carried out by the gene bank staff (characterisation),
> b) that which is largely or wholly carried out by plant breeders, phytopathologists, entomologists etc., either in other sections of the institute in which the gene bank is situated, or in other institutes ('in depth evaluation'),
> – that the 'in depth' evaluation ... be regarded as of equal or greater importance than 'characterization',
> – that if evaluation priorities are needed they should be set out approximately as follows, after consultation with the breeders:

> *Highest*:
> Resistance or tolerance to pests or diseases
> Agronomic characters
> Adaptation or tolerance to stress conditions such as cold, drought, salinity, etc
> Biochemical characters, where appropriate
> *Lowest*:
> Morphological/botanical characters
>
> This sequence may vary from crop to crop and from country to country. If materials in the gene bank show promise for some of the characters mentioned above but possess poor agronomic qualities (because they are landraces, wild species, etc) the gene bank manager should encourage pre-breeding (germplasm enhancement) of the material whenever and wherever possible so as to make it more immediately attractive and available to breeders.

The level of evaluation of most cereal collections in European gene banks falls far below the stated objectives and needs and much input will be needed to improve this situation.

'In depth' evaluation of disease resistance ultimately implies identification of its genetic base. Adaptation to stress conditions can only be measured if suitable methods are available. For the time being this may be possible for frost resistance but not for tolerance to drought or salinity. Finally, improvement in yield will have to be measured more in terms of combining ability (Frankel & Brown, 1984) than in terms of actual yield or excessive formation of individual yield components.

In all cases the outcome will be that evaluation needs to be much more strongly supported than before, and the restricted resources in labour and in money will make it necessary to concentrate 'in depth' evaluation on more promising accessions and also on pre-breeding activities.

This poses the important question of the identification of promising accessions.

From our experience in collaborative work with the University of Tel Aviv it is evident that *Hordeum spontaneum* collected in Israel harbours a wealth of genetic diversity for mildew resistance (Fischbeck *et al.*, 1976). Similar results have been obtained for resistance to crown rust in *Avena sterilis* used in the US oat breeding programme (Frey *et al.*, 1977) by Negassa (1986), who found 27 per cent of 293 entries of Ethiopian wheats to be resistant to mildew in Scandinavia. Only small numbers of *Hordeum spontaneum* accessions were held in the European barley collections until recently. But our screening for mildew resistance among single head progenies randomly chosen from natural stands of *Hordeum spontaneum* at 78 locations in Israel turned out to be much more efficient compared with earlier efforts devoted to about equal numbers of accessions of cultivated barley maintained in our gene bank. Using field observations of mildew attack in Israel and Germany, 42 of the large number of *H. spontaneum* lines collected in Israel have been subjected to 'in depth' evaluation of mildew resistance. Of these, 39 turned out to be different from sources for mildew resistance described in the literature including the 11 sources already in use by European barley breeders. Thirty-two new genes/alleles have been determined so far (Jahoor, 1986), certainly enough to achieve substantial progress in current efforts to broaden the genetic base of mildew resistance in European barley breeding. Landraces of barley only recently collected in co-operation with the International Center for Agricultural Research in the Dry Areas (ICARDA) in Jordan and in Syria provide another example for the extraction of 'promising' accessions for 'in depth' evaluation. The European Barley Data Base mentions only 129 accessions from Syria and Jordan (Knüpfer *et al.*, 1986). Also, the ICARDA barley collection which has been derived from the United States Department of Agriculture (USDA) originally contained only few accessions from these countries. New collections sampled the landraces which are still widely distributed in the area (Weltzien, 1986). Field evaluation under conditions of drought and salinity stress revealed definitely higher levels of adaptation to stress conditions compared with available standard cultivars, together with a rather large range of genetic variability immediately amenable to future barley breeding in the ICARDA region. Taking into consideration the increased interest of European barley breeders in drought resistance, the more promising accessions derived from these tests should be subjected to 'in depth' evaluation of potential use in European breeding programmes.

Both examples demonstrate that, in spite of large numbers of accessions stored in the barley collections in European gene banks, there is no certainty that a truly representative sample of genetic diversity within this crop has already been collected (Chapman, 1984).

Many more cases may be expected where there is no immediate access to or knowledge about more promising sectors for genetic variation of certain characters, most certainly not for yield or for general combining ability able to improve it. In such cases it will not be possible to provide useful 'in depth' evaluations unless 'core collections' are formed from existing gene bank materials, as was suggested by Frankel & Brown (1984), which are made up from five to ten per cent of the total number of accessions in existing collections (see also Brown, this volume). If a high amount of representation is reached, the results of 'in depth' evaluation for certain traits obtained with the core collection should be highly indicative for spotting limited sectors within the reserve collection containing 90 to 95 per cent of the total number from which more promising accessions are to be expected.

For this reason, research efforts to develop reliable measures of genetic diversity within gene bank materials should assist in the development of representative core collections. They should accelerate meaningful evaluation of the cereal collections in European gene banks and eventually speed up the introgression of genetic diversity into the working collections of European cereal breeders.

References

Bares, I., Vlasak, M. & Schmalova, J. (1985). Research, methods and results of rust resistance evaluation in wheat collections. In *Proceedings of the Eucarpia. International Symposium on Evaluation for the Better Use of Genetic Resources Materials*, pp. 13–24. V. Rogalewicz (ed.), Prague.

Bartos, P., Strichlikova, E., Bares, J. & Vlasak, M. (1985). Methods and results of rust resistance evaluation in wheat collections. In *Proceedings of the Eucarpia. International Symposium on Evaluation for the Better Use of Genetic Resources Materials*, pp. 25–34, V. Rogalewicz (ed.), Prague.

Chapman, C. G. D. (1984). On the size of a gene bank and the genetic variation it contains. In *Crop Genetic Resources: Conservation and Evaluation*, pp. 102–19, Holden, J. W. H. & Williams, J. T. (eds.). Allen and Unwin, London.

Dambroth, M. (1986). Verfügbarkeit genetischer Resourcen für die Pflanzenzüchtung in der Bundesrepublik Deutschland. *Vorträge Pflanzenzüchtung*, **10**, 18–31.

Eucarpia. (1986). Bulletin Nr. 15, Section Genetic Resources, pp. 32–33, Wageningen.

Fischbeck, G., Schwarzbach, E., Sobel, S. & Wahl, I. (1976). Types of protection against barley powdery mildew in Germany and Israel selected from *Hordeum spontaneum*. In *Barley Genetics III*, pp. 412–17, H. Gaul (ed.). Thiemig, München.

Frankel, O. H. & Brown, A. H. D. (1984). Plant genetic resources today: a critical appraisal. In *Crop Genetic Resources: Conservation and Evaluation*, pp. 249–59, Holden, J. W. H. & Williams, J. T. (eds.). London: Allen and Unwin.

Frey, K. J., Browning, J. A. & Simmonds, M. D. (1977). Management of host resistance genes to control disease loss. *Annals of the New York Academy of Sciences*, **287**, 255–74.

Jahoor, A. (1986). *Mehltauresistenz aus israelischen Wildgersten – Resistenzspektrum, Vererbung und Lokalisierung*. Dissertation, Technical University, Munich.

Knüpfer, H., Lehmann, Chr. O. & Scholz, F. (1986). Barley genetic resources in European gene banks – The European Barley Data Base. Contrib. to *V. International Barley Genetics Symposium*, Okayama, Japan.

Lehmann, Chr. O. (1984). Germplasm evaluation at Gatersleben: the relationship between genebank and breeder. In *Crop Genetic Resources:*

Conservation and Evaluation, pp. 202–6, Holden, J. W. H. & Williams, J. T. (eds.). Allen and Unwin, London.

Lekes, J., Krystof, Z., Machan, F. & Zezulova, P. (1985). Using of genetic resources in creation of new cereal cultivars in Czechoslovakia. In *Proceedings of the Eucarpia. International Symposium on Evaluation for the Better Use of Genetic Resources Materials*, pp. 85–94, V. Rogalewicz (ed.), Prague.

Negassa, M. (1986). Estimates of phenotypic diversity and breeding potential of Ethiopian wheats. *Hereditas*, **104**, 41–8.

Nover, I. & Lehmann, Chr. O. (1973). Resistenzeigenschaften im Gersten- und Weizensortiment Gatersleben. 19. Fortgesetzte Prüfung von Sommergersten auf ihr Verhalten gegen Mehltau (*Erysiphe graminis* DC. f. sp. *hordei* Marchal). *Kulturpflanze*, **21**, 275–94.

Schwarzbach, E. & Fischbeck, G., (1981). Die Mehltauresistenzfaktoren von Sommer- und Wintergerstensorten in der Bundesrepublik Deutschland, *Zeitschrift für Pflanzenzüchtung*. **87**, 309–18.

Seidewitz, L. (1986). Stand und Verfügbarkeit der Sammlung pflanzengenetischer Resourcen im Institut für Pflanzenbau und Pflanzenzüchtung der FAL. *Vorträge Pflanzenzüchtung*, **10**, 32–36.

UNDP/IBPGR (1986). *Report of a Barley Workshop*. European Cooperative Programme for the Conservation and Exchange of Crop Genetic Resources. IBPGR, Rome.

Walther, U. & Lehmann, Chr. O. (1980). Resistenzeigenschaften im Gersten- und Weizensortiment Gatersleben. 24. Prüfung von Sommer- und Wintergersten auf ihr Verhalten gegenüber Zwergrost (*Puccinia hordei* Orth.). *Kulturpflanze*, **28**, 227–38.

Weltzien, E. (1986). Anpassungsfähigkeit nahöstlicher Landgersten an marginale Wachstumsbedingungen. Dissertation, Technical University, Munich.

Wienhues, F. & Gieben, J. E. (1957). Die Abstammung europäischer Weizensorten. *Zeitschrift für Pflanzenzüchtung*, **37**, 217–30.

13

Evaluating the germplasm of groundnut (*Arachis hypogaea*) and wild *Arachis* species at ICRISAT

J. P. MOSS, V. RAMANATHA RAO AND R. W. GIBBONS

Introduction

The genus *Arachis* consists of the cultivated groundnut *A. hypogaea* L. and many wild species. Groundnut is grown throughout the tropical and warm temperate regions of the world, between 40 °N and 40 °S. It is an important oil and protein crop in Asia, Africa and the Americas. There is considerable variability in the genus, but Norden (1980) considered that some collections were poorly organised and few had been adequately evaluated. An extensive germplasm base, fully evaluated and properly documented, and the ability to make sufficient crosses to produce large segregating populations, are necessary for the germplasm to be properly exploited.

The genus has been subdivided into seven sections, distinguished on morphology and cross-compatibilities (Gregory & Gregory, 1979). Sections have been further subdivided into series. Cultivated groundnut, *A. hypogaea*, is a tetraploid, $2n = 4x = 40$, in series Amphiploides of section *Arachis*. It has two subspecies, each further subdivided into two botanical varieties (Table 13.1). Also in Amphiploides is *A. monticola*, a wild tetraploid species which is fully compatible with *A. hypogaea*; it has been suggested that it is a wild subspecies (Smartt & Stalker, 1982). These tetraploids all have the genomic formula AABB and they comprise the primary gene pool. The crossability between subspecies is such that they can easily be used in a breeding programme. Interspecific crossability with *A. monticola* is also fairly good. The hybrids have good fertility and produce large enough populations so that selections can be made; Spancross, an American cultivar, is derived from such a cross (Hammons, 1970).

The other two series in section *Arachis* consist of diploid species. The Annuae and Perennes are distinguished on morphological characters, but this division does not correlate with the occurrence of genomes, based on cytological analysis. All species of Perennes studied to date have the A genome, as do most of series Annuae, but *A. batizocoi* has the B genome, and *A. spiniclava nom. nud.* the D genome (Stalker, 1985). The A and B genome species, two of which are probably the ancestors of *A. hypogaea*, can be crossed with *A. hypogaea*. Although the hybrids produced are triploid and effectively sterile, some progeny is produced. Various ploidy manipulations that use different routes have been used to overcome this hybrid sterility (Moss, 1985). Triploid hybrids have been treated with colchicine, and the resulting fertile hexaploids backcrossed to *A. hypogaea* to reduce the chromosome number to the tetraploid level. Diploid species and hybrids have been brought to tetraploid level and these autotetraploids and amphidiploids have been crossed with *A. hypogaea* and fertile hybrids produced and selfed or backcrossed with *A. hypogaea*.

Table 13.1. *Classification of the genus* Arachis L.

- Section *Arachis*:
  - Series *Amphiploides* $2n = 4x = 40$; annual
    - *A. hypogaea* L.: cultivated groundnut
      - Subsp. *hypogaea* Krap. & Rig.
        - var. *hypogaea* Virginia type
        - var. *hirsuta* Kohler Peruvian type
      - Subsp. *fastigiata* Waldron.
        - var. *fastigiata* Valencia type
        - var. *vulgaris* Harz Spanish type
    - *A. monticola*, Krap & Rig.
  - Series *Annuae* $2n = 2x = 20$; Usually annual; 6 described species
  - Series *Perennes* $2n = 2x = 20$; Usually perennial; 6 described species
- Section *Erectoides* Krap. & Greg. *nom. nud.*: $2n = 2x = 20$.
  - Series *Trifoliolatae*: mostly with trifoliate leaves
    - 2 described species
  - Series *Tetrafoliatae*: tetrafoliate leaves
    - 3 described species, 1 *nom. nud.*
  - Series *Procumbensae* mostly prostrate;
    - 2 described species
- Section *Caulorhizae* Krap. & Greg. *nom. nud.*:
  - 2 described species
- Section *Rhizomatosae* Krap. & Greg. *nom. nud.*:
  - Series *Prorhizomatosae* $2n = 2x = 20$; 1 described species
  - Series *Eurhizomatosae* $2n = 4x = 40$; 2 described species
- Section *Extranervosae* Krap. & Greg. *nom. nud.*:
  - 4 described species
- Section *Ambinervosae* Krap. & Greg. *nom. nud.*: $2n = 2x = 20$.
  - No species described
- Section *Triseminalae* Krap. & Greg. *nom. nud.* $2n = 2x = 20$.
  - 1 species described

Selection for fertility, productivity and the desired character in succeeding generations have produced valuable lines through all these routes (Moss, 1984). Thus these diploid species, most of which have one genome in common with *A. hypogaea*, though differing in ploidy level, can be considered the secondary gene pool, and can be utilised using necessary cytogenetic manipulations. Priorities at The International Crops Research Institute for the Semi-Arid Tropics (ICRISAT) for evaluating these species are high, as ICRISAT has the expertise to use them and has produced valuable material from them.

Although some intersectional hybrids have been produced, there are no confirmed reports of any species other than those in section *Arachis* having been successfully crossed with *A. hypogaea*. Use of hormone treatments, and ovule and embryo culture, has resulted in hybrid *A. hypogaea* × *Arachis* sp. (Rhizomatosae) callus and shoots *in vitro* (Mallikarjuna & Sastri, 1985) but these have not been transferred to soil. Although Smartt & Stalker (1982) infer that there may be genomes common to *A. hypogaea* and species in other sections, with our present capability all sections except *Arachis* must be considered as the tertiary gene pool.

Subrahmanyam *et al.* (in review) have analysed the geographic origin of 88 germplasm accessions with resistance to rust and/or late leafspot. Of these, 84 per cent originated in South America, and 74 per cent in Peru. Of the 304 accessions from Peru in the ICRISAT groundnut germplasm collection, 182 have been screened and 47 per cent of these are resistant to rust and/or late leafspot. This clearly indicates that resistances to these two diseases have evolved in the Peruvian region, where these diseases are common. Resistances probably arose as mutations which had a selective advantage in the presence of the pathogens.

Evaluation

The ICRISAT Legumes Program includes groundnut breeders, cytogeneticists, pathologists, entomologists and physiologists. Germplasm is evaluated by the germplasm botanist, who is a member of the Genetic Resources Unit, in collaboration with the relevant scientists. Screening techniques are developed, adapted, and improved to screen large numbers of accessions.

The groundnut germplasm collection totals 11,548 accessions of *A. hypogaea* from 89 countries including 207 accessions of wild species. ICRISAT and IBPGR collaborated to develop a list of descriptors and their definitions and published them as 'Groundnut Descriptors'. These were revised in 1985 by IBPGR (IBPGR & ICRISAT, 1985). More than

10,000 accessions have been evaluated for various descriptors. Passport data for 11,400 ICRISAT accessions have been entered in the computer. The simple ICRISAT Data Management and Retrieval System (IDMRS), developed by ICRISAT Computer Services, is being used for this purpose. The computer file forms the base for a live catalogue. Future plans are to use the live catalogue to publish passport and evaluation catalogues.

Collection, maintenance and conservation are important for studies of taxonomic and evolutionary relationships between and within species, but the main justification for genetic resource conservation is for use in crop improvement (Rao, 1980) and evaluation is a necessary prerequisite. At ICRISAT, the 'preliminary evaluation' is carried out for 41 different morphoagronomic characters. Generally three four-metre long rows of each accession are sown and all the observations are recorded on the central row (Fig. 13.1) though the intergenotypic effects are not quantified in groundnuts. The screening of groundnut germplasm for reaction to diseases and insect pests that occur at ICRISAT, for reaction to drought, and for other important characteristics, is carried out in collaboration with scientists in various disciplines of the Legumes Program of ICRISAT. Details of evaluation for specific factors are provided in this paper and the results are summarised in Table 13.2. ICRISAT scientists have been

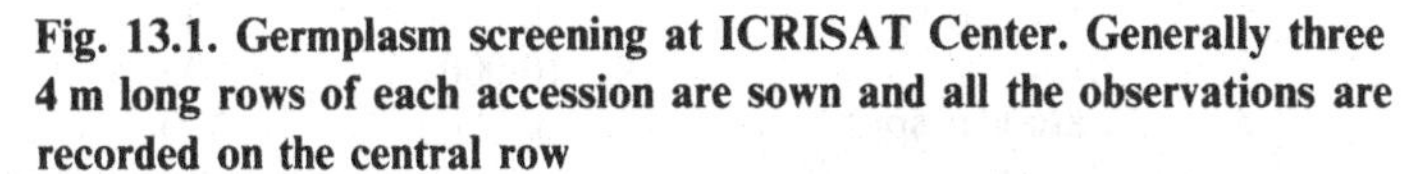

Fig. 13.1. Germplasm screening at ICRISAT Center. Generally three 4 m long rows of each accession are sown and all the observations are recorded on the central row

unable to evaluate germplasm for reaction to a number of groundnut pests and diseases. These include web blotch (*Phoma arachidicola* Marasas, Pauer and Boerema), cylindrocladium black rot (*Cylindrocladium crotolariae* (Loos) Bell and Sobers), bacterial wilt (*Pseudomonas solanacearum* E. F. Sm.) and spider mites (*Tetranychus* spp.) all of which are important pests in one or more groundnut producing countries. Some screening has been done by other workers and some germplasm accessions that have known reaction to these constraints are available in the ICRISAT gene bank.

In the future, groundnut germplasm will also be evaluated for such useful attributes as oil quality and resistance to other pests and diseases. Multi-locational testing is also envisaged. With the expansion of ICRISAT's programmes in southern and West Africa, more efforts for such a multi-locational evaluation are needed to fully exploit the

Table 13.2. Arachis *germplasm evaluation at ICRISAT, 1976–86*

Character		Number of accessions screened	Number of desirable accessions
Disease resistance			
1	Late leaf spot	10,000	44
2	Rust	10,000	95
3	Late leaf spot +rust	10,000	30
4	Bud necrosis	7,400	22
5	Peanut mottle virus	800	4
6	Seed invasion by *A. flavus*	582	11
7	Pod rot	3,222	24
Insect resistance			
1	Thrips	5,000	20
2	Jassids	6,500	25
3	Termites (pod scarification)	520	20
4	Leaf miner	930	18
5	Aphids	300	4
Multiple resistance to pests and diseases		3,400	56
Drought tolerance		578	17
Efficiency of utilisation of iron		225	5
Biological nitrogen fixation ability		342	3

available diversity. Since the production of groundnuts in the post-rainy season is on the increase in India, germplasm accessions are also characterised in the post-rainy season.

Fungal diseases

Techniques of screening for foliar disease resistance

An infector row technique is used for field screening at the ICRISAT Center, usually with one infector row of a susceptible genotype for every four test rows, but there can be up to 10 test rows. The infector row is inoculated with spores, or pots of a susceptible genotype heavily infected with the relevant pathogen are placed along the infector rows. When there is sufficient material, entries are replicated, but usually disease pressure is adequate for single row screening to give reliable results. Rust, *Puccinia arachidis*, and late leaf spot, caused by *Phaeoisariopsis personata* (Berk. & Curt.) v. Arx (= *Cercosporidium personatum* (Berk. & Curt.) Deighton) usually occur together during the rainy season. To avoid any interaction between rust and late leaf spot, rust is controlled by selective fungicides, or trials are grown at locations in India, for example, Bhavanisagar, where rust does not occur, or develops late in the season. However, these techniques are more often used on some advanced breeding material than for screening large numbers of germplasm accessions. The presence of rust and late leaf spot allows identification of sources of multiple resistance. Screening for a single pathogen can also be achieved by glasshouse testing of whole plants or laboratory testing of detached leaves. A nine-point scale is used for scoring, where 1 = immune and 9 = susceptible.

Laboratory testing is done on detached leaves. This has the advantage over field screening that there is not interference from other foliar pathogens, and that it is done under controlled conditions (Subrahmanyam *et al.*, 1983*b*). It is a good means of confirming the field resistance and has the additional advantage that it can be done at any time of the year and when test materials or inoculum of the pathogen is in short supply.

Leaf spots

Late leaf spot is usually the most predominant and economically important leaf spot in India, the People's Republic of China and some areas of West Africa, whereas early leaf spot predominates in most of southern Africa. However, in many locations both leaf spots occur and their incidence and severity differ from season to season.

Late leaf spot is most important during the rainy season at the

ICRISAT Centre, when field screening is possible. To date, 44 germplasm accessions resistant to late leaf spot have been identified and 39 of these are also resistant to rust.

Resistance or immunity to *P. personata* has also been found in wild *Arachis* species (Subrahmanyam *et al.*, 1985). As many wild species differ markedly from *A. hypogaea* in growth habit, season length, etc both field screening and whole plant and/or detached leaf screening is done, preferably with plants or leaves at different ages. Particularly important are the species in section *Arachis* as they can be crossed with *A. hypogaea*. Tetraploid wild species derivatives resistant to *P. personata* have been developed and deposited in the ICRISAT gene bank. Where resistance is identified, components of resistance are studied (Subrahmanyam *et al.*, 1982). This is valuable information which can be used in conjunction with findings on the genetics of resistance to further understand the different sources of resistance in the gene pool.

Although early leaf spot, caused by *Cercospora arachidicola* Hori, occurs at ICRISAT Center, its severity varies from year to year and the natural intensity of the disease is rarely sufficient for field screening. Spores can be collected from artificially inoculated susceptible plants grown in the glasshouse and used to screen a limited number of accessions in the laboratory or glasshouse.

Early leaf spot occurs regularly in southern Africa, and germplasm is being screened in the field by the ICRISAT Regional Groundnut Program for southern Africa in Malawi. So far no resistant accessions have been identified in *A. hypogaea* despite reports of resistance in the USA (Sowell *et al.*, 1976; Moraes & Salgado, 1979). Three diploid wild species, have shown promise and will be utilised in cytogenetics crossing programmes to make this resistance available to breeders.

Rust

Rust, caused by *Puccinia arachidis* Speg., occurs in most groundnut growing areas of the world. It normally occurs with leaf spots. Rust and late leaf spot together have caused yield losses of 70 per cent at the ICRISAT Center.

Fourteen rust-resistant lines, Tifrust-1 to Tifrust-14, have been registered in co-operation with scientists in the United States (Hammons *et al.*, 1982*a*; 1982*b*; 1982*c*; 1982*d*). Most of these, and other rust-resistant accessions, are *fastigiata* type of Peruvian origin. Many wild *Arachis* species are resistant or immune to rust (Subrahmanyam *et al.* 1983*b*), including a number of accessions in section *Arachis* that have been used as sources of resistance at ICRISAT (Moss, 1985). The resistance in wild species is dominant, that in *A. hypogaea* is recessive (Singh *et al.*, 1984), and these two sources of resistance are being combined.

Seed and seedling diseases

Seed rots and seedling diseases caused by soil fungi, particularly *Aspergillus flavus* Link ex Fr. and *Aspergillus niger* van Tieghem are common and serious in the semi-arid tropics (SAT). Two important seedling diseases are collar rot and 'aflaroot'. Seed dressing with fungicides can provide limited control. Little is known of resistance to these diseases. Limited germplasm screening has indicated that some genotypes may possess resistance to seed rot and seedling diseases caused by *Aspergillus* species (Mehan *et al.*, 1981).

Pod rots

Pod rot diseases are widespread in the SAT and are known to cause severe damage in a few countries (Mercer, 1977; Mehan *et al.*, 1981). A complex of species of *Fusarium oxysporum*, *Fusarium solani*, and *Rhizoctonia solani* have been identified as causing pod rots. Pod rots cause problems at different levels. Severely damaged pods are destroyed with direct loss of yield. In some cases damage may be less severe and partially rotted seeds may get into the harvested yield, which greatly reduces the quality of the produce and can act as a means of carry-over of the diseases. At the ICRISAT Center, 3,500 accessions have been screened and a few of these have consistently shown low percentages of rotted pods.

Aflatoxin contamination

The major economic importance of *A. flavus* is when it infests the developing pod, causing mouldy pods and seeds, but most importantly producing aflatoxins, extremely toxic and carcinogenic metabolites. The amount of aflatoxin produced is more critical than the degree of infestation by *A. flavus*. Many countries ban the import of groundnuts in which aflatoxin can be detected at 20 ppb.

Resistance to *A. flavus* depends on the integrity of the seed coat; two genotypes (PI 337394F and PI 337009) were reported resistant only if the seed coat was intact (Mixon & Rogers, 1973). Similarly UF 71513 was reported resistant by Bartz *et al.* (1978). Of the 850 germplasm accessions screened at ICRISAT, five were identified as resistant by inoculating undamaged stored seeds, rehydrated to 20 per cent moisture, with a spore suspension of 4×10^6 conidia ml^{-1} (Mehan *et al.*, 1981).

Aflatoxin can be extracted from groundnut seeds and their products by organic solvents and detected and quantified by using HPLC, TLC and mini-column techniques. In addition, large numbers of samples have been screened at ICRISAT using enzyme-linked immunosorbant assay (ELISA). Antibodies to aflatoxin have been developed and aflatoxin-detecting kits are also available. More than 500 genotypes were tested for the amount of aflatoxin they produced following infection with *A. flavus*, and two accessions, ICG 4286 and ICG 7101, have been identified as low toxin-producers (Mehan *et al.*, 1986). The two traits of resistance to infection and low toxin production are being combined (Mehan, 1985; Mehan *et al.*, 1986).

Viral diseases

Bud Necrosis Disease (BND)

BND is a serious problem in India, is economically important in Texas, USA, and parts of Thailand, and occurs in Southeast Asia, northern USA and Australia. The causal agent, tomato spotted wilt virus (TSWV), infects a wide range of host plants and is transmitted by thrips, predominantly *Frankliniella* spp. on groundnut. Although early and close sowing can reduce the incidence (Reddy *et al.*, 1983) cultivars combining resistance to both virus and thrips are required. Genotypes resistant to thrips have been identified (see page 223), but no *A. hypogaea* germplasm has been found that is resistant to the virus. Some accessions have a lower incidence of BND under field conditions, and their yields are less reduced (ICRISAT, 1985).

Some wild species had been screened under field conditions when disease incidence was high, and by caging viruliferous thrips on plants. In preliminary tests *A. chacoense* showed no symptoms of BND and no viral antigens were found in its leaves.

Peanut mottle

Peanut mottle disease caused by peanut mottle virus (PMV) occurs in all groundnut growing countries. Although symptoms are inconspicuous and often overlooked by the untrained eye, yield losses can be significant.

The virus is transmitted by aphids, *Aphis craccivora* Koch and *Myzus persicae* and is seedborne. Infection occurs after a feeding period of just a few minutes so resistance to the vector may not be of value unless it prevents any probing or feeding. Priorities are therefore to find resistance to the virus and to identify genotypes where no seed transmission occurs. Three such accessions have been identified (Bharathan *et al.*, 1984). Even if infected plants occur in the seed production crop, healthy seed can be produced if there is no seed transmission. Combining low seed transmission with resistance or tolerance to the virus would minimise yield losses. Field screening using an air-brush technique has identified one tolerant accession, ICG 5043.

The ELISA technique is used to test seed for the presence of the virus. At ICRISAT it is possible to ELISA-test 15,000 seeds in one day. Germplasm and advanced breeding lines are screened for frequency of seed transmission and PMV tolerance. All the germplasm planted for seed increase is also tested by removing a small portion of cotyledon to ensure that the seed is PMV-free.

While screeing wild species, a thick cuticle may prevent entry of the virus into the cells and the resistance observed may be due to the inadequacy of techniques developed for *A. hypogaea*. All wild species accessions are therefore being tested using the air-brush technique (Fig. 13.2) which ensures that virus particles penetrate the plant cells.

Peanut Clump

There are reports of severe stunting from India and West Africa caused by Peanut Clump Virus (PCV). The virus has been characterized and its occurrence is strongly associated with a fungus, *Polymyxa graminis* Ledingham (ICRISAT, 1985).

Fig. 13.2. Using a compressor and air brush to spray Peanut Mottle Virus into groundnut leaves. This technique ensures that the virus particles penetrate the plant cells

Several hundred germplasm accessions have been tested in plots infested with PCV in fields near Bapatla, Andhra Pradesh and at Punjab Agricultural University Farm, Ludhiana, India. Nine germplasm lines showed low percentages of infected plants. An accession GK 30036 did not become infected (ICRISAT, 1985).

Groundnut Rosette (GRV)

Groundnut Rosette is the most important virus disease of groundnut in Africa south of the Sahara. GRV is transmitted by *Aphis craccivora*, but only in the presence of groundnut rosette assistor virus. Mechanical inoculation has been achieved (Demski & Kuhn, 1985).

Resistance sources were identified in germplasm from the Ivory Coast and Burkina Faso (Sanger & Catherinet, 1954) and have been used successfully in breeding programmes.

Insect pests

***Aphids** – Aphis craccivora* Koch

Although ten aphid species have been listed as occurring on groundnut (Smith & Barfield, 1982) *Aphis craccivora* is by far the most serious pest. Even though aphid feeding activity can depress yield, they are far more important as vectors of virus diseases. *A. craccivora* is the vector of GRV which is one of the most significant constraints to yield throughout Africa.

Aphids do not occur regularly at the ICRISAT Center in sufficient numbers for field screening to be effective. Glasshouse screening is done by rearing aphids on susceptible genotypes, and then transferring them to test plants. One genotype with marked antibiotic activity to *A. craccivora* has been found. Aphid fecundity was reduced on three wild species (*A. villosa* and *A. chacoense* in section *Arachis*, and *A. glabrata* in section *Rhizomatosae*) (Amin, 1985).

***Jassids** – Empoasca* spp.

Jassids are found on groundnuts throughout the world. They are polyphagous so other crops and weeds act as reservoirs. Groundnut leaves are damaged by the toxic saliva injected into the leaf by these sap-sucking pests. The damage varies from leaf yellowing to stunting of plants when severe attacks occur early in the season.

Empoasca kerri Pruthi is the jassid which infests groundnuts at ICRISAT. About 6,000 accessions have been screened in field conditions and 25 with resistance have been identified (Amin *et al.*, 1985). Some have a hairy leaf surface, a thick cuticle and a high tannin content in the leaves (Amin & Singh, 1983).

Leaf miner – *Aproaerema modicella* Dev.

Leaf miner is a problem in India, Sri Lanka, Indonesia and Thailand. It is more serious under dry conditions, when an attack may result in retardation of growth or even death of plants. About 1,500 accessions have been screened at ICRISAT in the post-rainy season; 18 showed varying degrees of resistance (ICRISAT, 1985).

Thrips – *Scirtothrips dorsalis* Hood and *Frankliniella schultzei* Trybom

A total of 17 species of thrips has been reported on groundnut. Yield increases of 37 per cent in USA have been recorded after applying insecticides to thrips-infected crops, but how much of the yield loss could be attributed to thrips damage alone is questionable. Thrips are of major economic importance because they are vectors of TSWV which causes BND, an important groundnut disease in India (cf. page 220). As there is no known resistance to TSWV in *A. hypogaea* emphasis at ICRISAT is being placed on using 14 thrips-resistant accessions of *A. hypogaea* (Amin *et al.*, 1985). This resistance has been transferred into high-yielding breeding lines. Wild species have also been screened and six have been found resistant (Amin, 1985). Five of these are in section *Arachis* and compatible with *A. hypogaea.* The mechanism of resistance is not known, but the survival and fecundity of thrips cultured on wild *Arachis* species is less than on susceptible cultivars (Amin, 1985).

Termites

Termites are a serious problem in many groundnut-growing countries of the SAT. Termites can kill the plant or damage the pods by scarification. At present chemical control is the only effective method of controlling these pests, but large-scale chemical control is beset with many problems. Entomologists at ICRISAT have identified 20 germplasm accessions that have shown resistance to scarification by *Odontotermes* sp. (ICRISAT, 1983; Amin *et al.*, 1985). Resistance was transferred to progeny when ICG 2271 and ICG 5040 were used as resistant parents. However, these efforts are in a preliminary stage and need much more work before progress can be made towards developing termite resistant groundnut cultivars.

Components of resistance

Some accessions identified as resistant by field screening have been studied further in the laboratory. Resistance in the field is assessed on a scale which reflects the amount of damage to the crop, and is assessed

Table 13.3. *Reaction of some wild* Arachis *species to* Phaeoisariopsis personata *at the ICRISAT Center (from Subrahmanyam* et al., *1985).*

Species	USDA PI no.	ICRISAT groundnut accession	Components of resistance to *P. personata*: Infection frequency (lesions/cm)	Defoliation (%)	Lesion diameter (mm)	Sporulation index[1]
A. duranensis	219823	8123	8.0	35.0	0.49	1.8
A. spegazzinii	262133	8138	12.7	75.0	0.79	3.0
A. correntina	262137	8133	15.9	5.0	0.23	1.0
A. stenosperma	338280	8126	19.4	30.0	0.16	1.0
A. chacoense	276235	4983	17.4	32.6	0.26	1.0
A. hypogaea (Control TMV 2)	—	221	19.1	100.0	1.96	5.0

1. Extent of sporulation scored on a five-point scale where 1 = no sporulation and 5 = extensive sporulation.

at flowering and at maturity. Laboratory studies have shown that field resistance can be due to one or more factors, or components of resistance. Accessions resistant to a fungal disease can differ in respect to infection frequency, incubation period, pustule diameter, degree of spore production and viability of spores (Table 13.3) (Subrahmanyam *et al.*, 1983*a*; 1985). Thus different resistant parents can be used by breeders to produce genotypes with more than one component of resistance, which will be more stable.

Abiotic stresses

Drought

Drought is one of the important yield reducers of groundnut, and mid- and/or end-season drought plays a major role in reducing yields in the SAT countries. In addition to causing substantial yield losses, drought can result in poor seed quality (Davidson *et al.*, 1973) decreased germinability (Pallas *et al.*, 1977) and increased invasion by *A. flavus* (Diener & Davis, 1977). At ICRISAT emphasis has been placed on developing groundnut cultivars with tolerance of or resistance to drought. A series of drought-screening experiments conducted by ICRISAT physiologists, using a line-source irrigation system to impose different degrees of drought stress at different stages of growth, resulted in the identification of 17 tolerant accessions among about 1,000 accessions tested (ICRISAT, 1985). None of these accessions belongs to the important Spanish group (var. *vulgaris*), but they are of diverse origin and represent both subspecies. Four accessions, ICG 1697, 3657, 4728 and 4790, have been used in the hybridisation programme and the segregating material is under test. Selected lines have been tested under rainfed conditions at a drought-prone location, Anantapur, Andhra Pradesh, India, where some lines produced significantly higher yields than released cultivars. Attempts are being made to develop short-duration cultivars for regions where end-season drought is the major problem. Many of the drought-resistant lines, although they give acceptable yields under drought stress, under good conditions are lower yielding than released cultivars.

Nutrient stress

Although most cultivated soils contain large populations of *Rhizobia*, it is possible to increase nitrogen fixation by manipulation of the *Rhizobium* strain, host genotype and environment (Nambiar *et al.*, 1982). Over the past six years many accessions have been screened for higher biological nitrogen-fixing (BNF) ability. In general, Spanish (var. *vulgaris*) genotypes were found to fix less nitrogen, but one Spanish accession

Table 13.4. *Groundnut germplasm with more than one desirable attribute*

Resistance or tolerance to												
ICG No.	Synonym	Els[1]	LLS[2]	Rust	BND[3]	PMV[4]	Thrips	Jassids	Termites	LMN[5]	Aphids	Drought
A. hypogaea var *hypogaea*												
156	M 13						X	X		X	X	
2271	NC Ac 343				X		X	X	X	X	X	
2306	NC Ac 2142				X		X	X	X			
2307	NC Ac 2144				X		X					
2320	NC Ac 2462						X	X				
2741	Gujarat Narrow Leaf				X			X				
5030	NC Ac 1741				X			X	X			
5040	NC Ac 2214						X	X			X	
5041	NC Ac 2230						X	X	X			
5042	NC Ac 2232				X		X	X	X			
5043	NC Ac 2240				X	X	X	X	X		X	
5044	NC Ac 2242				X		X	X	X			
5045	NC Ac 2243						X	X	X			
6317	NC Ac 17888				X			X	X			
6323	RMP 91		X		X							
6764	NC Ac 1705							X	X	X		
7237	RMP 40								X	X		
7890	PI 323526		X	X								
A. hypogaea var *fastigiata*												
405	NC Ac 2663							X				
1660	NC Ac 2666							X				X
1697	NC Ac 17090			X						X		X
1703	PI 275750		X	X								X
1704	NC Ac 17129		X	X								
1705	NC Ac 17130		X	X								
1707	NC Ac 17132		X	X								

1710	NC Ac 17135		X	X				
2716	EC 76446 (292)		X	X	X			
4747	PI 259747		X	X				X
4790	KU No. 24		X	X				X
4995	NC Ac 17506		X	X				X
6022	NC Ac 927		X	X				
6280	NC Ac 17124		X	X				
6340	PI 350680		X	X				
7013	NC Ac 17133 (RF)		X	X	X			
7881	PI 215696		X	X			X	
7884	PI 351879		X	X				
7885	PI 381622		X	X				
7886	PI 390593		X	X				
7887	PI 390595		X	X				
7888	PI 393516		X	X				
7892	PI 393527B		X	X				X
7894	PI 393641		X	X				
7897	PI 405132		X	X				
Arachis species								
4983	*A. chacoense*	X		X				
8124	*A. batizocoi*			X		X		
8126	*A. stenosperma*	X	X	X				
8131	*A. pusilla*		X	X				
8132	*A. correntina*		X	X		X		
8133	*A. correntina*		X	X				
8140	*A. correntina*			X		X		
8149	*A. glabrata*		X	X				
8165	*A. glabrata*		X	X				
8916	*Arachis* sp.		X	X				
8938	*A. glabrata*		X	X				
8959	*A.* sp. 30085	X		X				

1 = Early leaf spot 2 = Late leaf spot 3 = Bud necrosis disease 4 = Peanut mottle virus 5 = Leafminer

(ICG 1561) was found to possess high nitrogenase activity indicating good nitrogen-fixing ability (Nambiar *et al.*, 1982). This accession is being used to improve the BNF ability of Spanish types. Among Virginia (var. *hypogaea*) types ICG 2405 and ICG 4969 were found to have higher nitrogenase activity and a few progenies of crosses involving these accessions were found to be high yielding (ICRISAT, 1983).

Calcium deficiency is common in many groundnut-growing regions and gypsum application is recommended in such regions. Studies at ICRISAT have indicated significant genotype drought gypsum interactions. However, the varietal differences in calcium uptake are yet to be conclusively demonstrated before any practical use can be made of such genotypes.

Chlorosis induced by the non-availability of iron is common in groundnuts grown in heavier soils. In preliminary tests five germplasm accessions have been identified as more iron-efficient than some of the released cultivars in India (ICRISAT, 1985). Whether such germplasm could be used for genetic enhancement of groundnut cultivars to be grown in heavier soils remains to be seen.

Multiple resistance

The final goal for a groundnut breeder interested in developing resistant cultivars is to combine resistance to one or more insect pests and/or diseases along with other desirable attributes. This is expected to have a stabilising effect on yields of groundnut in any given region. At ICRISAT 56 accessions with multiple pest and disease resistance and/or tolerance to drought have been identified (Table 13.4). This indicates that it is possible to identify a groundnut germplasm accession with more than one desirable attribute that could be used in groundnut improvement. It is interesting to note that no accession belonging to var. *vulgaris* has yet been identified with multiple resistance. There is also a need to screen more accessions of *Arachis* species since utilisation of wild species germplasm in groundnut improvement has proved to be a practical approach and is on the increase.

Conclusions

Evaluating more than 10,000 accessions is a major undertaking, but a concerted and co-operative effort by scientists at ICRISAT has identified many valuable accessions. This has been possible due to the commitment of ICRISAT, and to the development and adaptation of screening techniques to handle large numbers of accessions and to give reliable results.

Table 13.5. *ICRISAT Entries in All India Coordinated Research Project on Oilseeds (AICORPO) Trials*, 1986

Trial	Number of entries		Remarks
	A. hypogaea	Wild species	
Rainy Season			
1. Adaptive trials	3		Includes 1 Foliar Disease Resistant Selection (FDRS)
2. National elite trial	10		Includes 3 FDRS
3. Co-ordinated varietal trial	13		Includes 1 FDRS
4. Initial Evaluation trial	13	2	Includes one pest resistant selection and one multiple pest resistant accession
5. Foliar disease resistant varietal trial	5	2	
6. Hand picked selection varietal trial	16	1	
7. Bud necrosis disease varietal trial	8		Includes 1 FDRS
Post-rainy Season			
1. National elite trial	1		
2. Co-ordinated varietal trial	5		Includes 3 FDRS
3. Initial evaluation trial	5		

After assessment at the ICRISAT Center and other appropriate locations, ICRISAT groundnut material is submitted to the All India Coordinated Research Project on Oilseeds (AICORPO) for inclusion in its testing system. In 1986, 81 potential varieties of *A. hypogaea* material are being tested in AICORPO trials (Table 13.5). Derivatives of crosses between *A. hypogaea* and wild species (Fig. 13.3) have been selected for high seed yield, disease resistance, large seed for confectionary use, and for high haulm yield, and some of these entered into AICORPO trials. Germplasm lines, and segregating material from crosses at ICRISAT, are available to breeders worldwide.

The potential impact of these lines on groundnut growing in the SAT is considerable. Disease- and pest-resistant lines reduce or eliminate the cost of pesticides. Where the resource-poor farmer has not been using pesticides they result in substantial yield increases. The high haulm yield of the wild species is of great value, both in economies relying on animal power for transport or cultivation, and where animals produce milk or meat.

Summary

The groundnut germplasm collection at ICRISAT consists of accessions of cultivated groundnut and a number of wild species. Of the

Fig. 13.3. Derivatives of crosses between *A. hypogaea* and wild species are being screened for high seed yield, diesease resistance, large seed for confectionery use, and for high haulm yield

latter, the close relatives of groundnut are considered the secondary gene pool as genes from these wild species have been transferred to *A. hypogaea*.

Priorities at ICRISAT are to evaluate the primary and secondary gene pools as an essential prerequisite to their use in the breeding programme.

Accessions are screened for resistance or tolerance to the major yield reducers. These are diseases, drought, insect pests and nutritional factors. Accessions are also screened for yield potential, seed size, oil content and time to maturity. Large-scale screening in the field or laboratory has been possible for some of these constraints, but this is not possible for all because of environmental or biotic factors.

ICRISAT scientists have identified three wild species resistant to early leaf spot; 54 accessions resistant to late leaf spot; 95 resistant to rust; 11 resistant to *Aspergillus flavus* invasion, two of which produce only small amounts of aflatoxin; one accession tolerant to peanut mottle virus; one with antibiotic activity to *Aphis craccivora*; 25 resistant to jassids; 18 resistant to groundnut leaf miner; and 20 resistant to thrips.

In addition, 17 accessions tolerant to drought, one with good nitrogen fixing ability, and five accessions with better growth in heavy soil where iron is less available, have been identified. Fifty-six accessions, including 12 wild species accessions, with resistance to more than one pest have been identified. Many of these have been used in breeding programmes; and ICRISAT has 84 lines, including five derivatives from crosses of *A. hypogaea* with wild species, undergoing testing in India for release as cultivars.

Acknowledgements

This paper (Conference paper no. 316 of ICRISAT) summarises the work of many scientists of the ICRISAT Legumes Program, and we acknowledge their co-operation over many years, and their help in preparing this manuscript.

References

Amin, P. W. (1985). Resistance of wild species of groundnut to insect and mite pests. In *Proceedings of an International Workshop on Cytogenetics of Arachis*, pp 57–60. ICRISAT, India.

Amin, P. W. & Singh, K. N. (1983). *Studies on Host Plant Resistance in Groundnut to Jassid* (*Empoasca kerri Pruthi*). Progress Report, Groundnut Improvement Program. ICRISAT, India. (Limited distribution).

Amin, P. W., Singh, K. N., Dwivedi, S. L. & Rao, V. R. (1985). Sources of resistance to the jassid (*Empoasca kerri* Pruthi), thrips (*Frankliniella schultzei* Trybom) and termites (*Odontotermes* sp.) in groundnut (*Arachis hypogaea* L.). *Peanut Science*, **12**, 58–60.

Bartz, Z. A., Norden, A. J., LaPrade, J. C. & DeMuynk, T. J. (1978). Seed tolerance in peanuts (*Arachis hypogaea* L.) to members of the *Aspergillus flavus* group of fungi. *Peanut Science*, **5**, 53–6.

Bharatan, N., Reddy, D. V. R., Rajeswari, R., Murthy, V. K., Rao, V. R. & Lister R. M. (1984). Screening peanut germplasm lines by enzyme-linked immunosorbant assay for seed transmission of peanut mottle virus. *Plant Disease*, **68**, 757–8.

Davidson, J. M., Garten, J. E., Mathock, A. S., Schwab, D., Stone, J. F. & Tripp, L. D. (1973). Irrigation and water use. In *Peanuts Culture and Uses*, pp. 361–82. American Peanut Research and Education Association, Stillwater, OK.

Demski, J. W. & Kuhn, C. W. (1985). The peanut collaborative research support program (CRSP) – project on rosette virus disease. In *Collaborative Research on Groundnut Rosette Virus*. Summary proceedings of a consultative group meeting, 13–14 April 1985, Cambridge, UK. ICRISAT, India.

Diener, V. L. & Davis, N. D. (1977). *Aflatoxin formation in peanut by Aspergillus flavus*. Auburn University Agricultural Experiment Station Bulletin 493, Auburn.

Gregory, M. P. & Gregory, W. C. (1979). Exotic germplasm of *Arachis* L. interspecific hybrids. *Journal of Heredity*, **70**, 185–93.

Hammons, R. O. (1970). Registration of Spancross peanuts. *Crop Science*, **10**, 459.

Hammons, R. O., Branch, W. D., Bromfield, K. R., Subrahmanyam, P., Rao, V. R., Nigam, S. N. & Gibbons, R. W. (1982*a*). Registration of Tifrust-14 peanut germplasm (Reg. No. GP31). *Crop Science*, **22**, 697–8.

Hammons, R. O., Subrahmanyam, P., Rao, V. R., Nigam, S. N. & Gibbons, R. W. (1982*b*). Registration of eight peanut germplasm lines resistant to rust (Reg. No. GP22 to GP29). *Crop Science*, **22**, 452–3.

Hammons, R. O., Subrahmanyam, P., Rao, V. R., Nigam, S. N. & Gibbons, R. W. (1982*c*). Registration of peanut germplasm Tifrust-1 to Tifrust-4 (Reg. No. GP18 to 21). *Crop Science*, **22**, 453.

Hammons, R. O., Branch, W. D., Bromfield, K. R., Subrahmanyam, P., Rao, V. R., Nigam, S. N., Gibbons, R. W. & Goldin, E. (1982*d*). Registration of Tifrust-13 peanut germplasm (Reg. No. GP30). *Crop Science*, **22**, 697.

IBPGR & ICRISAT. (1985). *Descriptors for Groundnut* (*Revised*). International Board for Plant Genetic Resources, Rome and International Crops Research Institute for the Semi-Arid Tropics, Patancheru, A. P. 502 324, India.

ICRISAT (1983). Groundnut. In *Annual Report* 1982, pp. 189–229, ICRISAT, India.

ICRISAT (1985). Groundnut. In *Annual Report* 1984, pp. 195–244, ICRISAT, India:

Mallikarjuna, N. & Sastri, D. C. (1985). *In vitro* culture of ovules and embryos from some incompatible interspecific crosses in the genus *Arachis* L. In *Proceedings of an International Workshop on Cytogenetics of Arachis*, pp. 153–8. ICRISAT, India.

Mehan, V. K. (1985). The aflatoxin problem in groundnut-approaches to prevention and control. In *XXVII Annual Oilseed Workshop of Groundnut*. ICAR, Nagpur, India.

Mehan, V. K., McDonald, D. & Ramakrishna, N. (1986). Varietal resistance in peanuts to Aflatoxin production. *Peanut Science*, **13**, 7–10.

Mehan, V. K., McDonald, D. & Rao, V. R. (1981). Pod rot disease of peanut at ICRISAT. In *Proceedings of American Peanut Research and Education Society* **13(1)**, 91 (Abstract).

Mercer, D. C. (1977). A pod rot of peanuts in Malawi. *Plant Disease Reporter*, **61**, 51–5.

Mixon, A. C., & Rogers, K. M. (1973). Peanut accessions resistant to seed infection by *Aspergillus flavus*. *Agronomy Journal* **65**, 560–2.

Moraes, S. A. & Salgado, C. L. (1979). Evaluation of the resistance of groundnut (*Arachis hypogaea* L.) to *Cercospora arachidicola*. *Fitopathologia* **14**, 65–72.

Moss, J. P. (1984). Wild species in crop improvement. In *Biotechnology in International Agricultural Research*, pp. 199–209. IRRI, Los Banos.

Moss, J. P. (1985). Breeding strategies for utilisation of wild species of *Arachis* in groundnut improvement. In *Proceedings of an International Workshop on Cytogenetics of Arachis*, pp. 9309. ICRISAT, India.

Nambiar, P. T. C., Dart, P. J., Nigam, S. N., & Gibbons, R. W. (1982). Genetic manipulation of nodulation in groundnut. In *Biological Nitrogen Fixation for Tropical Agriculture*, pp. 49–56. Graham, P. H. & Harris, S. C. (eds.). CIAT Cali, Colombia.

Norden, A. J. (1980). Breeding methodology for groundnuts. In *Proceedings of the International Workshop on Groundnuts*, pp. 58–61. ICRISAT, India.

Pallas, J. E., Stansell, J. R. & Bruce, R. R. (1977). Peanut seed germination as related to soil water regimes during pod development. *Agronomy Journal.* **69**, 381–3.

Rao, V. R. (1980). Groundnut genetic resources at ICRISAT. In *Proceedings of the International Workshop on Groundnuts*, ICRISAT, India.

Reddy, D. V. R., Amin, P. W., McDonald, D. & Ghanekar, A. M. (1983). Epidemiology and control of groundnut bud necrosis and other diseases of legume crop in India caused by tomato spotted wilt virus. In *Plant Virus Epidemiology*, eds. Plumb, R. T., & Thresh, J. M. (eds.). Blackwell Scientific Publications, Oxford.

Sanger, L. & Catherinet, M. (1954). New observation on chlorotic rosette of the peanut and bred strains. *Annales du Centre de Recherches Agronomiques de Bambey au Senegal.* Bulletin No. 11, pp. 204–16. INRA, Bambey.

Singh, A. K., Subrahmanyam, P. & Moss, J. P. (1984). The dominant nature of resistance to *Puccinia arachidis* in certain wild *Arachis* species. *Oléagineux*, **39**, 535–7.

Smartt, J. & Stalker, H. T. (1982). Speciation and cytogenetics in *Arachis.* In *Peanut Science and Technology*, pp. 21–49, Pattee, H. E. & Young, C. T. (eds.). American Peanut Research and Education Society, Yoakum, Texas.

Smith, J. W. & Barfield, C. S. (1982). Management of preharvest insects. In *Peanut Science and Technology*, pp. 250–3259. Pattee, H. E. & Young, C. T. (eds.). American Peanut Research and Education Society, Yoakum, Texas.

Sowell, G., Smith, D. H. & Hammons, R. O. (1976). Resistance of peanut plant introductions to *Cercospora arachidicola*. *Plant Disease Reporter*, **60**, 494–8.

Stalker, H. T. (1985). Arachis spiniclava, a D Genome species of section *Arachis*. In *Proceedings of American Peanut Research and Education Society*, **17**, 23 (Abstract).

Subrahmanyam, P., McDonald, D., Gibbons, R. W., Nigam, S. N. & Nevill, D. J. (1982). Resistance to rust and late leafspot diseases in some genotypes of *Arachis hypogaea*. *Peanut Science*, **9**, 6–10.

Subrahmanyam, P., McDonald, D., Gibbons, R. W. & Subba Rao, P. V. (1983*a*). Components of resistance to *Puccinia arachidis* in peanuts. *Phytopathology*, **73**, 253–6

Subrahmanyam, P., Moss, J. P. & Rao, V. R. (1983*b*). Resistance to peanut rust in wild *Arachis* species. *Plant Disease*, **67**, 209–12.

Subrahmanyam, P., Moss, J. P., McDonald, D., Subba Rao, P. V. & Rao, V. R. (1985). Resistance to *Cercosporidium personatum* leafspot in wild *Arachis* species. *Plant Disease*, **69**, 951–4.

Subrahmanyam, P., Rao, V. R., McDonald, D., Moss, J. P. & Gibbons, R. W. (In review). Origins of resistance to rust and late leaf spot in peanut (*Arachis hypogaea* L.)

14

Practical considerations relevant to effective evaluation

J.T. WILLIAMS

Introduction

The description of accessions and the recording of the information in databases are one aspect of genetic resources work which has progressed slowly. The large number of samples now conserved in gene banks makes their comprehensive description and evaluation a formidable task. Over the past decade the IBPGR alone has organised the collection of more than 130,000 primitive accessions of diverse crops which are, in the main, population samples. Other genetic resources programmes have developed in scores of countries so that curators have been faced with an unprecedented management problem. It is conservatively estimated that the number of accessions in germplasm collections now totals several million due to extensive duplication.

While breeders have never had so much material available for use, it would be nonsense to conclude that they should, or will, start systematically evaluating samples for numerous traits. Only a clearly perceived, short-term benefit to their breeding programmes will lead to breeders taking on additional work. Instead curators of germplasm collections must first organise them and then initiate programmes for their evaluation by breeders. Such an undertaking, when coupled with limited funding and manpower, requires rational strategies defining priorities and procedures.

However, not all the constraints relate to the availability of financial resources. It has taken, and is taking, time to clarify concepts and procedures, and to identify sufficient centres with appropriate expertise to do the work, especially for the less important crops. Additionally the division of labour between curators holding similar materials, and between the breeders and other scientists who use the materials, needs to be explored.

Proposing anything more than guidelines for the development of evaluation is senseless. Genetic resources centres vary widely in the crops they conserve, their standard of work and their level of sophistication. In some cases centres (particularly international centres) are involved with in-depth evaluations; others, notably the base collections, are concerned with conservation only. Some are small special-purpose collections; some hold global collections of numerous crops. They may be parts of international, national or university programmes. However, a number of general recommendations can be made, based on the axiom that the evaluation of accessions should be facilitated so that their use can be accelerated.

During the past decade, experience has demonstrated the importance of the following points in relation to genebank management and the development of evaluation:

1. Gene bank curators have tended to be passive, waiting for samples and information to arrive. The careful acquisition of passport data (and the pursuit of those which are missing), the assessment of gaps and their filling through purposeful collecting have rarely occurred;
2. Characterisation and evaluation of accessions, prerequisites for their use in breeding, have lagged behind expectations and, in some cases, have scarcely begun;
3. Few gene banks have developed bilateral agreements with others holding the same or similar materials, in order to partition responsibility for characterisation and evaluation, and to make arrangements for safe duplication;
4. Few gene banks have developed clear links with breeders, although some notable successes are the international centres of the Consultative Group on International Agricultural Research (CGIAR); ZIGUK, Gatersleben, GDR; and the Plant Germplasm System of the United States Department of Agriculture (USDA). Indeed, there are a number of cases of the establishment of gene banks in countries and areas where there are no breeders to use the conserved materials. Hopefully national authorities have in mind the establishment of plant breeding programmes able to exploit the resources so carefully conserved!
5. Similarly, few gene banks have developed links, national or international, with laboratories specialising in related scientific research.

These considerations, and others relevant to effective evaluation, are discussed below.

Management and documentation of collections

The description of accessions and databases

The ability of curators to respond most helpfully to the requests of breeders for material clearly depends on the adequate description of accessions and the ability to manipulate the information in the computer database. This is essential and, without this ability, gene banks, in the longer term, will be of limited usefulness.

The International Board for Plant Genetic Resources (IBPGR) has long stressed the importance of passport data: the unique identifiers and basic data relating to the origin of accessions. Additionally, the full description of accessions requires their 'characterisation' by scoring a limited number of morphological traits which have a high heritability (i.e. whose expression is unaffected by the environment) and also the 'evaluation' of more variable traits of interest to the users.

Hitherto, insufficient emphasis has been placed on recording the passport data, and their absence is a major hindrance to curators in assessing the range of variation in their collections, and in identifying gaps which should be filled. The users also require this information so that they can make rational choices on which accessions to examine. Curators must therefore put a high priority on registering all available passport data and on obtaining missing data on altitude, latitude and longitude for landraces and wild material, or on the origin of obsolete varieties by reference to published information. Subsequent associations between passport and characterisation data can then provide an overall picture of the range of diversity in the collections and will also provide a much-needed service to users. This is of such importance that the IBPGR might well consider drawing attention to those who have taken initiatives in this area in the directories of germplasm collections it publishes. Such information would go a long way towards opening up to breeders those collections which had hitherto seemed disorganised and from which the rational selection of material for evaluation was not possible.

In addition, characterisation data can indicate the genetic uniformity of a sample. The present collections largely consist of landraces with intrinsic genetic diversity within samples. Procedures should be designed to take account of this variation, and databases designed to accommodate the information obtained. These heterogeneous population samples should be conserved as such (Erskine & Williams, 1980), a point which has important implications for regeneration procedures.

For most crops the language of characterisation has been standardised through descriptor lists, but not without much difficulty. Representative groups of breeders have been asked to agree on lists of traits for use in systematic characterisation; but in many cases when they are published, users, both curators and breeders, have decided that not all are required. Hence the IBPGR has experienced a sequence of crop-specific descriptor lists being developed, then revised, and subsequently adapted when used. From a practical point of view this has been disquieting and caused inevitable delays in the description of accessions so that a huge backlog remains which still requires basic characterisation.

Although the traits considered under evaluation are more variable in expression, they are generally those of most interest to breeders. The list of such traits is virtually open-ended. It is beyond the means of most collections to do all this work and, ideally, most evaluation should be carried out in co-operation with breeders or other scientists.

When breeders select materials for screening for specific characteristics they usually follow a well-defined sequence of steps: first, by searching their working collections of locally adapted 'improved' material (which breeders have traditionally kept relatively small); second, by searching among landraces locally adapted; then, among exotic landraces and/or wild species (which may or may not be easily available from collections). Of these it is the last two groups that yield results of value for genetic resources work and these should be incorporated into the collection's database. Breeders are, however, notoriously bad at providing such data to the curator.

In addition, because of environmental sensitivity, evaluation data may have little relevance outside the circumstances in which they were obtained. The detailed recording of the test conditions can aid the interpretation and application of the results. Nevertheless, genotype × environment interaction is a major problem which has yet to be properly addressed (Westcott, 1985) for meaningful interpretation of data in databases.

Well-organised databases are clearly essential and the data they contain should meet the needs of curators, breeders and research scientists but resource limitations require that the data be kept to an essential minimum (Marshal & Brown, 1981; Frankel & Brown, 1984). However, it should be borne in mind that information needs vary between users and over time. In practice, curators have developed documentation systems from which much information is missing.

Implementing core collections

The collections of landraces of many major crops have become so large (e.g. wheat, rice, barley, sorghum and pearl millet) that there is a need to help breeders avoid the full screening of the large 'world collections'. Haddad (1984) stressed that it is impracticable even for many national breeding programmes to do so.

One approach for improving the usefulness of gene banks to breeders could be through the establishment of 'core collections' which will include a representation of genetic diversity in the collection as a whole, and upon which work would be concentrated (see Frankel & Brown, 1984; Brown, this volume, for detailed consideration). However, the selection of a 'core' does require quality passport and characterisation data. Ecogeographic data would clearly be of great value, so too would be any analysis of the range of variability within the collection. Yet often data are only recorded during the multiplication process and many gene banks have large proportions of recently accessed materials for which grow-out is necessarily some years hence. In this case the selection of a core would rely simply on passport data. For older collections of accessions lacking comprehensive passport data, e.g. historical collections of old varieties in Europe, the selection of a core would have to rely on pedigree and characterisation data.

Although comprehensive quality databases would be the ideal starting point, it would be relatively easy for small groups of experts to examine existing databases – imperfect though they might be – and provide interim criteria for the delimitation of the core. Then, following evaluation of this core, users will have a much more secure basis for selecting material for breeding, and curators will be assisted in the management and maintenance of their collections as work would tend to be concentrated on this part. Further, by concentrating on a reduced number of accessions, it would be possible to contemplate the detection of new yield-promoting germplasm by tests for general combining ability (Frankel & Brown, 1984).

Co-operation between collections

The serious lack of attention to passport data has been mentioned above but experience has shown that the IBPGR's publicising of this point has done little to rectify the situation. Co-operative action may be a solution.

The UNDP/IBPGR ECP/GR programme for six important crops or crop groups in Europe has given much attention to registering data and has enabled numerous collections in the various countries to assess which accessions are unique and which are redundant duplicates. Collections are now being rationalised and gaps in both passport and characterisation data identified and filled (Perret, this volume).

It had been expected by the IBPGR that the centres holding the base collections would, between them, maintain comprehensive regional or global databases for their crops. However, this has rarely happened. One of the reasons is that data, other than passport data, are generated at numerous active centres and their systematic transfer to a central database is dependent on understandings such as those devised and operated in the ECP/GR. It seems likely that, as a general rule, any regional crop database will require prompting, and perhaps funding, from an international body such as IBPGR. Experience gained in the European programme has shown that it is essential to establish such databases at centres both holding a major collection and possessing experienced computer personnel and scientists capable of transforming the data on old accessions, recorded in various formats and codes, into the agreed format for present use.

It has also become clear over the past few years that the soundest basis for initiating a planned evaluation strategy is by using the crop databases spanning several gene banks. Peeters & Williams (1984) pointed out that the information now provided by most individual gene banks is inadequate. Information from a broader crop database is likely to enable a breeder to make a more discriminating choice among available accessions.

Linkage between gene banks, breeders and scientists

In developing evaluation strategies, the input from gene banks is largely related to the initial description, ordering and sorting of accessions. Beyond this the gene banks alone are rarely able to go. Nonetheless, co-operation with other scientists can yield results of interest and use to both parties (Burdon, this volume). There are, for example, major gaps in the knowledge of the extent and content of the wider crop gene pools, and often a poor understanding of species relationships and patterns of variation. Curators need to be aware of these problems in order to estimate how comprehensive their collections are.

Techniques have been available for some time which may be used to describe patterns of variation, e.g. numerical taxonomic ordination and the analysis of isozyme patterns through electrophoretic surveys (Simpson & Withers, 1986). More recently newer techniques have emerged which show promise, e.g. restriction fragment length polymorphisms. Few, if any, gene banks are staffed or equipped to apply these methods to studies of variation and much of the published work of value stems from scientific interests in the origins and evolution of cultivated plants rather than the description of the variation in genetic resources collections.

Carefully planned research projects jointly organised by gene banks and scientific laboratories could produce results which: 1) enable the curator better to understand his collection: to know how much variability is present and to what extent it is representative of that in the field, and 2) could improve the quality of the databases and help in the selection of materials for further evaluation.

Examples which may be quoted include – for numerical analysis – Burt *et al.* (1971) on *Stylosanthes*; Hamon and van Sloten (this volume) on *Abelmoschus*; Burt, Reid & Williams (1976) on pasture collections; Small & Lefkovitch (1985) on the *Medicago* gene pool; – for isozyme analysis – Nevo *et al.* (1979) and Nevo *et al.* (1986) on wild barley; – for molecular techniques – Tanksley & Bernatsky (this volume). In some cases simple grouping techniques, used to determine the distinctness of varieties, can be used and these encompass a mix of qualitative and quantitative characters (Baltjes *et al.*, 1985).

This approach could well run in parallel with evaluation of specific traits by conventional methods of screening collections for the 'most sought' characters. This typifies the approach of the USDA Plant Germplasm System guided by its crop committees, or the routine activities in several of the (CGIAR) international centres.

In developing an evaluation programme, attention will have to be paid to the nature of the breeding effort on that crop, a point which underlines the need to involve breeders in evaluation work to ensure its relevance to practical problems. For minor crops in many countries straightforward introduction and selection is sufficient. Crops at this stage of development and not yet properly subject to breeding are unlikely to be candidates for the implementation of an enhanced evaluation strategy.

By contrast, there are major staples which are the object of advanced breeding techniques. These are, therefore, amenable to inputs from breeders from international centres, country programmes and private concerns. Many of these crops are served by specialised scientific communities and are the subject of regular symposia. Hence there is a vast pool of diverse talent which may be tapped and no scientific constraints are envisaged in guiding the work although there may be organisational problems.

The evaluation of wild species is subject to a number of special problems. Breeders are often not the best people to handle this type of material, even if they do use it from time to time. Harlan (1984) has stressed that breeders are rarely familiar with wild species, are often misled by inappropriate or inept taxonomies, and may be deterred by problems of sterility and deleterious genes. All too often the centres of expertise for breeding are located in places which for ecogeographical reasons, can maintain only a portion of the wild species gene pool. As a result, evaluation of this type of material is best conducted by appropriate specialists working in conjunction with breeders and the gene banks.

It has been proposed that gene bank collections would become more useful if stocks were developed which incorporate desirable genes from landraces and related wild species in an improved background. Whether such programmes of parental line breeding (or pre-breeding) are justified depends on the crop and the needs of breeders. While this is the normal work of breeders, the gene banks have a role in obtaining and conserving stabilised hybrid segregates (often produced with much labour), introgressed and weedy forms, *and* clearly identify them in the databases. Similarly when a bridging-species has been identified, this too should receive appropriate conservation and documentation in the gene banks. In general funds are lacking for such long-term strategic breeding work of a speculative nature, but they are more likely to be found when common problems with a major crop prompt co-operative international action.

Lastly, evaluation must interface with the newer techniques of molecular biology, where gene transfer techniques are being developed. This could, in part, close the gap between gene banks, traditional breeders and the newer technologies which are often located in laboratories away from the genetic resources community. In any case the newer technologies are going to rely for their raw materials on well-documented germplasm collections (Peacock, 1984; this volume). These raw materials will comprise, in the main, identified genes and gene sequences and evaluation will need to encompass work of this detail. As a concept this will present few problems in those crops where, historically, there have been communities of scientists engaged in basic genetic research, e.g. maize, wheat, barley and pea, but the genetic basis of many important agronomic traits remains obscure.

Limiting resources: costs v. competence

Estimates are available for the average costs of growing out and evaluating an accession. These are usually derived from breeders' screening trials and costs are high. There are no figures available to estimate the initial costs of putting a collection in order, nor until tested widely, the costs of delimiting a core collection on which to do detailed work.

It would be wrong to suppose that the constraints which are currently hindering progress can be removed simply by increasing funding to existing organisations. As gene banks move towards their central role in the promotion of characterisation and evaluation, different skills and scientific expertise will be required. Here competent staff are more important than expensive buildings and equipment. This point has been missed by many of the pressure groups arguing for enhanced funding. Political manoeuvring, often linked to questions of ownership of materials, will not lead to the provision of trained and efficient staff.

Conclusions

A number of practical considerations have been raised in this paper through the definition and discussion of constraints. Although the IBPGR is addressing some of them, its own resources are limited, and it is the purpose of this paper to bring them to the attention of the wider community of gene bank managers and their funding authorities for action.

Effective evaluation and more extensive use of germplasm collections will follow when:

1. The gene bank collections are well managed and documented.
2. There are comprehensive crop databases which combine national collections into regional or – perhaps later – world groups.
3. Strategies are determined by the collaboration of experts knowledgeable on the crop, perhaps to delimit a core collection, certainly to guide the evaluation work.
4. There are clear and good working relationships between the gene bank managers, breeders and other scientific users.

If these practical considerations receive attention then it can be expected that the breeding of crops to meet agricultural needs could be advanced significantly, and the principal purpose of germplasm conservation, namely the utilisation of those genetic resources for the common good, be achieved.

References

Baltjes, H. J., Geltink, D. J. A. Klein, Nienhuis, K. H. & Luesink, B. (1985). Linking distinctness and description of varieties. *Journal of the National Institute of Agricultural Botany*, **17**, 9–19.

Burt, R. L., Edye, L. A., Williams, W. T., Grof, B. & Nicholson, C. H. L. (1971). Numerical analysis of variation patterns in the genus *Stylosanthes* as an aid to plant introduction and assessment. *Australian Journal of Agricultural Research*, **22**, 737–57.

Burt, R. L., Reid, R. & Williams, W. T. (1976). Exploration for, and utilization of, collections of tropical pasture legumes. 1. The relationship between agronomic performance and climates of origin of introduced *Stylosanthes* spp. *AgroEcosystems*, **2**, 293–307.

Erskine, W. & Williams, J. T. (1980). The principles, problems and responsibilities of the preliminary evaluation of genetic resources samples of seed-propagated crops. *Plant Genetic Resources Newsletter*, **41**, 19–33.

Frankel, O. H. & Brown, A. H. D. (1984). Plant genetic resources today: a critical appraisal. In *Crop Genetic Resources: Conservation and Evaluation*, pp. 249–57, Holden, J. H. W. & Williams, J. T. (eds.), Allen and Unwin, London.

Haddad, N. (1984). Utilization of genetic resources in a national food legume program. In *Genetic Resources and their Exploitation – Chickpeas, Faba beans and Lentils*, pp. 85–94, Witcombe, J. R. & Erskine, W. (eds.), M. Nijhoff/Dr. W. Junk, the Hague for ICARDA and IBPGR.

Harlan, J. R. (1984). Evaluation of wild relatives of crop plants. In *Crop Genetic Resources: Conservation and Evaluation*, pp. 212–22, Holden, J. H. W. & Williams, J. T. (eds.), Allen and Unwin, London.

Marshall, D. R. & Brown, A. H. D. (1981). Wheat Genetic Resources. In *Wheat Science – Today and Tomorrow*, pp. 21–40, Evans, L. T. & Peacock, W. J. (eds.), Cambridge University Press, Cambridge.

Nevo, E., Beites, A. & Zohary, D. (1986). Genetic resources of wild barley in the near East, structure, evolution and application in breeding. *Biological Journal of the Linnean Society*, **27**, 355–80.

Nevo, E., Zohary, D., Brown, A. H. D. & Haber, M. (1979). Allozyme-environment relationships in natural populations of wild barley in Israel. *Evolution*, **33**, 815–53.

Peacock, W. J. (1984). The impact of molecular biology on genetic resources. In *Crop Genetic Resources: Conservation and Evaluation*, pp. 268–76, Holden, J. H. W. & Williams, J. T. (eds.), Allen and Unwin, London.

Peeters, J. P. & Williams, J. T. (1984). Towards better use of gene banks with special reference to information. *Plant Genetic Resources Newsletter*, **60**, 22–32.

Simpson, M. J. A. & Withers, L. A. (1986). *Characterisation of Plant Genetic Resources using Isozyme Electrophoresis: A Guide to the Literature.* International Board for Plant Genetic Resources, Rome.

Small, E. & Lefkovitch, L. P. (1986). Relationships among morphology, geography, and interfertility in *Medicago*. *Canadian Journal of Botany*, **64**, 45–52.

Westcott, H. B. (1985). Some methods of analysing genotype-environment interaction. *Heredity*, **56**, 243–53.

15
Principles and strategies of evaluation

O.H. FRANKEL

Introduction: the curator and the breeder

In the light of 20 years' experience, evaluation of genetic resources in germplasm collections is due for reappraisal. It has been a widely accepted axiom in earlier genetic resources literature that evaluation is an essential preliminary to utilisation (Frankel & Bennett, 1970). Evaluation was seen as an organised and institutionalised activity resulting in information which comes to the user (the breeder) as standardised and computerised documentation (Finlay & Konzak, 1970). Evaluation was the responsibility of the curators of germplasm collections. The characteristics to be evaluated were nominated by specialists in the various crops (Hyland, 1970). Essentially this remains the basic strategy. It is the gene bank manager who is responsible for planning and executing the evaluation process which is to make collections accessible to the breeders (Peeters & Williams, 1984).

Breeders, as is often stated, look to the gene banks or genetic resources centres (GRCs) for information on 'agronomically useful traits' (Duvick, 1984). Peeters & Williams (1984) conducted a survey on the utilisation of germplasm resources. They concluded that genebanks are not being used very extensively by plant breeders, the main reason being the scarcity of information that was of use to the individual breeder. It seems, then, that in the opinion of authoritative judges the prevailing strategy for evaluation – enshrined in the 'genetic resources dogma' (Frankel, 1986) of the 1960s – has not been altogether successful. And the proposed cure, more of the same, scarcely inspires confidence.

This may be an unduly pessimistic view of the present situation. Germplasm collections of some of the international agricultural research centres (IARCs) of the Consultative Group on International Agricultural Research, or the Small Grains Collection of the United States Department of Agriculture, have been purposefully evaluated and extensively used; and, as is evident from the surveys already quoted, a significant proportion of breeders turn to germplasm collections when their own do not provide the genes they require. All the same, the opinion polls also indicate that there is a need for change if breeders are fully to exploit the genetic resources that are now available to them as they have never been before.

It seems that a lead to more effective evaluation comes from institutions where there is a close organisational and personal contact between curator and breeder, and where the breeding objectives are reflected in the evaluation programme. An example is GEU, the Genetic and Utilisation Programme of the International Rice Research Institute (IRRI), in which curator, breeder and all the relevant specialists participate. The contact with breeders is carried beyond the confines of the international institute to national breeding institutions. Thanks to the outreach programme, collection–user contact is a two-way traffic of information and materials between the collection and the breeding programmes. Much as breeders desire a hand-out of ready-usable information that is meaningful for their breeding programmes, 'evaluation by proxy' is marked more by its limitations than its successes. It seems inevitable that breeders take an active part in the process, especially when breeding objectives are diverse and breeders competitive, as tends to be the case in developed countries. But in any circumstance it is becoming apparent that breeders need to be directly involved. No one but the individual breeder can determine the genes or characters he wants to introduce into his breeding material. However, these are not necessarily the 'most sought' characters as suggested by Peeters & Williams, 1984; evaluation according to opinion polls could stifle innovation. Breeding aims change rapidly, hence evaluation needs to be adaptive. And only the breeder can recognise which characters need to be evaluated, or at least checked, in his own environment.

There is also need for re-examining the responsibilities of the curator in gathering and disseminating information. In my view he has five important tasks. First, he is responsible for obtaining information on the origin of accessions (the 'passport data') from collector, breeder or other sources. Second, he is responsible for characterisation. Third, on the basis of passport data and characterisation, he may proceed to rationalise the collection, so as to reduce the task of evaluation to a manageable scale; a 'core collection' is one possibility (see Brown, this volume). Fourth, he should organise and co-ordinate the participation of relevant specialists such as plant pathologists, entomologists, etc. Fifth, he is – or should be – responsible for co-ordinating and making available the information that results from the evaluation process. But he should not be directly involved in evaluation, nor in processes designed to modify or adapt germplasm for use in plant breeding (pre-breeding).

These ideas on the respective functions of curator and breeder in generating germplasm information are in general agreement with the procedures suggested in the most recent IBPGR publications (e.g. IBPGR, 1985, page 1). In the sections which follow, the steps which constitute the characterisation and evaluation process are examined more closely in the light of the general principles outlined above. The terminology is that used in the widely distributed publications of the IBPGR, e.g. IBPGR (1985).

Biological status and mode of utilisation

To avoid misunderstanding and repetition, some basic distinctions on the botanical status and the mode of use need to be made. The status of an accession, whether wild or domesticated, interacts with the mode of utilisation, whether plant introduction (i.e. for direct use) or gene introduction (i.e. as donor of genetic components). The chapter in this book by Hamon & van Sloten is concerned with plant introduction, those by Gill and Smith & Duvick with gene introduction.

Fig. 15.1. shows that, broadly speaking, status and mode interact and sort out the different groups of crops. Introductions of pasture and forest plants are mainly wild species used directly; those of field crops are mainly used in recombination, with some important exceptions such as the introduction of improved cultivars in less developed regions, or of highly successful and adaptable cultivars anywhere. Each of the four states – wild, domesticated, plant and gene introduction – makes specific demands

Fig. 15.1. Biological status and mode of ultilisation – interactions

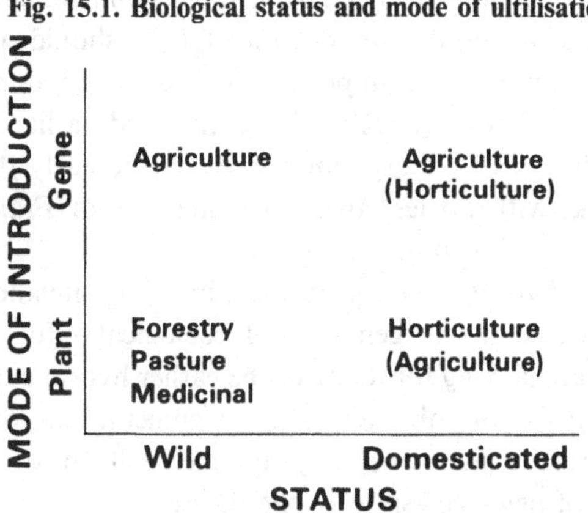

on some phase of the information process. If these interactions fail to be appreciated, one or other of the components is likely to be overlooked.

Origin (passport data) (Table 15.1:1)

Information on collecting sites of accessions is important for all phases of genetic resources work. It is therefore regrettable that many of the older accessions in collections lack even the most elementary – and perhaps most valuable – information, data which identify the point of origin on a map, and/or by latitude, longitude and altitude.

Recent site descriptor lists (e.g. in IBPGR, 1985) are streamlined by comparison with earlier versions and should be a light burden for collectors. But the sins of omission of past collecting efforts cannot be undone.

Site identification is perhaps the most significant evidence available to the curator for designating a core collection and a great help in identifying duplicates. It is essential information for ecobiological, evolutionary or population genetic research and for planning further collecting. Records on topography or soil characteristics can be valuable for plant breeders concerned to improve adaptation to particular conditions or tolerance of edaphic or climatic stresses. There is increasing emphasis around the world on seeking genes for tolerance to specific adverse environmental factors such as waterlogging, soil acidity, specific mineral deficiencies or toxicities, in addition to climatic stresses which should be ascertainable from climate tables.

In addition, for material likely to be used for plant introduction, a record of the associated flora, the hydrology and the land use system could be of interest.

Site documentation for wild material should be a great deal more descriptive of the environment than is relevant for gene introduction of domesticates. For wild material, site documentation should include ecogeographic information, which is important for assessing redundancy and representativeness and for research fields mentioned earlier in this section. Site description was strongly emphasised in the early days of genetic resources work, with no less than four chapters of Frankel & Bennett (1970) largely devoted to it.

For accessions with a hybrid origin, genealogy, breeding site and areas of cultivation can provide useful genetic and ecological information. Regrettably such data are lacking for many of the earlier hybrid cultivars, but curators should insist on obtaining full documentation on new accessions of this nature which, one may assume, will constitute an increasing proportion of new accessions in the future.

Collection data that can only be obtained in the field are the natural responsibility of the collectors. But they should not be made responsible for climate data as required, for example, in the revised descriptor list for sorghum (IBPGR, 1984*a*), since they might be more difficult to obtain on the spot than from climate tables or data banks. It should be the responsibility of the curator of the recipient collection to see that passport data are as complete as can be contrived.

Characterisation (Table 15.1:2)

Characterisation, according to a recent IBPGR (1985) definition, 'consists of recording those characters which are highly heritable, can be easily seen by the eye, and are expressed in all environments'. At a basic level characterisation is an account of the plant's morphology, either throughout its development, or only at maturity. Its primary user is the curator. Characterisation data are useful in management measures such as the designation of a 'core', as discussed in the preceding section: 'characterization should provide a standardized record of readily assessable plant characters which, together with passport data, go a long way to identify an accession' (Frankel, 1986). On the other hand, to the plant breeder, this level of characterisation is of only limited interest since the characters it covers are, as a rule, no more than subsidiary to those that are his main reasons for using alien germplasm. Yet information on characters such as seed colour or plant height might be useful in selecting accessions which meet some of the requirements a breeder may have to satisfy in a new cultivar.

Characterisation for the purpose of classification can usefully be complemented by characters that cannot be 'seen by eye', such as isozyme data.

Of greater interest to the plant breeder are physiological characters which can be examined in a neutral environment, i.e. one in which a broad representation of a crop can be grown, as is usually the case for major collections. In IBPGR descriptor lists they are included under the headings 'further characterisation' or 'preliminary evaluation'. Characterisation by curators can be of considerable help to the breeder through providing a preliminary account of physiological baseline data such as vernalisation requirement, tillering, times of flowering and maturity, which would help to narrow the selection of potential breeding stocks. Reference varieties should effect a degree of comparability across latitudinal or climatic diversity, obviating the need for multi-site tests except in more extreme environments. Characterisation is generally regarded as a responsibility of curators. Whether it is confined to one central collection, such as one of the base collections in the IBPGR system, or shared by two or more, may depend on the diversity of the crop – three centres are responsible for the three major cultivated subspecies of rice – and on environmental or organisational circumstances. To what extent this responsibility is discharged, resulting in published data, varies between crops. Extensive documentation is available for many crops within the mandates of the IARCs, for some national systems like the Plant Gene Resources Programme of Canada, and some regional ones like some of the national collections linked to the IBPGR Southeast Asia Committee, and for many individual crops in a number of countries. But, as is often pointed out (e.g. by Peeters & Williams, 1984) regrettably, there are many gaps.

One would wish to see this large effort widely utilised, not only by curators, but by plant breeders who, as we have seen in the first section and in the chapters by Smith & Duvick and Gill in this book, ask for more characterisation. Curators would need to know which of the descriptors commonly included – and others that might be – help plant breeders to identify accessions for further evaluation or for direct use in their programmes. This, from the breeder's point of view, would surely be the main purpose of characterisation.

Having presented the case for characterisation on the grounds of its usefulness for curators and breeders, one is tempted to ask *how essential is the information* derived from it? It is perhaps a sobering thought for those insisting on complete characterisation of collections that germplasm collections had existed and had been used near on a century before characterisation and evaluation had been thought of and formalised. Collections were used for both plant and gene introduction and some of them were of considerable size. Selection for resistance or tolerance can be carried out on a large collection provided the pathogen or stress are present; I selected for large size of grain in a collection of some 3,000 wheats; and discrimination on the basis of origin results in a preliminary selection for adaptation. The point of these remarks is not that characterisation is unnecessary but that *its absence does not prevent the utilisation of a collection* as long as there is a clear idea of objectives, and the initiative to seek out and apply unfamiliar, and even exotic, gene resources.

Evaluation

In the widest sense, the evaluation of genetic resources is a multi-dimensional endeavour, involving scientific fields as diverse as crop cytogenetics and evolution, physiology, pathology and agronomy. They all contribute information that bears on the choice and utilisation of genetic resources by the breeder, the emphasis on particular fields varying with the status of the material. Cytogenetic information is essential for the use of many wild relatives of crops; genetics of host–parasite interactions is equally essential for the choice of resistant genotypes of any status. Contributions such as these must be constantly kept in view as basic not only for an understanding of, but for the evaluation and exploitation of genetic resources.

The level of evaluation which is the subject of this chapter is closer to the working face of the plant breeding station. At this level also, evaluation draws on interacting or collaborating disciplines. Primarily involved are genetics, plant physiology, plant pathology, entomology, biogeography, with others like biochemistry frequently called in, interacting with the ecology and agronomy of a defined environmental range. Multi-disciplinary participation is essential, but this is generally understood. But two of the participants hold the keys to the system: *the breeder*, to point the direction and to provide the final and crucial test, and *the curator*, to steer participation and to guide the resulting information.

While the identity of the curator is unambiguous, that of the breeder needs to be defined. In this chapter it is used to identify an individual breeder (or institution) rather than a group joining in an opinion poll. In defining evaluation targets the former expresses specific requirements to which his programme is committed. An opinion poll is likely to result in divergent views (Peeters & Williams, 1984) with priorities to which no one is committed. Nor is a group necessarily as committed, as the individual breeder is likely to be, to tests in his own environment, although group testing does occur (Smith & Duvick, this volume). The curator of the Australian Winter Cereal Collection has a co-operative evaluation programme with a number of Australian breeders. These indications will suffice to outline the framework of evaluation. We must avoid the error of some discussions on evaluation which confine themselves either to generalities or to specific cases. I shall attempt to deal with the elements of evaluation, the characters – whether defined in genetic, physiological, agronomic or any other terms – which serve as the descriptive elements of evaluation.

In this discussion of evaluation I confine myself to gene introduction, hence to the evaluation of accessions as potential donors of characters or genes. This justifies an analytical versus a holistic approach which would be in place for plant introduction (see above). The evaluation of plant introductions, especially of forest and pasture species, has been discussed many times in the last 25 years.

The characters

In a paper called 'the theory of plant breeding for yield' (Frankel, 1947) I suggested that the characters of concern to plant breeders were, broadly speaking, of two kinds:

(a) Observable or strongly expressed characters. These can be identified in single plants or their immediate progenies. They are expressed under growing conditions which are normal for the crop or they may require special conditions for expression such as a specific parasite or environmental stress. They are simply, or, if polygenic, strongly inherited and can be readily selected for in hybrid generations.

(b) Variable or complex characters. These are subject to environmental variation. They are largely responsible for differences in yield or adaptation. Inheritance is polygenic. Evaluation of accessions may require replicated tests in multiple sites. The relevance of such tests is discussed below.

One can recognise four categories of characters, the first three conforming to the concept of observable characters.

Observable characters

Morphological, physiological or biochemical characters relating to survival, productivity or quality. (Table 15.1:2)

As we have seen in the preceding section, some characters in this category may, and as a rule should, be included in the characterisation by curators. But many of them would need to be re-examined in the breeder's own environment where also characters of particular concern need to be examined in appropriate tests. For example, to score for high-level resistance to lodging in wheat, accessions with apparent resistance could be tested further by providing extra water and nitrogen. Under conditions where grain shedding is liable to occur, a propensity to shedding can be diagnosed, even in the absence of strong winds, by a gentle rubbing of the mature inflorescence – all too familiar to a former New Zealand wheat breeder.

Characters (or, when known, compounds) involved in the quality of the products or the presence of toxic substances, may involve in the evaluation process not only the curator and the breeder, but biochemists and technologists. The initiative may come from the breeder or the curator, but it may fall to the curator to organise the appropriate action.

Resistance to diseases and pests. (Table 15.1:3)

Resistances to parasites are some of the most sought after contributions of germplasm collections. Appropriate combinations of pathogen availability and environment make such resistances 'observable'. As a rule, the evaluation of pathogen resistance is made possible by extensive research by plant pathologists and entomologists into the genetics and ecology of host–pathogen relations.

At the International Rice Research Institute (IRRI) the results of extensive research into the diseases and pests of rice are applied in the evaluation programme (GEU) of the germplasm collection, to the effect that a wide range of resistances has been obtained and studied. At national level an example is the Australian National Rust Control Programme (NCRP) at the University of Sydney. It has a continuing research and survey programme into the currently occurring biotypes of cereal rusts and examines accessions of the Australian Winter Cereal Collection and breeding lines of Australian plant breeders. Another extensive university programme of evaluation is carried out by the Wheat Genetic Resources Center (WGRC) at Kansas State University which explores the genetic potential of wild species related to wheat. A representative collection of accessions with a known site of origin has been assembled and is being classified – in co-operation with other institutions – for reactions to biotypes of the main diseases, and for a range of other characters. Genetic analysis and chromosome mapping provide information on the interrelations between resistance genes, and transfers to cultivated material facilitate further examination and utilisation.

These few examples are indicative of the diversity of research programmes which are basic to the evaluation of genetic resources. The question needs to be asked to what extent this largely spontaneous research network covers the diversity of crops and environments, especially in the tropics where, compared with the temperate zone, there are fewer scientists to work on a much greater number of species. As the number of breeders of tropical crops increases – the present number in some developing countries being pitifully small – the demand for pathogen resistances is bound to stimulate research, as had been the case in the most successful research programmes such as those on the rust diseases of wheat.

Thus the plant breeder shares with scientists in allied fields the responsibility for initiating the process of research and exploration that is to lead to resistant cultivars. Yet his task includes also a more direct participation. Before committing resources to a long-term programme of resistance breeding he will need to check in his own environment the validity of tests conducted in the laboratory and/or in other environments. This will be the more necessary, the less the genetic diversity and the heredity–environment interactions of the pathogen have been explored.

Tolerance of adverse conditions or stresses (Table 15.1:4) such as high and low temperature, drought, winter killing, frost, waterlogging, soil acidity, salinity, nutritional deficiencies or toxicities.

Such tolerances have features which place them between 'observable' and 'variable' characters. They are mostly polygenic, against the prevailingly oligogenic pathogen resistances. Though laboratory tests may be helpful for some stress identifications, the crucial test is under appropriate field conditions.

Stress tolerances have one feature in common with pathogen resistances. Both are environment dependent in their expression: parasite resistance on the parasite, stress tolerance on the stress. The conditions for expression are specific and can be induced. Tests require specific, not multiple, environments as is sometimes claimed. So both resistances and tolerances differ from complex characters like yield which require tests in multiple sites and seasons to attain significance.

In the evaluation for stress tolerance, the breeder plays a central role. Being familiar with the environment and its challenges he is likely to take the initiative in the search for tolerant breeding materials. He has not only the interest but he is likely to have the physical conditions for the evaluation of a range of accessions. The site descriptions (see above) should provide preliminary information on accessions that should be worth testing.

Variable or complex characters (Table 15.1:5)

For a paradigm it is appropriate to turn to yield as the most representative and important complex character. Clearly it is a breeder's, not a curator's character, although it is included in some IBPGR descriptor lists (e.g. IBPGR, 1984*b*). It is hard to see what information on yield capacity a small evaluation plot could contribute, even if enlarged to a replicated state; nor what significance it would have for the breeder's environment. But it has been suggested that a close site description of such tests would facilitate comparisons with the breeder's environment and thus validate the test data obtained elsewhere. Indeed, IBPGR descriptor lists prescribe such site data, and their importance was emphasised in a recent survey of plant breeders who indicated 'the importance of either quantifying the environmental conditions in which the data are being recorded or of doing multi-site evaluation.... If no way is available to quantify the impact of environmental conditions...the interpretative value of the results is very severely limited.' (Peeters & Williams, 1984). While test site descriptions may have indicative value, as a rule they will do little to replace tests in the breeder's own environment.

Inevitably, evaluation for complex characters such as yield is expensive in time and resources. It also presents problems that do not apply in the evaluation of observable characters. The latter results in information that is directly applicable in breeding for resistance or tolerance, whereas in the evaluation for yield capacity the information derived from yield tests is not directly applicable. The information that is relevant to the breeder is not the yield performance of accessions as such – though it may be suggestive – but the interaction between accessions and representative local cultivars. Accordingly, there are two approaches to evaluate germplasm accessions for their capacity to raise yield levels above those of locally successful cultivars. The first assesses the prospect of success by the yield performance of accessions themselves. One can call this the indirect method, since the expectation of positive interactions with adapted cultivars is based on tests of the parents rather than of their interactions. A good example of a consistent testing and development programme by the breeder – in this case by a group of collaborating breeders – is reported in the chapter by Smith & Duvick in this volume and may be briefly summarized here. Sifting a large collection for desirable traits, multi-site yield tests and partial conversion to adapted germplasm resulted in a number of lines with satisfactory yields and no less than 25 per cent exotic germplasm. These are to be used in the quest for genetic diversification, hopefully resulting in raising yields beyond levels that are approaching a yield plateau.

The second approach, which can be called the direct test, examines, as already outlined, the interactions between accessions and locally adapted cultivars. The relevant test is one for combining ability, which is difficult in normally self-fertilised plants. However, using chemical hybridising agents, which are becoming available for wheat and are likely to be developed for other crops, and using male testers with good general combining ability and a capacity for pollen dispersal, it should be possible not only to test large numbers of accessions but to produce seed for multi-site F_1 tests (Bhatt *et al.*, 1984).

Extensive tests by either of the two methods could scarcely be contemplated on a large germplasm collection. However, starting with a core collection (Brown, this volume) they clearly become feasible. If results are encouraging it should be possible to attempt surveys of the propensities for yield improvement in a wide and representative range of the world's germplasm.

Pre-breeding

It is a frequently voiced demand by plant breeders that valuable genes be transferred from poorly adapted landraces into adapted germplasm, i.e. cultivars with a performance approximating that of current cultivars. This applies to an even greater degree to gene transfers from related wild species which may have to overcome hybrid incompatibility. As Smith & Duvick point out in their chapter, pre-breeding may be a necessity for the evaluation of some characters. There can be little doubt that pre-breeding would be a convenience for those who could take advantage of it. The questions are whether agreement could be reached among interested parties on the choice of donor and recipient stocks, and who would do the work.

Pre-breeding is the early phase of any breeding programme utilising exotic germplasm. For the most part, such programmes have been carried out by public rather than private institutions in both developed and developing countries. Recently, with a rapid increase in the size and sophistication of private breeding companies in developed countries, and with the growing interest in the largely unexplored potential of wild germplasm, the demand for gene transfers to more manageable resources has become more frequently voiced.

It is clear enough who should *not* be involved – the curators of germplasm collections, since they are not only fully occupied with tasks only they can perform, but may lack the essential expertise (see also J.T. Williams, this volume). Hence it is left to public and private breeders, and there are reasons why either or both should play their part. I can see every reason why large companies – where they exist – should contribute their share of research and development as companies in other industries are accustomed to do. Indeed, there can be little doubt that in this highly competitive industry, individual initiative will prevail in exploring new pathways for enriching the pool of available gene resources. Diversity of enterprise is more likely to result in the genetic diversity which everyone regards as essential, than would be possible through reliance on public institutions alone.

Breeders lacking the human and capital resources directly to exploit exotic germplasm may have to rely on pre-breeding by public breeders, or on using cultivars which incorporate such germplasm. Such a procedure which is not restricted by variety rights regulations has been the standard approach of many breeding institutions for the best part of a century.

With few or no private breeders, most developing countries are not faced with such a choice. Pre-breeding, if it is to be performed, must be a function of international or national institutions. This has the advantage that its products will become generally available without delay, but the drawback that it is unlikely to be responsive to local requirements. With the inevitable increase in plant breeding activities, pre-breeding will be an important contribution to the use of genetic resources in developing countries.

Summary and conclusions

1. The germplasm resources available to plant breeders tend to be of two kinds: a working collection consisting of cultivars, breeding lines and various genetic stocks, with which they are familiar; and cultivars or wild species from germplasm collections selected on the basis of characterisation and evaluation which breeders expect to be carried out by GRCs.
2. When curator and breeder belong to one institution, as is the case in the IARCs, or breeders have close working relations with GRCs, the characterisation–evaluation system can work well because the breeders can influence and participate in all phases, from characterisation to pre-breeding.
3. In general, however, distance and/or environmental disparities intervene between GRC and breeder. There is, therefore, a need for a collaborative system based on consultation and labour-sharing between curator and breeder, each playing the principal role at opposite ends of the process (see 5 and 8 below). Clearly, evaluation at GRCs without breeder participation has not been entirely successful.
4. Evidence of origin, 'passport data', help the curator in dealing with duplicates and in designating a core collection. For the breeder they provide information on daylength, climate and ecology and/or the genetic affiliations of accessions.
5. Characterisation serves the same ends. It adds information on readily identified characters which may range from simple morphology to phenology, physiology and agronomy. Characterisation is the responsibility of the curator, but it would gain from consultation with breeders.

Table 15.1. *The roles of curator, breeder and specialists in characterisation and utilisation.*

Subject	Fact-finding		Utilisation	
	Action	Responsibility	Action	Objective
1. Origin Passport data	Collector Curator	Collector Curator	Curator Breeder	Ecological characterisation
2. Characterisation of morphological and physiological characters	Curator (Breeder)	Curator	Curator Breeder	Morphology Phenology Physiology
3. Resistance to parasites	Entomologist Pathologist Breeder	Breeder (Curator)	Breeder Entomologist Pathologist	Resistance breeding Pathogen research
4. Tolerance of stress	Ecologist Physiologist Breeder	Breeder	Breeder	Resistance breeding
5. Variable or complex characters	Breeder		Breeder	Adaptation Performance
6. Information	Curator	Curator Databank	All participants	Information

6. Basically, evaluation draws on many sciences, from evolution and biochemistry to agronomy. At the working face of plant breeding it is multi-disciplinary and needs to be collaborative, with the breeder and the curator as lead operators, the former to define objectives, either of them to organise participation by specialists, and the curator to collate the information.
7. Evaluation identifies characters (or genes) for incorporation in adapted genotypes. Characters can be classified as 'observable' (i.e. highly heritable), or 'variable' and complex. Observable characters include (a) morphological, physiological or biochemical characters which are usually evaluated at a GRC; (b) resistances to diseases and pests, usually evaluated by specialists; (c) stress tolerance, best evaluated by breeders in stress environments.
8. Variable or complex characters have yield as the paradigm. They can be meaningfully evaluated only in the breeder's environment. The capacity to raise yields above current levels can be assessed either indirectly, by yield testing of accessions, or directly and more meaningfully, by compatibility tests with locally adapted cultivars. Both are expensive and time consuming and are feasible only on small collections or on core collections.
9. Pre-breeding is an essential concomitant of the utilisation of exotic, and especially of wild germplasm. It cannot be the responsibility of curators, but should devolve on those who profit most from the introduction of new germplasm – the large private companies. Where they do not exist, as in most developing countries, pre-breeding clearly is an important responsibility of international and national centres.
10. The role of characterisation and evaluation in the utilisation of genetic resources is now recognised, but the responsibilities need to be more clearly understood (Table 15.1). The curator's responsibilities are to maintain and characterise a collection, to assemble the information resulting from all parts of the process, and to transmit it to the data banks (J.T. Williams, this volume). The breeder's participation is essential, from the definition of objectives, to the checking of evaluation results and the test for the capacity to raise yields. This needs to be more widely understood by both GRCs and breeders. In the last resort, it is the breeder who makes the decision what resources to use: germplasm collections were used for a

century before the advent of characterisation, thanks to plant breeders' initiative and imagination.

Acknowledgements

I am grateful to Drs A. H. D. Brown and D. R. Marshall for many stimulating discussions and for critical comments, and to Dr B. S. Gill for information on the work of the Wheat Genetic Resources Center at Kansas State University.

References

Bhatt, G. M., Ellison, F. W. & Marshall, D. R. (1984). A case for unreplicated plots for multi-site yield testing in wheat. *Australian Journal of Agricultural Research*, **35**, 197–214.

Brown, W. L. (1987). The exchange of genetic materials: a corporate perspective on the internationalization of the seed industry. In *Seed and Sovereignty*. Kloppenburg, J. Jr. (ed.) Duke University Press (in press).

Duvick, D. N. (1984). Genetic diversity in major farm crops on the farm and in reserve. *Economic Botany*, **38**, 161–78.

Finlay, K. W. & Konzak, C. F. (1970). Information storage and retrieval. In *Genetic Resources in Plants – their Exploration and Conservation*, pp. 461–5.

Frankel, O. H. (1947). The theory of breeding for yield. *Heredity*, **1**, 109–20.

Frankel, O. H. (1986). Genetic resources – museum or utility? In: *Proceedings of Plant Breeding Symposium DSIR 1986*. Department of Scientific and Industrial Research, Wellington (in press).

Frankel, O. H. & Bennett, E. (1970). *Genetic Resources in Plants – their Exploration and Conservation*. IBP Handbook No. 11. Blackwell Scientific Publications, Oxford.

Hyland, H. L. (1970). Description and evaluation of wild and primitive introduced plants. In: *Genetic Resources in Plants – their Exploration and Conservation*, pp. 413–19. Frankel, O. H. and Bennett, E., (eds.). IBP Handbook No. 11. Blackwell Scientific Pubications, Oxford.

IBPGR (1984*a*). *Revised Sorghum descriptors*. IBPGR, Rome.

IBPGR (1984*b*). *Descriptors for Soyabean*. IBPGR, Rome.

IBPGR (1985). *Oat descriptors*. IBPGR, Rome.

Peeters, J. P. & Williams, J. T. (1984). Towards better use of genebanks with special reference to information. *Plant Genetic Resources Newsletter*, **60**, 22–31.

Part V

Wild relatives of crops

16
Collection strategies for the wild relatives of field crops

C. G. D. CHAPMAN

Introduction

The cultivar collections of many major crops are now very large indeed, frequently running into several tens of thousands globally (Lyman, 1984). As a consequence, further general collecting is becoming less and less efficient since new accessions are increasingly likely to carry alleles and traits already to be found in existing collections.

The wild relatives of crop plants represent additional and novel gene pools of genetic resources. Table 16.1 lists some comparisons between crop species and their wild relatives based on isozyme studies. With respect to the loci assayed, wild species are generally more variable than the corresponding crop (despite often smaller sample numbers) and have larger and/or different spectra of alleles. Reasons for this include the founder effect of domestication and other genetic 'bottlenecks' in the evolution of the crop, and its isolation from other species – including its own progenitors – by geography, ecology, disruptive selection and crossability barriers (Ladizinsky, 1985).

The value and contributions of wild crop relatives to breeding have been discussed by Hawkes (1977), Stalker (1980), Harlan (1984), and Brown & Marshall (1986). The documented successes relate largely to the transfer of simply inherited characters, particularly disease resistance. However, it is evident that quantitative improvements, even in as complex a character as yield, can also be introduced (Frey *et al.*, 1984).

Attention is, therefore, shifting to the wild relatives to further enlarge the genetic base available to the breeder. The approach to collecting this material is somewhat different to that for the cultivars for a number of reasons:

1. Collecting landraces has been an emergency activity due to the rapid spread of new varieties. Some wild relatives are also

Table 16.1. *Isozyme polymorphisms in some crops and their wild relatives*

Crop and species	Cultivated (C) or wild (W)	Number of lines or populations	Origins	Number of polymorphic loci Overall	Number of polymorphic loci Each species	Number of alleles at all polymorphic loci	Number of alleles in wild species not in cultivated	References
AMARANTHS (*Amaranthus*)								
A. caudatus	C	51	Latin America, India and Nepal	8	7	20	—	Hauptli & Jain (1984)
A. cruentus								
A. hypochondriacus								
A. quitensis	W	21	Latin America and USA		8	20	4	
A. hybridus								
A. powellii								
A. retroflexus								
PEPPERS (*Capsicum*)								
C. pubescens	C	49	—	19	10	32	—	McLeod *et al.* (1979)
C. cardenasii	W	16	—		6	26	4	
C. eximium	W	36	—		11	32	8	
C. tovari	W	1	—		3	23	6	
WATERMELON (*Citrullus*) (*Citrullus*)								
C. lanatus	C	12	Israel and USA	13	0	13	—	Zamir *et al.* (1984)
C. colocynthis	W	31	Israel and Sinai		8	21	16	
BARLEY (*Hordeum*)								
H. vulgare	C	1,358	Global	4	4	29	—	Kahler & Allard (1981)
H. spontaneum	W	148	Israel and Turkey		4	30	3	
H. vulgare	C	1	Global composite	25	7	36	—	Nevo *et al.* (1979)
H. spontaneum	W	28	Israel		22	102	66 + ?	
LENTIL (*Lens*)								
L. culinaris	C	31	Medit. and S.W. Asia	13	6	22	—	Pinkas *et al.* (1985)
L. orientalis	W	19	Israel and Turkey		9	26	5	

L. odemensis	W	5	Israel and Turkey		5	19	3	
L. ervoides	W	9	Israel and Turkey		12	28	7	
L. nigricans	W	4	Spain and Turkey		4	18	7	
TOMATO (*Lycopersicon*)								
L. esculentum	C	178	USA, Latin America and Europe	9	6	19	—	Rick & Fobes (1975)
L. esculentum var. *cerasiforme*	W	98	Latin America, Old World tropics		8	22	6	
RICE (*Oryza*)								
O. sativa	C	468	Africa and Asia	33	25	—	—	Second (1982)
O. glaberrima	C	515	Africa		6	—	—	
O. brevigulata	W	965	Africa		22	—	—	
FOXTAIL MILLET (*Setaria*)								
S. italica	C	223	Europe, Africa and Asia	10	10	22	—	Jusuf & Pernes (1985)
S. viridis	W	45	China and France		8	23	5	
POTATO (*Solanum*)								
S. tuberosum								
group Stenotomum	C	11	Peru and Bolivia	12	7	23	—	Oliver & Martinez Zapater (1984)
subgroup Gonicalyx	C	3	Peru		7	21	—	
group Andigena	C	15	Peru, Bolivia and Colombia		9	29	—	
group Tuberosum	C	76	Northern hemisphere		9	27	—	
S. sparsipilum	W	3	Peru		7	20	1	
S. pinnatisectum	W	3	Mexico		5	17	6	
MAIZE (*Zea*)								
Z. mays	C	43	Mexico and Guatemala	19	19	95	—	Smith *et al.*
Z. mexicana	W	79	Mexico and Guatemala		19	133	58	(1984; 1985)

endangered and deserve priority in collecting e.g. *Zea diploperennis* which is confined to only a few sites (Iltis *et al.*, 1979). On the whole though, few of the wild relatives of crops are in great danger and many are, in fact, quite abundant. In the latter case, collecting is needed not to preserve genetic resources, but to make the variation available for use. There is no compulsion to collect anything and everything, only to secure samples sufficient and representative enough to meet the requirements of breeders and researchers in the near future.

2. Whereas any crop is usually represented by one or two species, it may have several or many wild relatives to be considered for collecting. It may be necessary to set priorities among these.
3. While cultivars can be highly mobile, being planted in different fields or even in different villages to that from which they were previously harvested, most wild species do not migrate far in any one generation and can develop highly localised patterns of gene distribution.

Hence there is a need to consider separate strategies for collecting wild material. Because of this diversity of species that crops and their relatives represent, this paper suggests broad principles that could be adapted to any one group of crop relatives.

Setting priorities by taxa

The taxonomic hierarchy used in classification and nomenclature can be a poor guide to the utilisation of genetic resources, and hence to what should be collected. The definition of each taxon is often fundamentally morphological rather than genetic. This can lead, on the one hand, to an abundance of fully inter-fertile species based on gross but genetically simple differences, and on the other, to the lumping together of distantly related species into genera on the basis of superficial similarities. For crop plants and their relatives, the situation is complicated by the proliferation of synonyms arising from the number of studies to which they have been subjected. The tetraploid wheat *Triticum turgidum* (L.) Thell., for example, has 14 other names.

A more useful system is the gene pool system devised by Harlan & de Wet (1971) for cultivated plants, which basically reflects the ease of utilisation. This proposes three categories:

Primary gene pool (GP-1); the true biological species including all the cultivated, wild and weedy forms of a crop species. Hybrids among these are more or less fully fertile and gene transfer to the crop is simple. A subdivision into (A) the cultivated types and (B) the wild forms is generally useful.

Secondary gene pool (GP-2); the coenospecies from whose members gene transfer is difficult, but possible. Hybrids may be weak or partially sterile, chromosomes may pair poorly and there may be differences in ploidy levels.

Tertiary gene pool (GP-3); from which moving genes is very difficult. Embryo culture of either the hybrid or its offspring may be needed or bridging crosses necessary to bring genes over. This represents the outermost limit for breeding by conventional means, but it is often ill defined due to lack of research.

Table 16.2 summarises the gene pools of a number of crop species, and shows that divisions do not necessarily coincide with traditional taxonomy. In some cases, members of the same genus as the crop fall outside the tertiary gene pool, e.g. all other species of *Vicia* for *faba* bean. At the other extreme, some crops (e.g. wheat) contain several genera within their secondary gene pools. The size of, and rankings in, the gene pool of a crop also reflect to some extent the amount of research carried out on interspecific hybridization. Improvements in crossing techniques, the discovery of genotypes with higher crossabilities or attempts at novel crosses will expand these gene pools in due course.

The availability of such information on gene pools enables priorities to be set for collecting. The general importance of a species, and the number of accessions that can be exploited, will be related to the ease with which it can be used. Thus the highest priority and most extensive collecting should be for species in GP-1B, with decreasing emphasis on GP-2 and GP-3 respectively. While it is desirable to have some accessions of every species available for research work, large-scale collecting of species from the lower order gene pools is pointless if there is little prospect of their being used.

Investigating which relatives lie in which gene pools may require an extensive literature search. So far as is known, this has been done thoroughly only for the legumes (Smartt, 1986). It is, nevertheless, an essential prerequisite when traditional taxonomy is so variable a guide. In most cases the literature will be found deficient in some respects, and decisions on the importance of some species may have to be taken based on what is known about their near relatives.

In the future this hierarchy of ease of use may need drastic revision. Developments in biotechnology promise the movement of genetic material to a crop species from almost any source. It is presently unclear whether,

Table 16.2. *The gene pools of some crop species*

Crop	Gene pools			References
	Primary	Secondary	Tertiary	
BARLEY *Hordeum vulgare*	A. cultivars B. *H. spontaneum*	*H. bulbosum*	other *Hordeum* species other Triticeae	Bothmer *et al.* (1983) Dewey (1984)
BREAD WHEAT *Triticum aestivum*	A. cultivars B. —	other *Triticum* species *Aegilops* *Secale* *Thinopyrum* *Haynaldia* *Leymus*	other Triticeae	Sharma & Gill (1983) Dewey (1984)
GROUNDNUT *Arachis hypogaea*	A. cultivars B. *A. monticola*	section *Arachis*	other *Arachis*	Smartt (1984) (see also Moss *et al.* this volume
COMMON BEAN *Phaseolus vulgaris*	A. cultivars B. var. *aborigineus*	*P. coccineus* *P. polyanthus*	other *Phaseolus* species	Smartt (1984)
FABA BEAN *Vicia faba*	A. cultivars B. —	—	?	Smartt (1984)
SUNFLOWER *Helianthus annus*	A. cultivars *B.* subsp. *annuus*	Mainly annual and diploid *Helianthus* species	Mainly perennial and polyploid *Helianthus* species	Thompson *et al.* (1981) Georgieva-Todorova (1984)

in those circumstances, genes from the recognised gene pools will integrate better with the crop genome than those from other species. If that is so, then gene pool species may still be the preferred sources of new genes. If not, then all plant species may come to be regarded as crop genetic resources (see also Peacock, this volume).

Collecting in the field

Choosing the target area(s)

Once the priorities for species have been determined, the target area(s) in need of collecting have largely been determined too, based on what is known about species distribution. Investigating this will involve a literature search unless a recent publication has reviewed the situation. Floras will be a major source, but recourse to journals and herbaria is also likely to be necessary. Should the taxonomy of the group be difficult, then familiarity with the material is highly desirable.

Even when this sort of work has been done there may still be surprises. Harlan & Zohary (1966) plotted the distribution of *Hordeum spontaneum* in some detail, showing it at its farthest west in Libya. It was subsequently found to be growing in the Atlas mountains of Morocco (Molina-Cano *et al.*, 1982), apparently as a long-standing weed. This sort of problem will recur time and again with less well-known species, and any long-term collecting plans must be adjusted to meet such discoveries.

With distributions determined, the next step is to review what is already in collections to see if any areas can be considered to be adequately covered for the present. For crop relatives stored in many gene banks, this may be a laborious undertaking. Nonetheless it is important, for experience with both cultivars and wild species has often shown representation within collections to be uneven and that this has not been appreciated previously by crop experts. The time and money saved in fielding missions only when they are necessary make this a very economic exercise.

The areas outstanding for collecting may still be vast, sometimes to be measured on a continental scale, however, it is possible to suggest two criteria for giving some regions a higher priority.

1. *Interspecific diversity*: When there are several species with similar priorities it may be useful to consider them together. Fig. 16.1 shows the result of one such exercise for the nearer relatives of wheat in the genera *Aegilops* and *Triticum*. Although spread across the Mediterranean and southwest Asia, there is a clear concentration of species in the 'fertile crescent'. Within this region also occur all the wild *Triticum*

Fig. 16.1. The broad distribution of the wild relatives of wheat (genera *Triticum* and *Aegilops*). The outer dotted line marks the absolute limits for all species of these genera; the area bounded by triangles marks the general area in which 15 or more species can be found.

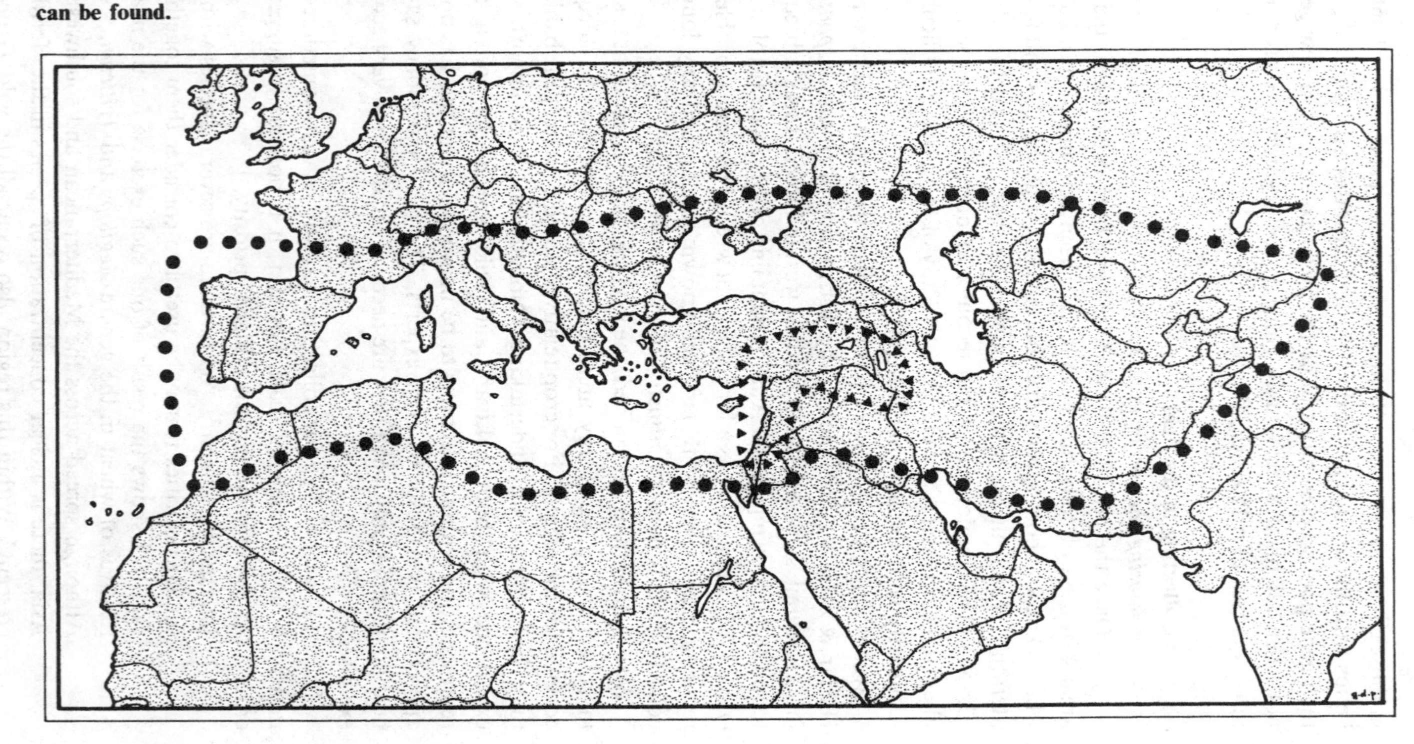

species and a part of the distribution of *Aegilops squarrosa*, the ancestral species of cultivated wheats. This region therefore, deserves a higher priority than regions to either the east or west (Chapman, 1985). Similar maps have been plotted for some other grass taxa (Hartley, 1961).

2. *Intraspecific diversity*: A second possibility for setting priorities by regions arises with weedy species which have spread with man. More will be said about the geographical distribution of variation below, but it does appear that in these cases diversity may be greatest in the region of the weed's origin. For example, *Lycopersicon esculentum* var. *cerasiforme* is most diverse in Ecuador and the Peruvian coast when compared to other regions of South America and the Old World (Rick & Fobes, 1975) and *Avena barbata* is more diverse in Israel than California (Kahler *et al.*, 1980). Thus, preference should go to collecting a species in those regions from which it appears to have originated.

Collecting within a target area

If alleles were dispersed at random among and within populations, or appeared to be so, then it would not be possible to make any recommendations about collecting practices. The amount of genetic variation secured would depend solely on the number of plants collected, regardless of how the work was done.

In fact, several studies demonstrate that the frequencies of many alleles vary significantly among populations (reviewed in Loveless & Hamrick, 1984). This has led Marshall & Brown (1975; 1981; 1983) to make two arbitrary yet useful divisions of alleles into those which are common versus rare, and those which are widespread versus local. They suggest a collecting strategy aimed at locally common alleles arguing that: widespread common alleles will inevitably be collected; widespread rare alleles will be included in relation to the total number of individuals regardless of strategies; and that locally rare alleles are more difficult to collect and not likely to be of interest to breeders (on this last point see below).

The basis of this strategy is, generally, to maximise the number of populations visited and not to spend an inordinate amount of effort collecting plants at each. Criticisms of this approach, basically that locally rare alleles will be overlooked, have been answered (Marshall & Brown, 1981) and it appears to be a robust strategy.

Choosing collecting sites

The next question concerns the choice of populations to be sampled. In some cases there appears to be no pattern to the variation among populations, as found by Hauptli & Jain (1984) for both wild and cultivated *Amaranthus* species within the Andean area. More commonly, and more usefully, some allele frequencies can be shown to vary with geography, e.g. clines in *Lycopersicon* and *Solanum* species on, the west coast of South America (Rick *et al.*, 1977; Rick & Tanksley, 1981) and/or with environmental factors, e.g., *Triticum dicoccoides* and *Hordeum spontaneum* is Israel (Nevo *et al.*, 1982; 1979). Such trends may be on a grand scale across the length and breadth of a country, or they may reflect changes that can occur in the space of a few metres such as waterlogging (Torres *et al.*, 1977), soil depth or shade (Nevo *et al.*, 1986*a*), or salt spray (Aston & Bradshaw, 1966). A further useful feature is that the most variable populations tend to be in the centre of a species' distribution, whether defined geographically or ecologically.

Lest these relationships be considered a panacea for guiding collectors, a number of points need to be made.

1. Different loci or suites of characters can show discordant patterns of variation. This has been reported in *Hordeum murinum* (Giles & Edwards, 1983; Giles, 1984), teosinte (Smith *et al.*, 1984) and *Lycopersicon pimpinellifolium* (Rick *et al.*, 1977). Collecting strategies designed around variation in one set of characters may be no better than random for others.
2. Although statistically significant, the proportion of the variation regarded as explicable by geographical and/or environmental association rarely exceeds 50 per cent. These higher correlations depend on particular multiple sets of factors and/or unusual environmental indices, e.g. mean day/night temperature differences (Nevo *et al.*, 1986*b* and references therein). To expect planning around the possibilities of such incomplete and/or subtle relationships is unreasonable. Should such relationships already be well documented and understood, it is probably an indicator that the target area is already adequately collected.
3. The relationships are invariably purely descriptive, and their adaptive significance unclear. McDonald (1983) cites three general problems in assessing whether a relationship is adaptive: (a) the necessary parameters for neutralist models may be imprecisely known; (b) hitchhiking responses from linkage of selected and observed loci; (c) observations in line

with a neutralist model never exclude a selectionist one. It follows that developing strategies based on one area or species for application to another is highly suspect. Only the most broadly based principles can be carried over in this way. Thus essentially parallel studies on wild barley from Israel and Turkey (Nevo *et al.*, 1979; 1986*d*) both show diversity at its maximum in the more favourable environments, but no significant correlation was common.

The prudent strategy remains, therefore, the one suggested by Marshall & Brown (1975): to collect from populations as well dispersed over space and environments as possible. If this is an early mission for a species then the best chance for gathering the maximum variability is to visit the apparent centre of its distribution, though a transect through it may give more information. Later missions should aim to cover the entire area in which it is found.

General route planning before a mission sets out can help ensure that diverse sites are likely to be visited. Temperature, rainfall, soil and vegetation maps adequate for this are likely to be found in most large libraries (e.g. World Atlas of Agriculture, 1969 *et seq.*). As far as is practical the entire range of any one variable should be visited, and in as many combinations as possible.

Within any one locality the selected sites should be as diverse as possible, based on what can be observed from the local topography, soils and vegetation. Marshall & Brown (1975) suggested that a clustered rather than dispersed sampling pattern of sites was more appropriate for wild species as it saved time and encouraged the search for diverse habitats. Certainly this is a practical approach for it enables collectors to work out of a centre for several days before packing up and moving on to the next.

Collecting at a site

At any one site Marshall & Brown (1975) recommend collecting 50 to 100 randomly selected plants with the objective of collecting all alleles with a frequency of five per cent or greater, i.e. those that are considered common. They argue against the selection of rare off-types as this is more time consuming, biases the collection against locally common alleles and may lead to the collection of diseased specimens.

For some species, such as wild cereals, the suggested 50 individuals per site may be collected in a matter of minutes. With others the local rarity of the plants, the labour involved in collecting them or in later handling, particularly from vegetative material, will make this an impractically large figure. Marshall & Brown (1983) suggest that in such circumstances as few as ten plants may constitute an adequate sample, which happens to be in line with the figures suggested by Yonezawa (1985). More alleles are likely to be collected by taking a few plants from many sites than many plants from a few sites.

Since micro-geographic differentiation does occur, plants from obviously different environments should be collected as separate samples, even if they seem to form part of one contiguous population. On the other hand, for any one sample, it is as well to disperse collecting within a site as much as practicable. Although pollen, seed and other ramets can be dispersed over long distances, most tend to be deposited near the parent plant so that plants near one another are quite likely to be related.

Digression on 'useful' variation

Some authors have suggested that certain classes of 'useful' variants be targeted in collecting, arguing that this makes for greater efficiency. Two broad categories seem to have been proposed, locally common alleles and types with important agronomic traits, e.g. disease resistance. Both appear attractive but respectively, are probably misleading and inefficient in the long run.

The argument for locally common alleles being useful is that, being common, they are therefore adaptive, whereas rare alleles are those at a selective disadvantage (Marshall & Brown, 1975). Be that as it may, Donald & Hamblin (1983) point out that much plant breeding success in recent years has depended on features which diminish competitiveness, e.g. dwarf stature, reduced tillering or branching, open canopy structure. Though of great interest to the breeder, alleles conferring these characteristics are likely to be rare in heterogenous populations. Even highly deleterious, and hence extremely rare, traits can be important for breeding, the major example perhaps being male sterility.

Conversely, common traits may not necessarily be advantageous to the fitness of a plant. Accessory and gametocidal chromosomes – and the alleles they carry – can reach high frequencies through their aberrant behaviour at or after meiosis (Jones & Rees, 1982; Endo, 1982). Though unadaptive, they are also of interest to breeders.

The point about locally common alleles is that, unlike other classes, an effective strategy can be devised to increase their chance of being collected, and they form a significant part of most species' genetic variation. It is far better, therefore, to target this class than to collect entirely at random.

Targeting important agronomic traits presents two difficulties. The first is that breeders' needs for some of these are ephemeral when compared to the time span from planning a mission to the release of a new variety based on the material collected. For example, two decades ago there was much interest in hybrid wheat, stimulating quests for sterilising cytoplasm and fertility restoration genes. Today, it is apparent that the yield advantages obtained do not justify the problems and costs of hybrid production, and less than one tenth of one per cent of the US wheat area was planted to hybrids in 1984 (Siegenthaler *et al.*, 1986). Additionally, plant variety rights legislation has removed one prime reasons for private industry to promote F_1 hybrids. The need to seek germplasm for this technology has largely disappeared.

Another difficulty is that some needs, though long-term, can lead to conflicting priorities. For example, in wheat *Triticum dicoccoides*, the maximum resistance to powdery mildew is to be found in central populations (Moseman *et al.*, 1984), whereas high protein is to be found in marginal areas (Nevo *et al.*, 1986*c*) both hot and cool. Thus useful characteristics may require different strategies (and missions) were they to be targeted separately.

Given the problems of reconciling today's breeding needs and satisfying those of tomorrow (ephemeral or not), the only wise course must be to maximise the genetic variation captured, whether it appears to be useful or not.

Conclusions

The broad strategy proposed for collecting wild crop relatives can be summarised as follows:

1. Choose the species to be collected. Give priority to those most easily utilised and, where necessary, those that are endangered.
2. Choose the target area(s) in which to collect, based on the species distribution(s). If the total area is too large, priorities may be set by considering where most species can be found or where species originated. If there is no better guideline, a transect through the centre of distribution may be the best option.
3. Collect from as diverse a set of sites, both geographically and ecologically, as possible.
4. Emphasise the number of sites at the expense of the number of individuals collected per site.

In considering the strategies suggested here, it is apparent that scientific planning is as much a part of organising the collecting of wild species as is the practical planning of organising finance, visas, vehicles, etc. Since fielding a mission is expensive and planning by comparison cheap, it is arguable that, unless a species is thought to be endangered, missions should be held in abeyance until comprehensive plans for crops or their relatives can be developed. If there are gaps in the information necessary for important decisions to be made, then the early missions should be organised to address these problems.

It is felt that this is not so much a job for committee as for an individual with time to do the necessary research in libraries and herbaria, and to gather information from experts and collections. Such planning will maximise the value of each mission and may thereby reduce the total quantity of material collected to the benefit of the quarantine organisations and collections which must handle it.

References

Aston, J. L. & Bradshaw, A. D. (1966). Evolution in closely adjacent plant populations. II *Agrostis stolonifera* in maritime habitats. *Heredity*, **24**, 349–62

Bothmer, R. von, Flink, J., Jacobsen, N. & Landström, T. (1983). Interspecific hybridization with cultivated barley (*Hordeum vulgare* L.). *Hereditas*, **99**, 219–44.

Brown, A. H. D. & Marshall, D. R. (1986). Wild species as genetic resources for plant breeding. In Plant Breeding Symposium DSIR, pp. 9–14, Williams, T. A. & Wratt, G. S. (eds.).

Chapman, C. G. D. (1985). *Genetic resources of Wheat. A Survey and Strategy for Collecting*. IBPGR, Rome.

Dewey, D. R. (1984). The genomic system of classification as a guide to intergeneric hybridization with the perennial Triticeae. In *Gene Manipulation and Plant Improvement*, pp. 209–79, Gustafson, J. P. (ed.).

Donald, C. M. & Hamblin, J. (1983). The convergent evolution of annual seed crops in agriculture. *Advances in Agronomy*, **36**, 97–143.

Endo, T. R. (1982). Gametocidal chromosomes of three *Aegilops* species in common wheat. *Canadian Journal of Genetics and Cytology*, **24**, 201–6.

Frey, K. J., Cox, T. S., Rodgers, D. M. & Brambel-Cox, P. (1984). Increasing yields with genes from wild and weedy species. In *Proceedings of the XV Congress of Genetics*, Volume IV, pp. 51–68, Chopra, V. L., Joshi, B. C., Sharma, R. P. & Bansal, H. C. (eds.).

Georgieva-Todorova, J. (1984). Interspecific hybridization in the genus *Helianthus* L. *Zeitschrift für Pflanzenzüchtung*, **93**, 265–79.

Giles, B. E. (1984). A comparison between quantitative and biochemical variation in the wild barley *Hordeum murinum*. *Evolution*, **38**, 34–41.

Giles, B. E. & Edwards, K. J. R. (1983). Quantitative variation within and between populations of the wild barley, *Hordeum murinum*. *Heredity*, **51**, 325–33.

Harlan, J. R. (1984). Evaluation of wild relatives of crop plants. In *Crop Genetic Resources: Conservation and Evaluation*, pp. 212–22, Holden, J. H. W. & Williams, J. T. (eds.), George Allen & Unwin, London.

Harlan, J. R. & de Wet, J. M. J. (1971). Toward a rational classification of cultivated plants. *Taxon*, **20**, 509–17.

Harlan, J. R. & Zohary, D. (1966). Distribution of wild wheats and barley. *Science*, **153**, 1074–180.

Hartley, W. (1961). Studies on the origin, evolution and distribution of the Gramineae IV. The genus *Poa*. *Australian Journal of Botany*, **9**, 152–61.

Hauptli, H. & Jain, S. (1984). Allozyme variation and evolutionary relationships of grain amaranths (*Amaranthus* spp.). *Theoretical and Applied Genetics*, **69**, 153–65.

Hawkes, J. G. (1977). The importance of wild germplasm in plant breeding. *Euphytica*, **26**, 615–21.

Iltis, H. H., Doebley, J. P., Guzman, R. M. & Pazy, B. (1979). *Zea diploperennis* (Gramineae): A new teosinte from Mexico. *Science*, **203**, 186–7.

Jones, R. N. & Rees, H. (1982). *B. Chromosomes*. Academic Press, New York.

Jusuf, M. & Pernes, J. (1985). Genetic variability of foxtail millet (*Setaria italica* P. Beauv.) Electrophoretic study of five isoenzyme systems. *Theoretical and Applied Genetics*, **71**, 384–91.

Kahler, A. L. & Allard, R. W. (1981). Worldwide patterns of genetic variation among four esterase loci in barley (*Hordeum vulgare* L.). *Theoretical and Applied Genetics*, **59**, 101–11.

Kahler, A. L., Allard, R. W., Krzakowa, M., Wehrhahn, C. F. & Nevo, E. (1980). Associations between isozyme phenotypes and environment in the slender wild oat (*Avena barbata*) in Israel. *Theoretical and Applied Genetics*, **56**, 31–47.

Ladizinsky, G. (1985). Founder effect in crop-plant evolution. *Economic Botany*, **39**, 191–9.

Loveless, M. D. & Hamrick, J. L. (1984). Ecological determinants of genetic structure in plant populations. *Annual Review of Ecology and Systematics*, **15**, 65–95.

Lyman, J. M. (1984). Progress and planning for germplasm conservation in major food crops. *Plant Genetics Resources Newsletter*, **60**, 3–21.

Marshall, D. R. & Brown, A. H. D. (1975). Optimum sampling strategies in genetic conservation. In *Genetic Resources for Today and Tomorrow*, pp. 53–80, Frankel, O. H. & Hawkes, J. G. (eds.), Cambridge University Press, Cambridge.

Marshall, D. R. & Brown, A. H. D. (1981). Wheat genetic resources. In *Wheat Science Today and Tomorrow*, pp. 21–40, Evans, L. T. & Peacock, W. J. (eds.), Cambridge University Press, Cambridge.

Marshall, D. R. & Brown, A. H. D. (1983). Theory of forage plant collection. *In Genetic Resources of Forage Plants*, pp. 135–8, McIvor, J. G. & Bray, R. A. (eds.), CSIRO, Melbourne.

McDonald, J. F. (1983). The molecular basis of adaptation: A critical review of relevant ideas and observations. *Annual Review of Ecology and Systematics*, **14**, 77–102.

McLeod, M. J., Eshbaugh, W. H. & Guttman, S. I. (1979). An electrophoretic study of *Capsicum* (Solanaceae): the purple flowered taxa. *Bulletin of the Torrey Botantical Club*, **106**, 316–33.

Molina-Cano, J. L., Gomez-Campo, C. & Conde, J. (1982). *Hordeum spontaneum* C. Koch as a weed of barley fields in Morocco. *Zeitschrift für Pflanzenzüchtung*, **88**, 161–7.

Moseman, J. G., Nevo, E., El Morshidy, M. A., & Zohary, D. (1984). Resistance of *Triticum dicoccoides* to infection with *Erysiphe graminis tritici*. *Euphytica*, **33**, 41–7.

Nevo, E., Beiles, A., Kaplan, D., Goldenberg, E. M., Osvig-Whittaker, L. & Naveh, Z. (1986*a*). Natural selection of allozyme polymorphisms: a microsite test revealing ecological differentiation in wild barley. *Evolution*, **40**, 13–20.

Nevo, E., Beiles, A., Kaplan, D. & Zohary, D. (1986*b*). Genetic resources of wild barley in the Near East: structure, evolution and application in breeding. *Biological Journal of the Linnean Society*, **27**, 355–80.

Nevo, E., Goldenberg, E., Beiles, A., Brown, A. H. D. & Zohary, D. (1982). Genetic diversity and environmental associations of wild wheat, *Triticum dicoccoides*, in Israel. *Theoretical and Applied Genetics*, **62**, 241–54.

Nevo, E., Grama, A., Beiles, A. & Goldenberg, E. M. (1986*c*). Resources of high-protein genotypes in wild wheat, *Triticum dicoccoides*, in Israel: predictive method by ecology and allozyme markers. *Genetica*, **68**, 215–27.

Nevo, E., Zohary, D., Brown, A. H. D. & Haber, M. (1979). Genetic diversity and environmental associations of wild barley, *Hordeum spontaneum*, in Israel. *Evolution*, **33**, 815–33.

Nevo, E., Zohary, D., Beiles, A., Kaplan, D. & Storch, N. (1986*d*). Genetic diversity and environmental associations of wild barley, *Hordeum spontaneum*, in Turkey. *Genetica*, **68**, 203–13.

Oliver, J. L. & Martinez Zapater J. M. (1984). Allozyme variability and phylogenetic relationships in the cultivated potato (*Solanum tuberosum*) and related species. *Plant Systematics and Evolution*, **148**, 1–18.

Pinkas, R., Zamir, D. & Ladizinsky, G. (1985). Allozyme divergence and evolution in the genus *Lens*. *Plant Systematics and Evolution*, **151**, 131–40.

Rick, C. M. & Fobes, J. F. (1975). Allozyme variation in the cultivated tomato and closely related species. *Bulletin of the Torrey Botanical Club*, **102**, 376–84.

Rick, C. M., Fobes, J. F. & Holle, M. (1977). Genetic variation in *Lycopersicon pimpinellifolium*: evidence of evolutionary change in mating systems. *Plant Systematics and Evolution*, **127**, 139–70.

Rick, C. M. & Tanksley, S. D. (1981). Genetic variation in *Solanum pennellii*: comparisons with two sympatric tomato species. *Plant Systematics and Evolution*, **139**, 11–46.

Second, G. (1982). Origin of the genetic diversity of cultivated rice (*Oryza* spp.): study of the polymorphisms scored at 40 isozyme loci. *Japanese Journal of Genetics*, **47**, 25–57.

Sharma, H. C. & Gill, B. S. (1983). Current status of wide hybridization in wheat. *Euphytica*, **32**, 17–31.

Siegenthaler, V. L., Stepanich, J. E. & Briggle, L. W. (1986). *Distribution of the Varieties and Classes of Wheat in the United States*. Estimates Division, Statistical Reporting Service, US Department of Agriculture. Statistical Bulletin.

Smartt, J. (1984). Gene pools in grain legumes. *Economic Botany*, **38**, 24–35.

Smartt, J. (1986). Exploitation of grain legumes. VI. The future – The exploitation of evolutionary knowledge. *Experimental Agriculture*, **22**, 39–58.

Smith, J. S. C., Goodman, M. M. & Stuber, C. W. (1984). Variation within teosinte III. Numerical analysis of allozyme data. *Economic Botany*, **38**, 97–113.

Smith, J. S. C., Goodman, M. M. & Stuber, C. W. (1985). Relationships between maize and teosinte of Mexico and Guatemala: Numerical analysis of allozyme data. *Economic Botany*, **39**, 12–24.

Stalker, H. T. (1980). Utilization of wild species for crop improvement. *Advances in Agronomy*, **33**, 111–47.

Thompson, T. E., Zimmerman, D. C. & Rogers, C. E. (1981). Wild *Helianthus* as a genetic resource. *Field Crops Research*, **4**, 333–43.

Torres, A. M., Didenhope, U. & Johnstone, I. N. (1977). The early allele of alcohol dehydrogenase in sunflower populations. *Journal of Heredity*, **68**, 11–16.

World Atlas of Agriculture (1969 *et seq.*). Edited by the Committee for the World Atlas of Agriculture. Novara Instituto Geografico de Agostini.

Yonezawa, K. (1985). A definition of the optimal allocation of effort in conservation of plant genetic resources, with application to sample size determination for field collection. *Euphytica*, **34**, 345–54.

Zamir, D., Navor, N. & Rudich, J. (1984). Enzyme polymorphism in *Citrullus lanatus* and *C. colocynthis* in Israel and Sinai. *Plant Systematics and Evolution*, **146**, 163–70.

17

Wild relatives as sources of disease resistance

J.J. BURDON AND A.M. JAROSZ

Introduction

Wild relatives of crop plants have proved to be fruitful sources of resistance genes for the control of serious crop diseases. Many examples of the extraction and utilisation of such genes have been documented in the literature and it is not our intention to repeat them here. On the other hand, less attention has been given to the fact that populations of wild relatives of crop plants can be viewed as representatives of a wide range of naturally co-evolved host–pathogen associations which, over long periods of time, have developed a variety of ways of ameliorating the worst effects of disease epidemics. These contrast with the relatively restricted range of disease control options that are widely used in agriculture. In the light of this difference, it is appropriate to consider whether wild relatives of crop plants may provide us with a source of novel strategies for disease control as well as of the traditionally used race-specific resistance genes. Should this be the case it will not only strongly influence the ways in which wild crop relatives are viewed by the scientific community in general but also the ways in which they are collected, stored and utilised as genetic resources. Initially to address this question we need to develop a basic understanding of the disease resistance mechanisms present in plant populations.

Disease resistance mechanisms in plants

A popular picture of disease resistance in agricultural crops is one of resistance that typically: (i) has major phenotypic effects; (ii) is controlled by single dominant genes; and (iii) is effective against only some races of a pathogen. Such resistance is generally seen as providing effective protection against a pathogen for only a short period of time. The same popular view of disease resistance mechanisms tends to see wild plant populations in an entirely different light. Such populations supposedly rarely suffer severe disease epidemics. This is assumed to result from the action of a battery of different resistance mechanisms particularly those that: (i) permit limited pathogen reproduction; (ii) are under the control of many genes that individually have very minor effects; and (iii) are expressed against all pathogen isolates.

Are these views a true reflection of the disease resistance mechanisms found in agricultural and wild plant populations? Do the perceived dissimilarities represent absolute differences in the basis of disease resistance or are they simply differences of degree? Moreover, if these differences are real, can approaches to breeding for disease control in agriculture be widened by their incorporation? In order to answer these questions it is necessary to consider just what disease resistance and reduction mechanisms occur in wild plants. That is, what are the plant-based factors which effect the duration, intensity and spatial scale of disease epidemics in plant populations?

Although this question has yet to be answered fully for most, if not all, plant–pathogen interactions, it is clear that these mechanisms may include those that operate at the level of the individual and those that operate at the level of the population. The former mechanisms have an immediate genetical basis while the latter result primarily from the demographic, spatial and temporal dynamics of wild plants and their pathogens. From the point of view of utilising wild relatives of crop plants as sources of resistance, many of these features, particularly those in the latter category, are inappropriate to a modern agricultural setting. However, recognition of their contribution to wild systems is extremely important in developing a realistic understanding of the level of disease control that may be expected when one or two other mechanisms are taken from such systems and placed in highly artificial agricultural environments.

Disease resistance at the level of the individual

Genetically based disease resistance in plants can be divided into two broad categories – passive and active. These reflect the degree of interaction occurring between host and pathogen. Passive resistance, encompasses mechanisms resulting from past interactions between host and pathogen. Through selection these have led to an avoidance of contact or to mechanisms whereby the detrimental effects of disease are at least partly absorbed by infected plants. Avoidance of contact may be achieved by either physiological or morphological means. The most obvious physiological characters are those that ensure that susceptible stages in the life cycle of the host do not occur at times of the year during which the incidence of the pathogen is high. Precocious germination, rapid 'hardening-off' of seedling tissues and variations in flowering time may all minimise contact between host and pathogen (Burdon, 1987*a*). Similarly, morphological characters like the closed-flower habit of many cereals may provide, for otherwise susceptible plants, protection against a range of pathogens.

By contrast, active disease resistance mechanisms occur as a result of continuing interactive responses between hosts and pathogens. In turn, these mechanisms may be further subdivided into resistance that is recognised by differential interactions between host and pathogen genotypes (race specific resistance) and that which is expressed generally against all pathogen genotypes (race non-specific resistance). (However, it should be recognized that some forms of resistance that are race non-specific are passive in nature.)

Race specific resistance (generally equivalent to 'vertical resistance' *sensu* van der Plank, 1963) is normally expressed as differences in the visual appearance of disease symptoms. In a susceptible host, a completely compatible interaction between host and pathogen results in disease symptoms characterised by large, profusely sporulating lesions. Resistant or incompatible reactions, on the other hand, may take on a variety of appearances. For example, a complete lack of macroscopically visible symptoms; non-sporulating lesions; or sporulating lesions with restricted fecundity. In general, race specific resistance has been found to be controlled by single, dominant genes that are inherited according to simple Mendelian principles. When two or more race specific resistance genes occur within the one plant they may function independently of one another or may interact (Day, 1974). Commonly, genes that condition higher levels of resistance are epistatic to those that condition lower levels of resistance. Race specific resistance may also be conditioned by resistance genes that are recessive or that show incomplete dominance. On average, however, less than 10 per cent of examples are of this type (Sidhu, 1975).

In contrast to race specific resistance, race non-specific resistance (generally equivalent to 'horizontal resistance' *sensu* van der Plank, 1963) may be expressed in a variety of ways and at various different stages in the development of the host. Features like (i) reductions in the frequency of successful host penetration; (ii) reductions in the survival rate of developing fungal colonies; (iii) increases in the length of the latent period; (iv) reductions in colony and lesion size; and (v) reductions in spore production and infectious period, are all manifestations of race non-specific resistance (Parlevliet, 1979). The genetic control of race non-specific resistance is poorly documented but it can be conditioned by one to many genes (Nelson, 1978). However, in the majority of cases it appears to be under polygenic control with each gene having a small phenotypic effect. Breeding experiments involving this type of resistance have commonly found F_2 segregating populations to display responses ranging continuously between, and sometimes beyond, those of the parental genotypes.

The procedures involved in the quantification of race specific and race non-specific resistance are very different. Race specific resistance can usually be determined on a single occasion by visually assessing the infection type response of plants to specific races of a pathogen. Typically, such determinations are made on seedling plants and the response thus detected is manifest throughout the life of the plant. Increases or decreases in race specific resistance may occur as the ontogenetic age of plants increases but such variations are comparatively rare. Quantification of race non-specific resistance, on the other hand, is severely complicated by its expression in features that require detailed microscopic examination (for example, establishment success) or that can only be determined by multiple monitoring (for example, latent and infectious periods).

Disease resistance at the population level

In contrast to resistance mechanisms that operate at the level of the individual, those that operate at the level of the population have, with the exception of mixtures and multilines, been ignored. Unlike crop plants, populations of most wild plants are patchily distributed through a habitat. Overall host plant density is low (relative to agricultural situations) and effectively this is further reduced by the patchy nature of plant distribution. Individual 'islands' of hosts are separated from one another by a heterogeneous 'sea' of non-hosts. Within patches, plant density may be reasonably high but, because patches may be quite small, each may contain only a few individuals. Between patches, the species of concern may be represented by few if any individuals. As a result, compared with a single uniform stand, the average patch may be expected to: (i) escape initial infection at the beginning of an epidemic for longer; (ii) fail to support the continued existence of the pathogen throughout the growing season; and (iii) sustain fewer pathogen individuals than would be present within a similar area in a uniform stand. Spatial heterogeneity and reduced plant density, especially when combined with genetic heterogeneity for resistance, will tend to impede the development of disease in a population as a whole.

Agricultural and wild plant populations also differ in a multitude of other ways, many of which inevitably have an effect on the occurrence of disease and its likely rate of increase. Differences in the age structure, nutritional status and phenology of plant populations all tend to affect disease development. Similarly, variations in the age structure of individual plant populations may lead to a lower *effective* density of susceptible plant material as variations in the ontogenetic age of plant parts may have a marked effect on their susceptibility to particular diseases. The overall effect of these features is often to reduce variations in disease levels in wild plant populations.

The relative importance of different types of resistance

What are the relative contributions that these forms of resistance make to disease control in agricultural and wild plant populations? The answer to this question, inevitably, depends on a number of factors including the extent of diversity in particular communities. Extensively managed rangelands and grazed pastures with the diversity of constituent species that many of them possess, may show many, if not all, of the resistance features found in undisturbed wild plant populations. Intensively managed cropping systems generally show considerably fewer although some of the control strategies practiced in such systems are only crude approximations to those occurring in wild systems, e.g. inter-field varietal or species diversification and intra-field strip-cropping (patchy distribution). Indeed, it is only in the use of multilines, varietal and inter-specific mixtures where attempts have been made in crop production systems to approximate the diversity of non-agricultural plant populations. In these systems, host density is effectively reduced by the simultaneous use of many host genotypes each carrying different combinations of race specific resistance genes.

Because of difficulties in their quantification, the relative importance of different forms of genetically based resistance is more difficult to assess. However, while race non-specific resistance is used in some crop-pathogen systems, race specific resistance dominates breeding for disease resistance in agriculture. Crops as diverse as wheat, flax, cotton, soybean, sunflower, lettuce and potato have all been protected from a variety of bacterial, fungal and viral pathogens through the use of race specific resistance characterised by single dominant genes.

An extreme view of this bias is perhaps found in the suggestion that gene-for-gene systems (typically characterised by differential interactions between hosts and pathogen and by resistance genes that have marked phenotypic effects) are artifacts of agriculture (Day, 1974; Day *et al*, 1983;

Barrett, 1985). By implication the claim is that disease resistance in natural host–pathogen systems is almost exclusively race non-specific.

Certainly race non-specific resistance is to be found in wild host–pathogen systems. However, so is race specific resistance! Broad-scale geographic surveys of individual species or more detailed studies of individual populations have shown the occurrence of both race specific and race non-specific resistance (e.g. *Hordeum spontaneum* challenged with *Erysiphe graminis*; Wahl *et al.*, 1978; Moseman *et al.*, 1984). Unfortunately, because of the difficulties associated with quantifying race non-specific resistance, such studies are severely limited in frequency and scale. In the case of race specific resistance, these limitations are far less severe and a range of studies encompassing a wide variety of host species has been carried out. These studies, some of which are used in the next section to illustrate aspects of the geographic distribution of disease resistance, show that race specific resistance is common in many wild host–pathogen systems. However, the relative contribution of these two forms of resistance to disease reductions in wild systems remains difficult to assess. It is likely to vary markedly between different host–pathogen associations and even between different populations of the same species.

Geographic distribution of disease resistance in plants

The interactions occurring between plants and their pathogens are quite unlike those occurring between plants and abiotic selective agents. In the latter situation plants are affected by the occurrence of the selective agent (for example, acidity, heavy metals, salinity) but they have little or no reciprocal effect on the selective agent. Host–pathogen interactions are, on the other hand, potentially very different. While pathogens may exert a marked selective effect on host populations, host populations in turn may have a marked effect on the size and genetic structure of their pathogens. The intensity of these reciprocal selective pressures is likely to show considerable temporal and spatial variation and to be affected by a wide range of factors.

At the most general level the suitability of the physical environment for the growth and development of the pathogen population is of ultimate importance in ensuring the possibility of an interaction between host and pathogen. On a geographic scale, environmental constraints are often important in limiting the range of a pathogen. However, as the geographic scale encompassed is reduced, a simple knowledge of gross climatic variables is not sufficient to provide an accurate and infallible predictive guide to the occurrence of disease resistance. Even within a pathogen's range, localised variation in the environment caused by a wide range of abiotic (for example, aspect, slope, soil type) and biotic (for example, associated species in the community) features can markedly affect the incidence and likely severity of disease epidemics and thus the selective pressure they may exert (Palti, 1981; Burdon, 1987*a*). Equally, differences in the ecology of host species (for example, whether they are weedy or non-weedy, annual or perennial) and their mating systems may influence both regional and local patterns of distribution of resistance.

Distribution of resistance on a continental scale

In searching for particular traits needed for the genetic improvement of crop plants, it has become axiomatic to concentrate collecting in areas within the primary and secondary centres of diversity of the species in question. Because of their antiquity these centres are generally regarded as the most likely area in which any particular trait may have arisen. In general, this argument is equally applicable to considerations of the geographic distribution of disease resistance. However, any tendency for resistance to be concentrated in a centre of origin not only reflects the antiquity of that centre (and hence the time available for a range of resistance genes to have arisen by mutation) but also the selective pressure exerted by the pathogen. Without this selective effect any appropriate resistance genes that may have arisen are likely to occur at extremely low frequencies and be virtually impossible to detect.

Indeed, the coincidence of host and pathogen distribution within collection areas is of fundamental importance. As crop species are continually refined and spread around the world and as modern transport spreads pathogens more widely, individual crop species will come into increasingly frequent contact with pathogens of generic or familial relatives. To some of these the crop may be completely susceptible. This possibility can be illustrated by reference to tobacco (*Nicotiana tabacum*) and the pathogen *Peronospora tabacina*, the causal agent of tobacco blue mould. Tobacco is believed to have evolved in South America (Gerstel, 1976), where *P. tabacina* does not occur naturally and resistance to this pathogen is unknown. *P. tabacina* is, however, native to Australia where it is part of the fungal flora associated with a number of native species of *Nicotiana*. Genes that confer resistance to specific races of *P. tabacina* have been found in several of these species (for example, *N. debneyi* and *N. goodspeedii*) and have subsequently been transferred to *N. tabacum* (Wark, 1970; 1975) and exploited in tobacco production.

Distribution of resistance on a regional scale

Even in regions where the distributions of hosts and pathogens overlap, disease resistance is not found uniformly in all populations of the host species. However, while the level of resistance possessed by adjacent populations may vary considerably, typically the overall frequency of resistant accessions is high in some areas and low in others. Such uneven regional distributions of resistance have been observed in a wide range of graminaceous species, for example the resistance shown by populations of *Avena sterilis* to *Puccinia coronata* (Dinoor, 1970), of *Hordeum spontaneum* to *Erysiphe graminis hordei* (Fischbeck *et al.*, 1976), and of *Triticum dicoccoides* to *E. graminis tritici* (Moseman *et al.*, 1984), in herbaceous, perennial species of *Glycine* (Burdon & Marshall, 1981) and *Phlox* (Jarosz, 1984) and in a range of gymnosperm trees (Hunt & van Sickle, 1984).

The actual causes of this patchy distribution have rarely been investigated in detail. However, in many cases, general parallels have been drawn between an increasing incidence of resistance and an apparently increasingly favourable environment for the development of particular diseases. Thus, for example, in Israeli populations of *Avena* and *Hordeum* the geographic distribution of resistance to *P. coronata* and *E. graminis hordei* respectively shows an association with rainfall patterns. Accessions derived from populations growing in the drier southern regions of the country are mainly susceptible while those growing in more mesic environments further north are more likely to be resistant (Dinoor, 1970; Fischbeck *et al.*, 1976). These perceptions have been supported by the studies of Nevo and his associates (1984), who found that the pattern of resistance in *Hordeum spontaneum* to *E. graminis hordei* was weakly correlated with gross climatic factors. Resistance was positively correlated with the number of dew nights and number of rainy days and inversely correlated with the number of days above 30 °C.

However, these same workers (Nevo *et al.*, 1985) found nearly the opposite pattern of resistance to *E. graminis tritici* in *Triticum dicoccoides*. It seems very unlikely that the two form species of *E. graminis* could have such widely different environmental requirements. One reason for this apparent conflict may lie in the small number of populations (10 to 15) that were used to describe the pattern of resistance. Correlations between resistance patterns and gross climatic features are usually weak because of two main problems inherent in the sampling procedure. First, the level of resistance within individual host populations is not expected to be static (Fig. 17.1). The pathogen will periodically enter a population and reduce the number or reproductive output of susceptible individuals. The net effect is that the population as a whole will appear to be more resistant after an epidemic (time 2) than before (time 1). In fact, the structure of resistance within crop multilines is known to change through a single growing season as a consequence of a pathogen epidemic (Murphy *et al.*, 1982). Secondly, climatic data based on fixed meteorological stations in the general vicinity of host populations can only provide a very generalised estimate of environmental suitability for pathogen infection. Localised variation in climate at the population site can either increase or decrease the propensity for pathogen infection relative to the average for the geographic area. This variation in both the host and local micro-climate reduces the precision and accuracy of any single population sample. To overcome this problem a greater number of populations must be examined in order to infer correctly any correlations with resistance.

Distribution of resistance on a local scale

Considerable variation may occur between adjacent populations in the frequency of resistant individuals. This was found in a survey of

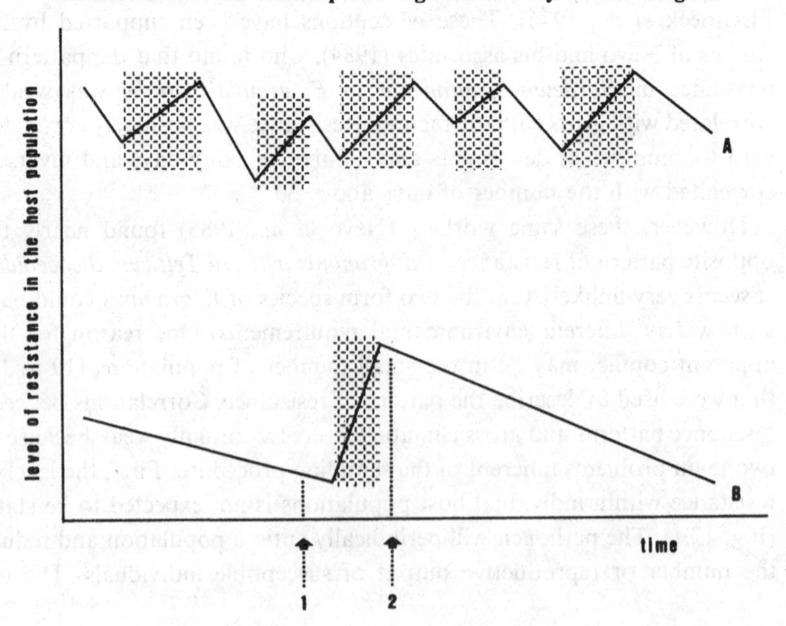

Fig. 17.1. Temporal change in the resistance structure of a population. Population A is in an environment that is near optimum for pathogen growth; population B is in a sub-optimum environment. Shading depicts the times during which the pathogen is present in the host population. The decrease in resistance in the absence of the pathogen is due to the assumed cost to the host of possessing 'unnecessary' resistance genes

more than 60 populations of *Phlox* spp. in the United States, where resistance to *Erysiphe cichoracearum* was best correlated with a habitat rating based on the aspect of the site and the amount of tree cover (Jarosz, 1984). Populations from shaded habitats tended to have greater levels of resistance than populations for exposed sites. Indeed, striking differences in the intensity of disease epidemics have been measured between populations of *Phlox* growing only 25 metres apart in shaded and exposed habitats. Differences on a similar spatial scale have also been noted in the level of resistance to *Erysiphe graminis hordei* found in plants of *Hordeum spontaneum* growing in the shade of trees or in the open (Dinoor & Eshed, 1984). The extent to which such local differentiation may occur has not been fully investigated but the lower limit would appear to be determined by the balance achieved between the rapidity with which selection pressures change across environmental gradients and the degree to which this is offset by gene flow.

The ultimate level of reduction of spatial scale that is possible, in the context of collecting and utilising wild relatives of crop plants as sources of disease resistance, is that of the individual population. Detailed studies of the resistance structure of individual populations are far less common than regional surveys but those available clearly show that even at this scale wild plant populations may be extremely variable. In a single Israeli population of *Avena* sp. challenged with six races of *Puccinia coronata*, Dinoor (1977) detected 18 different combinations of resistance and susceptibility. Some of these combinations occurred at relatively high frequencies (for example, 25 per cent of individuals were susceptible to all six races) but others were represented by single individuals only.

A similar diversity of response has been found in a study of the resistance of the Australian native legume *Glycine canescens* to the pathogen *Phakopsora pachyrhizi*. In that study the response of 22 different individuals to nine different races of the pathogen was determined. Twelve different multi-race resistance phenotypes were observed (Burdon, 1987*b*). Half of the individuals were resistant to all races of the pathogen, while others possessed resistance effective against from one to eight of the races. No individual was susceptible to all races. A detailed analysis of the genetic basis of these phenotypes has shown that any particular host individual carries one, two or, in some cases, three resistance genes, each conditioning a major phenotypic effect. This knowledge, combined with the observed phenotypic patterns, indicates that at least 12 different resistance genes are present in this one host population!

These examples should not be taken to imply, however, that all wild plant populations occurring in regions subject to marked pathogen activity will inevitably be variable in their resistance. This is not the case. Apparently uniformly resistant or susceptible host populations are also encountered in such environments in various host–pathogen combinations (for example *Avena* spp. – *Puccinia coronata*; Burdon *et al.*, 1983). Potentially this may reflect highly localised total absences of pathogen activity (susceptible populations); localised intense pathogen activity (resistant populations); or, perhaps more likely, 'founder effects' from the establishment of host populations from very limited initial seed inputs.

Guidelines for collecting disease resistance in wild relatives

Diversity is one of the most obvious features of most host–pathogen interactions regardless of the spatial scale of the association. At the macrogeographic scale, the causes of variation in the incidence of disease resistance are generally found in physical environmental factors that lead to the total or virtual exclusion of particular pathogens from given areas. As the spatial scale is reduced, the influence of these factors becomes modified by local topographic features and by aspects of the biology of individual host species. The effect of the latter aspects are easily overlooked and yet are undoubtedly of considerable importance in determining the final pattern of disease resistance within and between host populations. In this respect, differences in the life span of individuals (annual versus perennial), their ecological niche (colonisers versus non-colonisers) and their breeding system (inbreeding versus out-crossing) may all be important. As a consequence of these biotic effects it is very unlikely that any single approach to the sampling of wild plant populations will maximise the collection of disease resistance from all host–pathogen associations. Clearly, the intensity of collecting effort needed to obtain a comprehensive coverage of resistance must be affected by these factors and local habitat variations that affect the intensity of selection from one local population to another. Such increasing differentiation will be offset by features of plants like seed movement and pollen flow, and features of pathogens like efficient spore dispersal. Both these groups of features will tend to favour greater local similarities in host populations.

Despite the complications introduced by the numerous variables that affect the development and maintenance of resistance in wild plant populations, a generalised conservative strategy can be adopted that concentrates collecting efforts in areas where the environment is generally favourable for the growth and development of the appropriate pathogen (Harlan, 1977). These areas may represent part of the ancestral home of the species in question (for example, the earlier cited examples of the graminaceous genera *Avena*, *Hordeum* and *Triticum* in northern Israel) or may be in new environments where both host and pathogen have been introduced. The latter situation is exemplified by the distribution of resistance to *Puccinia coronata* in introduced weedy populations of *Avena* spp. occurring in New South Wales, Australia (Burdon *et al.*, 1983). Populations of *Avena* spp. growing in the northern half of the State were found to possess both a greater mean level of resistance and a greater overall diversity of resistant reaction types than those occurring further south. This reflects environmental conditions that allow a more prolonged and intense interaction between these hosts and their pathogen in northern New South Wales. Indeed, examination of the pathogen populations in these two areas shows a similar differentiation (Oates *et al.*, 1983). During the period 1975–79, the *P. coronata* population occurring in the north contained a significantly greater number of races. In addition the average pathogen isolate in that area was able to attack a wider range of host genotypes than isolates from the south.

For two reasons, however, a concentration of collecting effort on environmentally favourable areas should not be allowed to occur to the total exclusion of all other collecting. First, relatively few plant pathogens have precisely the same environmental optima and collection strategies that are designed specifically around the optimum environmental conditions of one pathogen species may well fail to cover those of other potentially important pathogens. Second, not all sources of disease resistance are restricted to host populations growing within the normally observed distributional range of the appropriate pathogen. Thus although most *Avena* populations growing on the edge of the Negev desert in southern Israel are susceptible to *Puccinia coronata*, some populations contain some resistant individuals (Dinoor, 1970; Wahl, 1970).

A more extreme case of this phenomenon is found in the distribution of resistance to *Phakopsora pachyrhizi* in wild *Glycine* species in Australia. At the present time, this pathogen is restricted to the mesic coastal regions of northern and eastern Australia and conventional logic would suggest that the appropriate resistance should be concentrated in species of *Glycine* growing in those areas. Conversely, other species like *G. canescens* that are restricted to the drier inland regions of the continent should be poor sources of resistance. As expected, species of *Glycine* growing in coastal regions are rich sources of resistance to *P. pachyrhizi* but, unexpectedly, so is *G. canescens*. Analysis of the genetic basis of disease resistance in six lines of *G. canescens* used to distinguish races of *P. pachyrhizi* have identified seven different resistance genes or alleles (Burdon & Speer, 1984; Burdon, 1986), while in the population study mentioned above, at least 12 different resistance genes appear to be present. Clearly a total concentration of collecting efforts in areas possessing environmental conditions particularly suitable for the development of *P. pachyrhizi* would have failed to detect, let alone exploit, the resistance present in *G. canescens*.

Indeed, it is possible to develop an argument which would support particular attention being given to areas which appear, on environmental grounds, to be marginal. The co-evolutionary interactions that are believed to lie at the heart of most host–pathogen interactions have been the subject of much theoretical speculation. While a variety of models have been developed (for example, Haldane & Jayakar, 1962; Jayakar 1970; Leonard, 1977), all recognise that balanced polymorphisms between resistance and susceptibility in the host, and virulence and avirulence in the pathogen (systems believed common in many wild host–pathogen interactions; Leonard, 1985) are only possible through the occurrence of fitness penalties that counterbalance the tendency of resistant host, and virulent pathogen, genotypes to dominate. If such fitness penalties are generally associated with resistance genes (and it should be noted that there is considerable controversy on this point alone, then it can be argued that resistance genes that are present at appreciable frequencies in areas where the appropriate pathogen no longer exerts a selective effect are likely to carry negligible fitness penalties. Those with higher cost would presumably have been reduced in the course of time to very low frequencies. In contrast, in areas where a pathogen continues to exert a strong selective pressure these costs would be hidden by the advantages conferred by resistance. As a result both 'expensive' and 'economical' resistance genes might be expected to be present.

Further light would be thrown on this intriguing possibility if detailed data were available concerning the resistance structure of populations of annual species that were completely or largely out-crossing. In such systems, selection against 'unnecessary' resistance would operate at the level of the individual gene. This would be a marked contrast to tightly inbreeding species like *G. canescens* where selection inevitably operates at the level of the phenotype of the whole plant. There, so long as resistance genes occur embedded in a background of average or above average fitness, their 'cost' will remain hidden.

This section must end on a cautionary note. In two recent papers, Nevo and his colleagues (Nevo *et al.*, 1984, 1985) have strongly argued the merits of using electrophoretic markers to detect resistance to *Erysiphe graminis* in *Hordeum spontaneum* and *Triticum dicoccoides*. Indeed, they propose that the use of electrophoretic markers will be more efficient than the methods currently used to detect disease resistance as they will: (i) 'identify multi-locus genotypes that are resistant to multiple diseases and are also agronomically superb', and (ii) 'may help us to discover new genes and/or alleles of resistance'. Some of these claims for the powers of electrophoresis appear to be quite extraordinary. Even should they be usable as a means of detecting resistance in the species and at the sites examined, it is important to realise that these are purely fortuitous associations resulting largely from the close inbreeding nature of both *H. spontaneum* and *T. dioccoides*. Furthermore, such associations only become apparent *a posteriori*. In many other host–pathogen combinations no consistent association between electrophoretic markers and disease resistance are likely to be detected (for example, resistance to *Phakopsora pachyrhizi* in *Glycine canescens*).

Future approaches to breeding for disease control

Over the past 50 years the use of simply inherited race specific resistance genes has been the mainstay of strategies for disease control in many agricultural crops. Certainly during that time the deployment of these genes has become more and more sophisticated so that today up to six different resistance genes may be incorporated into any one cereal variety. However, experience has shown that most pathogens are capable of responding to the evolutionary challenge implicit in such defences and eventually will be successful in overcoming them. In order to retard this cycle we must attempt to diversify the ways in which we protect our crops. Tentative steps have been taken in this direction through the increasing acceptance of varietal mixtures and the use of race non-specific resistances that reduce the rate of development of the pathogen population. It is vital that this trend continues and it is in this context that wild populations of crop relatives are likely to make two important contributions to disease control in agriculture in the twenty-first century.

Most immediately, wild relatives of crop plants will continue to provide sources of race specific resistance to protect agricultural crops in the short-term. Because of its relative ease of manipulation, this type of resistance is likely to retain, for the foreseeable future, a very important position in the plant breeders' armoury. In contrast, the use of wild relatives as a source of race non-specific resistance is likely to be very restricted despite the significant contribution it undoubtedly makes to disease reduction in wild plant populations. Because of its mode of expression, its multigenic, additive nature and the minor phenotypic effects that individual genes confer, such resistance is generally difficult to manipulate even in agronomically uniform backgrounds. In wide and often difficult crosses between crop plants and their wild relatives, such finely balanced combinations of genes are likely to be irrevocably broken up.

Second, and perhaps more excitingly, wild relatives may provide us with the insight needed to develop new and/or more reliable disease control strategies. Simultaneous studies of populations of wild relatives and their pathogens offer the possibility of the development of a better understanding of the reciprocal nature of plant–pathogen interactions. While such information is not important to the *ex situ* use of particular resistance genes, it is likely to provide a much better understanding of why and how resistance genes are distributed within and between plant populations and why and how the virulence structure of pathogen populations responds to selective pressures exerted by their hosts. With this knowledge we should be in a position to provide more reliable long-term protection for agricultural crops by developing practices that integrate a much broader range of resistance mechanisms than are now utilised.

References

Barrett, J. A. (1985). The gene-for-gene hypothesis: parable or paradigm. In *Ecology and Genetics of Host–Parasite Interactions*, pp. 215–25. Rollinson, D. (ed.), Academic Press, London.

Burdon, J. J. (1986). The potential of Australian native *Glycine* species as sources of resistance to soybean rust (*Phakopsora pachyrhizi*). In *New Frontiers in Breeding Researches*, pp. 823–32. Napompeth, B. & Subhadrabandhu, S. (eds.), Kasetsart University, Thailand.

Burdon, J. J. (1987*a*). *Diseases and the Population Biology of Plants.* Cambridge University Press, Cambridge.

Burdon, J. J. (1987*b*). Phenotypic and genetic patterns of resistance to the pathogen *Phakopsora pachyrhizi* in populations of *Glycine canescens*. *Oecologia (Berlin)*, **73**, 257–67.

Burdon, J. J. & Marshall, D. R. (1981). Inter- and intra-specific diversity in the disease response of *Glycine* species to the leaf-rust fungus *Phakopsora pachyrhizi*. *Journal of Ecology*, **69**, 381–90.

Burdon, J. J., Oates, J. D. & Marshall, D. R. (1983). Interactions between *Avena* and *Puccinia* species. 1. The wild hosts: *Avena barbata* Pott ex Link, *A. fatua* L. and *A. ludoviciana* Durieu. *Journal of Applied Ecology*, **20**, 571–84.

Burdon, J. J. & Speer, S. S. (1984). A set of differential *Glycine* hosts for the identification of *Phakopsora pachyrhizi*. *Euphytica*, **33**, 891–6.

Day, P. R. (1974). *Genetics of Host–Parasitic Interaction.* W.H. Freeman, San Francisco.

Day, P. R., Barrett, J. A. & Wolfe, M. S. (1983). The evolution of host–parasite interaction. In *Genetic Engineering in Plants*, Kosuge, T., Meredith, C. P. & Hollaender, A. (eds.), pp. 419–30. Plenum Press, New York.

Dinoor, A. (1970). Sources of oat crown rust resistance in hexaploid and tetraploid wild oats in Israel. *Canadian Journal of Botany*, **48**, 153–61.

Dinoor, A. (1977). Oat crown rust resistance in Israel. *Annals of the New York Academy of Sciences*, **287**, 357–66.

Dinoor, A. & Eshed, N. (1984). The role and importance of pathogens in natural plant communities. *Annual Review of Phytopathology*, **22**, 443–66.

Fischbeck, G., Schwarzbach, E., Sobel, Z. & Wahl, I. (1976). Types of protection against barley powdery mildew in Germany and Israel selected from *Hordeum spontaneum*. *Proceedings, Third International Barley Genetics Symposium, Barley Genetics*, pp. 412–17. Garching, 1975.

Gerstel, D. U. (1976). Tobacco, *Nicotiana tabacum* (Solanaceae). In *Evolution of Crop Plants*, Simmonds, N. W. (ed.), pp. 273–7. Longmans, London.

Haldane, J. B. S. & Jayakar, S. D. (1962). Polymorphism due to selection of varying direction. *Journal of Genetics*, **58**, 237–42.

Harlan, J. R. (1977). Sources of genetic defense. *Annals of the New York Academy of Sciences*, **287**, 343–56.

Hunt, R. S. & Van Sickle, G. A. (1984). Variation in susceptibility to sweet fern rust among *Pinus contorta* and *P. banksiana*. *Canadian Journal of Forestry Research*, **14**, 672–5.

Jarosz, A. M. (1984). *Ecological and evolutionary dynamics of* Phlox–Erysiphe cichoracearum *interactions*. PhD thesis, Purdue University.

Jayakar, S. D. (1970). A mathematical model for interaction of gene frequencies in a parasite and its host. *Theoretical Population Biology*, **1**, 140–64.

Leonard, K. J. (1977). Selection pressures and plant pathogens. *Annals of the New York Academy of Sciences*, **287**, 207–22.

Leonard, K. J. (1985). Population genetics of gene-for-gene interactions between plant host resistance and pathogen virulence. In *Genetics: New Frontiers*, Proceedings XV International Congress of Genetics, New Delhi, 1983.

Moseman, J. G., Nevo, E., El Morshidy, M. A. & Zohary, D. (1984). Resistance of *Triticum dicoccoides* to infection with *Erysiphe graminis tritici*. *Euphytica*, **33**, 41–7.

Murphy, J. P., Helsel, D. B., Elliott, A., Thro, A. M. & Frey, F. J. (1982). Compositional stability of an oat multiline. *Euphytica*, **31**, 33–40.

Nelson, R. R. (1978). Genetics of horizontal resistance to plant disease. *Annual Review of Phytopathology*, **16**, 259–78.

Nevo, E., Moseman, J. G., Beiles, A. & Zohary, D. (1984). Correlation of ecological factors and allozymic variations with resistance to *Erysiphe graminis hordei* in *Hordeum spontaneum* in Israel: Patterns and application. *Plant Systematics and Evolution*, **145**, 79–96.

Nevo, E., Moseman, J. G., Beiles, A. & Zohary, D. (1985). Patterns of resistance of Israeli wild emmer wheat to pathogens. I. Predictive method by ecology and allozyme genotypes for powdery mildew and leaf rust. *Genetica*, **67**, 209–22.

Oates, J. D., Burdon, J. J. & Brouwer, J. B. (1983). Interactions between *Avena* and *Puccinia* species. II. The pathogens: *Puccinia coronata* Cda and *P. graminis* Pers. f. sp. *avenae* Eriks. & Henn. *Journal of Applied Ecology*, **20**, 585–96.

Palti, J. (1981). *Cultural Practices and Infectious Crop Diseases*. Springer-Verlag, Berlin.

Parlevliet, J. E. (1979). Components of resistance that reduce the rate of epidemic development. *Annual Review of Phytopathology*, **17**, 203–22.

Sidhu, G. S. (1975). Gene-for-gene relationships in plant parasitic systems. *Science Progress (Oxford)*, **62**, 467–85.

van der Plank, J. E. (1963). *Plant Diseases: Epidemics and Control.* Academic Press, London.

Wahl, I. (1970). Prevalence and geographic distribution of resistance to crown rust in *Avena sterilis. Phytopathology*, **60**, 746–9.

Wahl, I., Eshed, N., Segal, A. & Sobel, Z. (1978). Significance of wild relatives of small grains and other wild grasses in cereal powdery mildews. In *The Powdery Mildews*, Spencer, D. M. (ed.), pp. 83–100. Academic Press, London.

Wark, D. C. (1970). Development of flue-cured tobacco cultivars resistant to a common strain of blue mould. *Tobacco Science*, **15**, 147–50.

Wark, D. C. (1975). The development of blue mould resistant cultivars of tobacco in Australia. *CSIRO (Aust.) Division of Plant Industry Annual Report*, pp. 31–3.

18

Ecological and genetic considerations in collecting and using wild relatives

G. LADIZINSKY

Plant breeders apparently will never use wild species if they can find the genetic diversity they need in the cultivated germplasm. Indeed, exploitation of the wild gene pool is not yet a common practice in breeding programmes, but this might be changed in the future. It is therefore, desirable to identify and to assess the major steps of this approach and to pinpoint problems and difficulties that might be associated with each phase.

Rational exploitation of wild relatives is based on, and conditioned by the following: 1) identification of the wild gene pool of the crop, 2) availability of sufficient material for screening and evaluation, 3) the use of appropriate methods for gene transfer.

Identification

Morphological similarity is the basis for the hierarchical arrangement of taxonomy. A taxon (taxonomic entity) is defined by a specific morphological trait or combination of traits; it is therefore not uncommon that taxonomic treatments by different authors yield different numbers of taxa, names and synonyms. The morphological diversity of cultivated plants and their wild relatives have been treated accordingly by classical taxonomists. The breeder, on the other hand, is more interested in the genetic affinities between the crop plant and its wild relatives and the extent to which their genetic diversity can be exploited. Seen in these terms, the wild relatives could be arranged in two groups: exploitable and unexploitable. The exploitable group includes the wild progenitor of the crop, and wild and weedy forms, which Harlan & de Wet (1971) denoted as the primary genepool of the crop. This group also includes the more distantly related but cross-compatible wild species whose hybrids with the cultigen are only partly fertile (the secondary gene pool, according to

Harlan and de Wet terminology). The unexploitable species correspond to the Harlan and de Wet tertiary gene pool and are of no direct value to the breeder, as long as methods for gene transfer from them are lacking.

Availability of wild relatives to the breeder

Existing collections

The aim of collecting wild relatives is to obtain genetic diversity which is absent from the cultivated germplasm. Over the years the wild gene pool has been an important source, not only of disease and pest resistance genes but also of characteristics such as stress tolerance, protein content and even increased total yield (Chang, 1985; Harlan, 1976; Stalker, 1980). The lack of genetic diversity in certain traits of the cultivated germplasm compared with that in the wild relatives, may stem from the 'founder effect' associated with domestication, low rate of gene flow from wild to cultivated populations and historical bottlenecks (Ladizinsky, 1985). Barriers to gene flow suggest further that wild species of the secondary gene pool possess genetic diversity not found in the cultigen. For example, wild tetraploid oats *Avena magna* and *A. murphyi* are rich in protein, containing about 30 per cent and 27 per cent respectively (Ladizinsky & Fainstein, 1977; Murphy *et al.*, 1968), while in superior hexaploid oat cultivars the protein content is only about 20 per cent. The green fruited wild tomato, *Lycopersicon minutum*, is noted for its high content of soluble solids (10–11 per cent), compared with 5.5 per cent in the main cultivar used for canning tomato in California (Rick, 1973). Wheatgrass, *Elytrigia pontica*, has an exceptionally high salt tolerance. While cultivated wheat seedlings do not live as long as 26 days at a concentration of 25 mM NaCl, all tested accessions of *E. pontica* showed high survival rates after 26 days at 750 mM NaCl (McGuire & Dvorak, 1981).

Plant breeders rarely collect wild relatives for use in breeding programmes. Instead, they rely on existing collections for screening and utilisation. Such collections presently contain only a limited number of accessions of most of the exploitable wild species and thus they hardly represent the main habitats or the distribution range of the wild relatives. Consequently, no ecological considerations can be applied in selecting specific accessions for screening. When large collections are available, screening for useful traits is sometimes more efficient if the origin and the habitat of the various accessions are taken into account. For example, germination and survival rates of *Lycopersicon hirsutum* seedlings at low temperature were found to be positively correlated with altitudinal origin of the accessions (Patherson *et al.*, 1978). Adaptation of *L. hirsutum* to low temperature is evident also by quicker pollen germination and pollen tube growth at 6–12 °C comparing the cultivated tomato (Zamir *et al.*, 1981). Resistance to barley yellow dwarf virus is controlled by a single gene (Rasmusson & Schaller, 1959). Highly resistant types in barley worldwide collections were found almost exclusively from Ethiopia and were dependent on altitude: in populations found below 7,000 feet only 3.8 per cent of the plants were resistant; 64.7 per cent were resistant in populations above 12,000 feet (Qualset, 1975). It has been suggested that higher altitudes are more favourable for the virus and the aphid vector.

Usually, however, the association between ecological conditions and genetic diversity is not obvious, possibly because of lack of knowledge or as a result of the adaptive neutrality of the trait in question. The systematic screening of the entire collections may then be required.

Considerations for obtaining additional collections

The potential of wild relatives as genetic resources can better be assessed when the collections available for screening represent the entire distributional range and the ecological niches where the wild species grow. Unfortunately this ecogeographic information is inadequate for the wild relatives of most crop plants. It can be said further that this is the main reason why in existing collections in gene banks, wild relatives are under-represented compared with cultivated germplasm. Wild relatives are more difficult to collect because they show specific and often restricted ecogeographic distributions. The wild oat *Avena canariensis* for example, is endemic to Lanzarote, Fuerteventura and Tanerife of the Canary Islands, and another wild oat, *Avena murphyi*, is known only from the southern tip of Spain and a small area near Tangier. In their homeland, wild relatives can be ecologically flexible or restricted to specific habitats and to undergo periodic cycles of abundance. *A. canariensis* grows in diverse habitats such as stony hillsides, the edges of roads and as a weed in abandoned fields. On the other hand, *A. murphyi* grows only in undisturbed habitats or grazing land on heavy alluvial soil. Another wild oat, *Avena longiglumis* is a psamophyte and is restricted to stabilised sand fields or sandy loam soils. In these particular habitats, it is found in the coastal belt of the Mediterranean sea and in adjacent areas. The ecological preferences of this species are so specific and amplitude is so narrow, that it is possible to predict its occurrence in specific areas according to the soil type. This exemplifies the ecological basis of the geographical distribution of many wild plants and the need of a combined knowledge of geology, pedology, phytogeography and plant communities for detecting wild relatives in their ecological niche. This ecological specificity is a major reason why wild relatives are more difficult to collect than cultivated species. Furthermore, many of the wild relatives are inconspicuous in nature, they grow in small populations and in mixed stands with other wild species. It is not uncommon that only tiny morphological differences separate the exploitable from the unexploitable species, and the two can easily be confused by an inexperienced collector.

Utilisation

An essential feature of exploiting wild species is the need for a pre-breeding phase during which desirable traits are transferred to a cultivated background and undesirable traits that have been contributed by the wild parent are removed. The length and complexity of this phase are conditioned by the genetic affinities and the nature of reproductive barriers between the crop plant and the wild species. The crossability of the parents, the fertility of the F_1 hybrids and hybrid derivatives, and linkage between desirable and undesirable traits, are the major factors determining the length of the pre-breeding phase. Barriers to gene transfer and methods to overcome them have been discussed in several reviews (Knott & Dvorak, 1976; Stalker, 1980) and only some points will be outlined.

Utilisation of the wild progenitor represents the simplest approach to gene transfer because, from the genetic point of view, the crop plant and its wild progenitor are members of the same species. They share the same chromosome number and arrangement, are fully inter-fertile and no special methods for gene transfer are required. Exploitation of more distantly related wild species occasionally depends upon sophisticated, time and labour intensive techniques aimed at improved crossability, enhancing hybrid fertility and inducing higher rates of recombination.

Improved crossability

More pollinations may be needed in order to obtain hybrid seeds with distantly related species than with the wild parent. As an example, the rate of hybrid seeds obtained per pollination is 0.07–0.32 when chickpea was crossed with its wild progenitor *Cicer reticulatum*, and only 0.004 when crossed with the more distantly related species *C. echinospermum* (Ladizinsky & Adler, 1976). When a large number of emasculations and pollinations are necessary to obtain hybrid seeds, it can be facilitated by using male sterile lines, provided that the pollen parent, usually the wild one, possesses a gene to restore fertility.

Lack of crossability between crop plant and its wild relative is occasionally unilateral. Both in pea (Ben Ze'ev & Zohary, 1973) and tomato (Rick, 1979), for example, hybrid seeds in specific combinations were obtained only when the pistillate parent was the crop plant. When the wild plant is the pistillate parent, the offspring might contain the wild parent cytoplasm capable of affecting various traits and specifically self-fertility (Frankel & Galun, 1977; Grun & Aubertin, 1965). Crossing the crop plant with a wild relative sometimes can be made only after doubling the chromosome number of one parent (Savitsky, 1975).

The breakdown of the hybrid embryo often acts as a barrier to interspecific hybridisation. The reasons for such a breakdown are yet poorly understood. When it results from endosperm breakdown or incompatibility between the embryo and the maternal tissue, explanting the fertilised ovule or the immature embryo in nutrient medium might allow the hybrid to develop normally. Embryo culture techniques have been successfully applied in interspecific crosses in *Cucurbita* (Wall, 1954), oats (Thomas & Rajhathy, 1967), *Phaseolus* (Braak & Kooistra, 1975) and lentil (Ladizinsky *et al.*, 1985). The application of embryo culture technique in wide crosses is limited by the specific nutrient requirements of different crops which are not always known and by the need to transplant fertilised ovules at an early stage when most of the common nutrient media have been proved inadequate.

Increased seed yield of interspecific hybrids

Low fertility is typical of interspecific hybrids. When production of hybrids is difficult because of strong reproductive barriers, vegetative propagation of the existing hybrid, by cuttings or top shoot culture, is the simplest way to obtain more plants (Braak & Kooistra, 1975).

Irregular chromosome pairing and segregation at meiosis are the main reasons for low fertility in interspecific hybrids, since they result in unbalanced gametes that usually abort. Low gamete viability occasionally leads to complete male sterility. In grasses, for example, pollen viability below 10 per cent usually fails to induce bursting of the anthers and hence the shedding of pollen. Even with low gamete viability, however, interspecific hybrids are still partially female fertile and can produce some seeds upon backcrossing with parental pollen. In wind-pollinated plants this backcross procedure can be facilitated by planting the hybrids between plants of the pollen donor (Ladizinsky & Fainstein, 1977; Vardi & Zohary, 1967).

When the F_1 hybrid is totally sterile and backcrossing is impractical, doubling of the chromosome number seems the only way to restore fertility. In many interspecific crosses the production of such amphiploids is an essential step in the backcrossing procedure (Johnson, 1974; Meyer, 1957; Sears, 1956; Thomas *et al.*, 1975).

Enhancement of recombination

Conventional methods for gene transfer are based on meiotic crossing over which follows chromosome pairing. The extent of chromosome pairing in interspecific hybrids is determined by the homology of the two chromosome sets. Differences, both gross and cryptic, in chromosome arrangement, are the major cause of reduced homology, but the lack of homology can also be genetically controlled. Reduced homology entails not only irregular chromosome association at meiosis, but also lower chance of crossing over and recombination. It is common that viable gametes of interspecific hybrids represent mainly the parental genomes and consequently cause significant deviation from expected segregation ratio in the offspring. Thus, besides being associated with sterility, hybridisation with distantly related species also requires larger populations of offspring to recover desirable combinations. Homology between sugarbeet, *Beta vulgare*, chromosomes and *Beta procumbens* chromosomes, which carry the gene for nematode resistance, is very low. In the addition line of sugarbeet containing that chromosome, 18 bivalents and one univalent were observed at metaphase. Following extensive backcrossing with pollen of diploid sugarbeet a single diploid nematode-resistant plant was recovered out of 8,834 progeny (Savitsky, 1975).

Irradiation to induce translocation and hence recombination with alien chromosomes was first attempted in wheat (Sears, 1956). *Aegilops umbellulata* ($2n = 14$) is resistant to leaf rust but this trait could not be transferred directly to bread wheat ($2n = 42$) since no viable hybrid seeds were produced. *A. umbellulata*, therefore, was crossed with tetraploid wheat ($2n = 28$) and the chromosome number of the F_1 hybrid was doubled to form an amphidiploid, which was then backcrossed to hexaploid wheat. It was then possible to select a leaf rust resistant plant that possessed the bread wheat chromosomes and an additional *A. umbellulata* iso-chromosome which carried the rust resistant gene. Following irradiation of the pollen of this plant prior to anthesis, translocation occurred between the wheat and the *Aegilops* chromosomes and among the progeny, resistant hexaploid plants were selected. Induced translocation was also utilised in transferring mildew resistance from *Avena barbata* ($2n = 28$) to the cultivated oat ($2n = 42$) (Aung *et al.*, 1977).

Genetic control of bivalent pairing has been demonstrated in several polyploid plants, but is best known in wheat where the Ph gene forces chromosomes to pair with their homologues and not with their homoeologues. This gene stabilises pairing at meiosis while at the same time preventing introgression from related species (see Sears, 1976 for more details). *Aegilops speltoides* possesses a gene that suppresses activity of the pH gene and thus allows pairing between chromosomes of the wheat genomes and homoeologous chromosomes in related species. Using the *A. speltoides* genome, Riley *et al.* (1968) manipulated the Ph gene in order to transfer stripe rust resistance from *A. comosa* to the bread wheat, *T. aestivum*, thus producing a new rust resistant cultivar, Campair. Regular bivalent pairing in hexaploid oats is also under genetic control (Gauthier & McGinnis, 1968). Accession cw 57 of wild diploid oat *Avena longiglumis* interferes with the bivalent pairing control of the hexaploid oats (Rajhathy & Thomas, 1972). Using the cw 57 genotype, Thomas *et al.* (1980) induced crossing over between chromosomes of the hexaploid oat and a chromosome of *A. barbata* carrying the gene for mildew resistance, and selected a mildew-resistant hexaploid plant from the offspring.

New techniques for gene transfer

Techniques for somatic hybridisation via the isolation and fusion of protoplasts and plant regeneration from hybrid calli, open up new possibilities for the exploitation of wild relatives that cannot by hybridised by conventional methods. However, two limitations to these techniques are already apparent: 1) In most plants it is extremely difficult to regenerate plants from hybrid calli. Success was reported for seven genera of three families, Solanaceae, Umbellifereae and Crucifereae (Evans & Wilson, 1984); 2) Somatic hybridisation might be useful when barriers to sexual hybridisation are of stylar origin or are due to failure of fertilisation, or to endosperm breakdown. It would be of little value, however, if the two genomes cannot function harmoniously in the fused cell because of chromosome elimination of one genome (Finch & Bennett, 1983), seedling mortality (Hollingshead, 1930; Ladizinsky & Porath, 1977) or complete sterility of the F_1 hybrid.

Plasmid-mediated gene transfer is potentially an attractive approach since it circumvents hybridisation barriers and shortens the time required to incorporate genes of wild relatives in the cultigen genome. However, as the genes are transferred in tissue culture, this technique may also encounter the problem of regeneration. Furthermore, the identification and selection of traits for plasmid-mediated transfer are carried out at the molecular level, where the gene product in the case of most agronomic traits, is not yet known. Thus, although asexual routes for gene transfer appear promising, further research is required for these techniques to yield pre-bred germplasm for the breeder.

References

Aung, T., Thomas, H. & Jones, I. T. (1977). The transfer of a gene for mildew resistance from *Avena barbata* (4*x*) into cultivated oat *A. sativa* by an induced translocation. *Euphytica*, **26**, 623–32.

Ben Ze'ev, N. & Zohary, D. (1973). Species relationships in the genus *Pisum* L. *Israel Journal of Botany*, **22**, 73–91.

Braak, J. P. & Kooistra, E. (1975). A successful cross between *Phaseolus vulgaris* L. and *Phaseolus ritensis* Jones with the aid of embryo culture. *Euphytica*, **24**, 669–79.

Chang, T. T. (1985). Germplasm enhancement and utilization. *Iowa State Journal of Research*, **54**, 399–424.

Evans, P. K. & Wilson, V. M. (1984). Plant somatic hybridization. *Advances in Applied Biology*, **10**, 1–57.

Finch, R. A. & Bennett, M. D. (1983). The mechanism of somatic chromosome elimination in *Hordeum*. In *Kew Chromosome Conference*, Brandham, P. E. & Bennett, M. D. (eds.), Allen and Unwin, London.

Frankel, R. & Galun, E. (1977). *Pollination Mechanism Reproduction and Plant Breeding*, Springer-Verlag, Berlin.

Gauthier, F. M. & McGinnis, R. C. (1968). The meiotic behaviour of a nulli-haploid plant in *Avena sativa*. *Canadian Journal of Genetics and Cytology*, **10**, 186–9.

Grun, P. & Aubertin, M. (1965). Evolutionary pathway of cytoplasmic male sterility in *Solanum*. *Genetics*, **51**, 399–409.

Harlan, J. R. (1976). Genetic resources in wild relatives of crops. *Crop Science*, **16**, 329–33.

Harlan, J. R. & de Wet, J. M. J. (1971). Toward rational classification of cultivated plants. *Taxon*, **20**, 509–17.

Hollingshead, L. (1930). A lethal factor in *Crepis* effective only in an interspecific hybrid. *Genetics*, **15**, 114–40.

Johnson, T. D. (1974). Transfer of disease resistance from *Brassica campestris* L. to rape (*B. napus* L.). *Euphytica*, **23**, 681–3.

Knott, D. R. & Dvorak, J. (1976). Alien germplasm as a source of resistance to diseases. *Annual Review of Plant Phytopathology*, **14**, 211–35.

Ladizinsky, G. (1985). Founder effect in crop plant evolution. *Economic Botany*, **39**, 191–9.

Ladizinsky, G. & Adler, A. (1976). Genetic relationships among the annual species of *Cicer* L. *Theoretical and Applied Genetics*, **48**, 197–203.

Ladizinsky, G. & Fainstein, R. (1977). Introgression between the cultivated hexaploid oat *A. sativa* and tetraploid wild *A. magna* and *A. murphyi*. *Canadian Journal of Genetics and Cytology*, **19**, 59–66.

Ladizinsky, G., Cohen, D. & Muehlbauer, F. J. (1985). Hybridization in the genus *Lens* by means of embryo culture. *Theoretical and Applied Genetics*, **70**, 97–101.

Ladizinsky, G. & Porath, N. (1977). On the origin of fenugreek *Trigonella foenum-graecum*. *Legume Research*, **1**, 38–42.

McGuire, P. E. & Dvorak, J. (1981). High salt-tolerance in wheatgrasses. *Crop Science*, **21**, 702–5.

Meyer, J. R. (1957). Origin and inheritance of D2 smoothness in upland cotton. *Journal of Heredity*, **48**, 249–50.

Murphy, H. C., Sadanaga, K., Zillinsky, F. J., Terrell, E. E. & Smith, R.T. (1968). *Avena magna*, an important new tetraploid species of oat. *Science*, **159**, 103–4.

Patherson, B. D., Paull, R. & Smillie, R. M. (1978). Chilling resistance in *Lycopersicon hirsutum* Humb. & Bonple, a wild tomato with a wide altitudinal distribution. *Australian Journal of Plant Physiology*, **5**, 609–17.

Qualset, C. O. (1975). Sampling germplasm in a center of diversity: an example of disease resistance in Ethiopian barley. In *Crop Genetic Resources for Today and Tomorrow*, pp. 81–96, Frankel, O. H. & Hawkes, J. G., (eds.), Cambridge University Press, Cambridge.

Rajhathy, T. & Thomas, T. (1972). Genetic control of chromosome pairing in hexaploid oats. *Nature New Biology*, **239**, 217–19.

Rasmusson, D. C. & Schaller, C. W. (1959). The inheritance of resistance in barley to the yellow dwarf virus. *Agronomy Journal*, **51**, 661–4.

Rick, C. M. (1973). Potential genetic resources in tomato species: clues from observations in native habitats. In *Genes, Enzymes and Populations*, pp. 255–69, Srb A. M., (ed.), Plenum Press, New York.

Rick, C. M. (1979). Biosystematic studies in *Lycopersicon* and closely related species of *Solanum*. In *The Biology and Taxonomy of Solanaceae*, pp. 667–78, Hawkes, J. G., Lester, R. N. & Skelding, A. D. (eds.), Academic Press, London.

Riley, R., Chapman, V. & Johnson, R. (1968). The incorporation of alien disease resistance in wheat by genetically induced homoeologous interference with the regulation of meiotic chromosomes synapsis. *Genetic Research*, **12**, 199–219.

Savitsky, H. (1975). Hybridization between *Beta vulgaris* and *B. procumbens* and transmission of nematode (*Heterodera schachtii*) resistance to sugarbeet. *Canadian Journal of Genetics and Cytology*, **17**, 197–204.

Sears, E. R. (1956). The transfer of leaf rust resistance from *Aegilops umbellulata* to wheat. *Brookhaven Symposium in Biology*, **9**, 1–22.

Sears, E. R. (1976). Genetic control of chromosome pairing in wheat. *Annual Review of Genetics*, **10**, 31–51.

Stalker, H. T. (1980). Utilization of wild species for crop improvement. *Advances in Agronomy*, **33**, 111–47.

Thomas, H., J. M. Leggett & Jones, I. T. (1975). The addition of a part of a chromosome of the wild oat *Avena barbata* (4*x*) into cultivated oat *A. sativa* ($2n = 42$). *Euphytica*, **24**, 717–24.

Thomas, H., Powell, W. & Aung, T. (1980). Interfering with regular meiotic behaviour in *Avena sativa* as a method of incorporating the gene for mildew resistance from *A. barbata*. *Euphytica*, **29**, 635–40.

Thomas, H. & Rajhathy, T. (1967). Chromosome relationships between *Avena sativa* (6*x*) and *A. pilosa* (2*x*). *Canadian Journal of Genetics and Cytology*, **9**, 154–62.

Vardi, A. & Zohary, D. (1967). Introgression in wheat via triploid bridge. *Heredity*, **22**, 541–60.

Wall, J. R. (1954). Interspecific hybrids of *Cucurbita* obtained by embryo culture. *Proceedings of the American Society of Horticultural Science*, **63**, 427–30.

Zamir, D., Tanksley, S. D. & Jones, R. A. (1981). Low temperature effect on selective fertilization by pollen mixture of wild and cultivated tomato species. *Theoretical and Applied Genetics*, **59**, 235–8.

Part VI

Technological or scientific innovations that affect the use of genetic resources

19

In vitro conservation and germplasm utilisation

L.A. WITHERS

Introduction

Problems with clonally propagated material and recalcitrant seeds have prompted examination of the storage of *in vitro* cultured plant parts (Withers & Williams, 1982; 1985*c*). However, four classes of subject can be identified that would benefit from *in vitro* handling. At one extreme are what might be termed 'obligate' *in vitro* systems wherein the storage of cultures provides the only means of establishing satisfactory conservation collections (especially for the long-term). Then there is material that can be conserved more efficiently or safely *in vitro* such as unique historic clones of seed-producing crops or germplasm that has been freed of pathogens. For a third group of subjects, *in vitro* culture might offer useful advantages at particular times such as when there is a need to propagate rapidly an exact genotype as part of a breeding programme or when a culture method for characterisation or evaluation is available. Such material would probably not be conserved *in vitro* except perhaps as a back-up to seed storage. Finally, consideration should be given to cultures involved in genetic manipulation and secondary product synthesis. For reasons of genetic instability, the latter subjects may, in fact, have the greatest need for *in vitro* conservation.

In vitro storage is only one component of an alternative conservation strategy which also embraces propagation and movement of germplasm. Furthermore, the partnership of *in vitro* culture and biochemical/molecular methodologies offers scope for improving other procedures relative to disease utilisation. *In vitro* related approaches to germplasm utilisation and conservation have great potential but should be treated circumspectly. Their relative benefits and drawbacks need to be weighed against each other. There is little point in imposing unnecessary and costly sophistication upon an already adequate practice. Individual judgements need to be made as to whether *in vitro* techniques are justified.

The *in vitro* genebank

The flow of material through an *in vitro* genebank and the relationship of the genebank to the field genebank and externally acquired material are shown in Fig. 19.1. (IBPGR, 1986). This plan is aimed specifically at the conservation of clones of vegetatively propagated material and populations of recalcitrant seed-producing species stored as shoots or embryos respectively, but provides a model for expansion or modification to embrace other subjects for storage as outlined above. The operational standards, initially of an *in vitro* active genebank, are currently being tested in a joint IBPGR/CIAT project.

The scheme can be broken down into a number of areas. These relate to: (i) germplasm acquisition; (ii) disease eradication, indexing and quarantine; (iii) culture establishment and multiplication; (iv) *in vivo* establishment and germplasm exchange; (v) storage by slow growth or cryopreservation; (vi) monitoring stability; (vii) characterisation and

Fig. 19.1. Relationships between accessed germplasm, the field genebank, material for distribution and exchange and the *in vitro* genebank

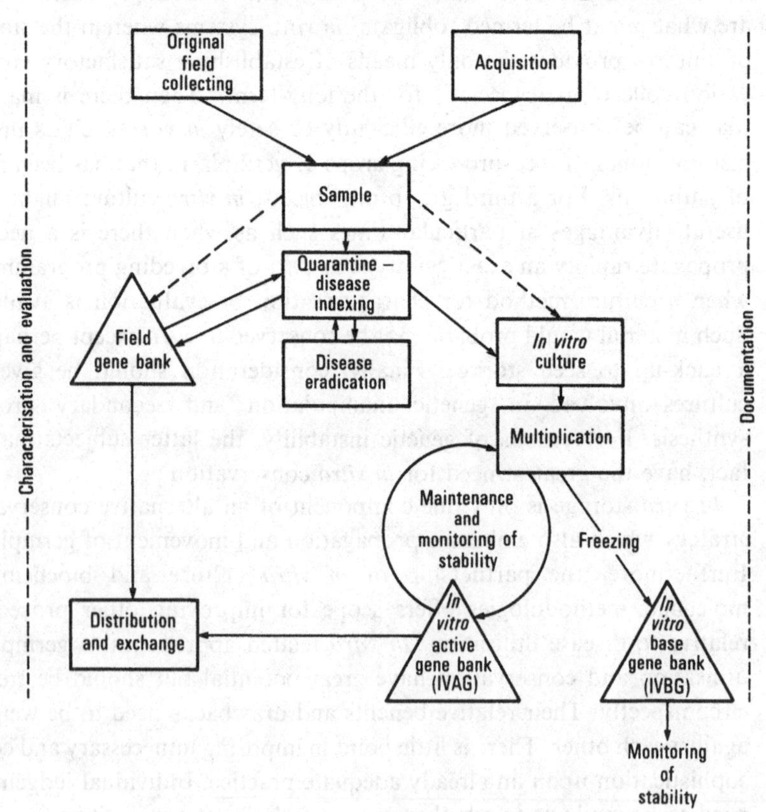

evaluation; and, ultimately, (viii) utilisation through *in vitro* breeding. Each of these areas is considered below.

Germplasm acquisition

Seed

Procedures for handling germplasm in the form of orthodox seeds are well established (Ellis *et al.*, 1985). These are designed to cope with mature seed amenable to drying and storage at ambient or sub-zero temperatures. However, there are instances with orthodox seed-producing species where mature seed is not available either for reasons of timing or because of biological problems. Germplasm represented by immature, damaged or aged seed or the products of wide crosses may be rescued by embryo culture. This procedure involves surface sterilisation of the seed, removal of the embryo or just the embryonic axis and inoculation into culture. One-to-one regeneration can ensue or, with the application of appropriate growth regulators, the production of secondary, asexual embryos. *In vitro* fertilisation is also being used for species which produce seed very infrequently in the wild and for which pollen is not available at the most appropriate time (Williams & Taji, 1986).

Recalcitrant seeds present more serious problems in that, on present knowledge, they lack a phase in their development at which they can be stored for any useful length of time. An additional practical problem is the frequently large size of the seed. Collection of zygotic embryos using a simplified *in vitro* propagation procedure is being tested for coconut (*Cocos nucifera*) in an IBPGR-sponsored project in the Côte d'Ivoire (see Withers, 1987; Withers & Williams, 1986*b*). With development, this approach should be useful for a wide range of recalcitrant seeds.

Vegetative material

It is necessary to collect vegetative material where seeds are either unavailable or inappropriate. The IBPGR, jointly with the Ghana Cocoa Growing Research Association, has been developing an *in vitro* method for collecting budwood of cocoa (*Theobroma cacao*). As microbial contaminants cannot be killed or excluded entirely, fungicides and antibiotics are incorporated into the culture medium. It appears possible at least to double the period of time for which budwood may be kept alive (Yidana *et al.*, 1987). Further processing of the collected tissues is problematic as there is, as yet, no good shoot culture method for cacao; grafting is currently the best approach.

The *in vitro* collecting technique is being or will shortly be tested for cassava (*Manihot esculenta*), *Citrus* and its relatives, breadfruit (*Arto-*

carpus spp.), temperate fruits and forage species, primarily in IBPGR projects (Withers, 1987). It will make available to breeders a much wider range of material than hitherto.

Disease eradication, indexing and quarantine

In vitro culture offers a double-edged sword to germplasm movement and quarantine. Undoubtedly, containment in a culture vessel is a powerful barrier against reinfection, but there is a danger that cultures will be assumed to be free of pathogens because they have passed through a surface sterilisation procedure. This misunderstanding could lead to devastating primary effects if a major pathogen were introduced into a previously uninfected region. Secondary effects upon the future credibility and acceptability of *in vitro* procedures would also be serious. Assumptions of cleanliness in the case of material that has been subjected to thermotherapy, chemotherapy and meristem-tip culture are better founded, but even then there is a variable probability of residual infection. Thus eradication can never be divorced from indexing for disease.

Disease eradication

Virus elimination by meristem-tip culture, first described more than 30 years ago, can now be applied to over 50 species (Kartha, 1986). A number of factors affect the efficiency of virus elimination. These include the size and location of the explant, *in vitro* culture conditions, the season, conditions of growth and physiological state of the donor plant, and the use of prior thermotherapy.

For subjects such as *Citrus* in which meristem-tip culture is difficult or impossible, Navarro (1981*a*) has pioneered the development of shoot-tip grafting. Success rates in grafting are moderate (30–50 per cent) but post-transplant survival is excellent (95 per cent) and success in virus elimination varies from 10–20 per cent (exceptionally) to 100 per cent. The technique is applicable to the majority of *Citrus* viruses and is a central procedure in the Citrus Variety Improvement Programme in Spain (Navarro *et al.*, 1981*c*).

Several chemicals, including antibiotics, have been used to suppress or inactivate plant viruses. Application in the culture medium used to grow excised meristem-tips, can dramatically increase the efficiency of recovery of pathogen-free plantlets. For example, the use of Ribavirin has been shown to raise the percentage elimination of PVM from the potato cv. Golden Wonder from 59 per cent to 100 per cent (Cassells & Long, 1982).

Disease indexing

A number of methods can be used for disease indexing including the inoculation of host or indicator plants, grafting and biochemical/ molecular assays. The latter are attractive as they involve tissue extracts (in some cases, non-infectious) and may, therefore, be carried out remotely. Several of the biochemical assay techniques such as ELISA (enzyme linked immunosorbent assay) and ISEM (immunosorbent electron microscopy), involve serological principles (Kartha,1986; Van Regenmortel, 1982).

Recently, progress in nucleic acid technology has led to the development of monoclonal antibody and nucleic acid hybridisation techniques. An example of the latter is the 'sap spot' test in current use to detect PVX in potato. It is applicable to other RNA viruses and viroids (Baulcombe *et al.*, 1984). The test involves synthesis of a DNA copy of the viral RNA, its replication in *E. coli*, radioactive labelling and use as a probe to identify the viral RNA in expressed sap. The procedure is more sensitive and, once developed, less complex than ELISA. It can also be used to reveal viruses that have been neither purified nor characterised.

Other potential indexing techniques include isozyme analysis in which disease is indicated by modification of enzyme banding patterns (Simpson & Withers, 1986), and two-dimensional electrophoresis of proteins (Celis & Bravo, 1984). Bidirectional electrophoresis using two different criteria of separation can yield a highly reproducible and characteristic pattern, in which the presence of pathogen- or symptom-related proteins is apparent.

Another broad approach to indexing is envisaged which has particular applications to *in vitro* material as it utilises the controlled interaction between host and pathogen in culture. First, segments of putatively infected material may be cultured and their performance observed. Shoot explants of *Citrus medica* infected with citrus exocortis viroid fail to root in culture, unlike uninfected controls (Navarro, 1981*b*). Callus of *Santalum album* infected with the mycoplasma causing spike disease requires GA_3 for normal morphogenic responses in culture (Gowda & Narayana, 1986).

Secondly, assays may be developed that involve the introduction of a pathogen into a test culture known initially to be free of the disease. Most appropriate is the situation which mimics natural interactions by challenging an *in vitro* shoot or plantlet with the pathogen or its extract. An unorganised culture (cell or callus) could be used but the interaction is inevitably more intimate and will by-pass mechanisms of resistance that depend upon a whole plant response. (Both approaches are developed further in the context of germplasm characterisation and evaluation in a later section.)

In *Citrus*, the bacterial disease canker caused by *Xanthomonas campestris* (Lopez & Navarro, 1981) and tristeza virus (Navarro, pers. comm.) can be detected by inoculating aseptically grown seedlings with a suspension of pathogen particles. In canker, the response is clearly specific to the pathogen under examination, no disease symptoms being elicited by either *Psudomonas syringae* or *Xanthomonas* pathogenic to non-citrus hosts. A similar approach has been used to study the interaction of *Xanthomonas ampelina* and grapevine (*Vitis vinifera*) (Lopez *et al.*, 1983).

In vitro cultured shoots of cultivars of apple (*Malus pumila*) susceptible to apple scab caused by *Venturia inaequalis* respond characteristically by becoming severely overgrown by mycelium (Yepes & Aldwinckle, 1986). A final example involving another fungal pathogen *Gremmeniella abiertina* and its coniferous hosts *Larix* spp. (Abdul Rahman *et al.*, 1986) reveals the dramatic reduction in the time that indexing may take *in vitro*. The pathogen normally takes 1–1.5 years to complete its life-cycle in nature but requires only four weeks in culture.

It is not a coincidence that the examples of *in vitro* indexing largely apply to bacterial and fungal diseases. Perhaps this is where *in vitro* techniques may find a genuine niche, biochemical/molecular approaches being the logical choice for viruses and viroids.

Quarantine

The movement of germplasm through quarantine is time consuming, often inefficient and may involve the incidental loss of valuable material whilst tests are under way. The choice of location of genebanks and quarantine stations is sometimes in this respect unfortunate being near centres of crop production. These factors, as well as frustrating germplasm introduction and breeding programmes may lead to breaches of quarantine regulations. Accordingly, it is proposed that the possibilities of *in vitro* movement of germplasm, contained indexing as envisaged above and *in vitro* propagation be brought together to provide a novel and safer quarantine system. No such system is in use as yet, although a number of existing practices include desirable features, e.g. the scheme for *Citrus* in Spain (see above), the Scottish Strawberry Certification Scheme (Harper *et al.*, 1986) and the Danish Seed Potato Programme (Kristensen, 1984).

The practical side of developing *in vitro*-based quarantine procedures must be given attention by scientists in the first instance but it is equally important that quarantine regulations be developed in parallel to facilitate implementation of new procedures.

Culture establishment and multiplication

A spectrum of culture systems exists, ranging from the unorganised protoplast, cell and callus to the organised shoot, embryo and plantlet. Some applications will dictate the use of unorganised systems but from the point of view of convenience and genetic stability it is preferable to use an organised system that multiplies by a non-adventitious route (i.e. one using pre-existing meristems) route. For general guidelines on culture procedures, reference should be made to texts on the species in question as culture medium and environmental requirements will vary. Advice is available on facilities needed for *in vitro* work (Withers, 1985*a* and references therein).

Rates of propagation will vary according to system and need. In general, material in storage should have as low a rate of growth and multiplication as possible. Some plant breeding applications will require a moderate rate, e.g. to bulk up parental lines for F_1 hybrid production.

Greater rates of multiplication are likely to be involved where *in vitro* propagation is the final stage before glasshouse/field growth of the crop. As stocks for future use would not derive from such material but a previous clonal generation, this stage could involve higher-risk propagation systems in a 'trade-off' between cost and genetic stability. The advantages of higher-risk systems lie particularly in the possibilities for automation, as discussed below.

Attempts have been made to estimate relative costs of *in vitro* and conventionally propagated plantlets. In some cases, the former cost some 2–10 times the latter; in others (e.g. hybrid broccoli as artificial seed; see below), the cost is comparable. Part of the equation is the relative performance of the two types of plant. For example, *in vitro* propagated tomato (*Lycopersicon esculentum*) plantlets fruit earlier on lower trusses than do seed-derived plants (Walkey, 1987).

***In vivo* transfer and germplasm exchange**

Transfer of rooted shoots or embryo-derived plantlets to *in vivo* growth involves 'hardening' them by exposure to a higher light intensity and lower humidity. Frequently, *in vitro* leaves lack on adequate cuticle and may have an abnormal internal structure. These conditions must be remedied during the hardening process to given transpirational control and photosynthetic competence. Shoots without preformed roots can be rooted either in an intermediate *in vitro* phase using appropriate growth regulators and physical supports (Short *et al.*, 1987), or *in vivo* using conventional propagation procedures.

For a crop such as potato, for example, the *in vivo* transfer procedure is carried out with relative ease. In Vietnam, clones provided from the International Potato Center (CIP) are multiplied rapidly by shoot culture in unsophisticated 'family tissue culture units' and then distributed to growers (Uyen, 1985).

Individual handling of *in vitro* shoots is both labour intensive and time consuming. Two alternatives are automated transfer coupled with artificial seed production and secondly *in vitro* formation of storage organs.

Shoot cultures could be chopped randomly and transferred to a medium that specifically induces further growth and rooting only of pre-existing shoot meristems. Alternatively, using more sophisticated robotics (De Bry, 1986), nodal cuttings could be produced for rooting. Embryogenic systems are particularly well suited for encapsulation in artificial seed. Considerable progress has been made recently in this area (Redenbaugh *et al.*, 1986). The progagule is encapsulated in a gel (e.g. sodium alginate) which is solidified and coated with a rigid shell and sown preferably by a fluid drilling technique. The author considers that there is scope for investigating the possibility of drying embryos as would occur in naturally maturing seed. Furthermore, if storage, particularly by cryopreservation could be introduced before drilling, the logistics of production, transport and field transfer would greatly be eased.

Turning to *in vitro* formed storage organs, a number of crop species such as potato (Wang & Hu, 1982) and yam (*Dioscorea rotundata*; Ng, 1986) that normally produce storage organs can be induced to form miniature ones in culture. Growers in the USA are now able to purchase microtubers commercially and the field performance of such propagules is quoted to be comparable to conventionally produced tubers (see Walkey, 1987).

The distribution of germplasm in the form of *in vitro* cultures is now used widely by international and national centres and commercial organisations. For example, potato is distributed world-wide from CIP. Microtubers are more tolerant of physical disturbance, environmental pertubations and delays in transit than are cultures. They are now in use for distributing germplasm of potato from CIP (Tovar *et al.*, 1985) and yam from the International Institute for Tropical Agriculture (IITA) (Ng, 1986).

Wider use of cultures for international transfer is recommended, particularly as part of a quarantine procedure. All that should be avoided is the use of unorganised cultures, liquid medium and lengthy periods in transit under unsuitable environmental conditions.

Storage of *in vitro* cultures

Storage of cultures under normal growth conditions is costly and carries risks of genetic change with time, contamination or loss through human error. Therefore, attempts are being made to develop methods that reduce the growth rate of cultures or suspend growth entirely by cryopreservation. The former, along with the cultures in normal growth, are active collections and parallel the *in vivo* field genebank, whereas the latter represent base storage.

Slow growth

Shoot cultures of many species can be stored under conditions in which growth is slowed down by use of a reduced culture temperature or the application of osmotica or growth regulators. Examples with a direct genetic conservation objective include banana and plantain (*Musa* spp.) stored at 15 °C (Banerjee & De Langhe, 1985), pear (*Pyrus* spp.) stored at 4 °C (Wanas *et al.*, 1986) and potato stored under a wide range of conditions. The latter is amenable to the application of abscisic acid, Alar (B-nine) or mannitol, with and without a temperature reduction (see Withers, 1986*a*). Other examples may be found in the literature and in the IBPGR International Data Base on *In Vitro* Conservation (Wheelans & Withers, 1984; Withers, 1985*b*; 1986*a*).

As there is evidence for instability in unorganised systems stored by slow growth, they should be avoided for genetic conservation purposes. Even in shoots and embryos maintained by slow growth, monitoring in storage should be strict.

Cryopreservation

The storage of germplasm in liquid nitrogen is widespread in microbiology and animal tissue culture. The application of cryopreservation to *in vitro* plant cultures has been under investigation for some twelve years and, despite a poor input of interest and resources, has yielded notable progress. A routine method is now available for the cryopreservation of cell cultures (Withers & King, 1980). All available evidence including data on chromosome number, secondary product synthesis, morphogenesis and antimetabolite resistance suggests that cryopreserved cell cultures are genetically stable (Withers, 1985*b*; 1986*a*).

Shoot and embryo cultures are more relevant to genetic conservation. A number of species have been cryopreserved successfully in such a form, e.g. pear (Wanas *et al.*, unpublished), shoot-tips of cassava (Kartha *et al.*, 1982), oilpalm, (Engelman *et al.*, 1985) although the situation is far from satisfactory. Problems are experienced in maintaining symplastic continuity in relatively large unit specimens and such material tends to regenerate after freezing by a route that involves disorganised growth. The latter is partly due to the magnitude of physical damage experienced, and the consequent need to apply growth regulators to induce a high percentage regeneration.

Although research is attempting to overcome such difficulties (e.g. Benson & Withers, 1986; Kartha *et al.*, 1982; Withers, 1979), it is clear that there may have to be a 'trade-off' between quantity and quality in cryopreservation. Quality must always predominate in genetic conservation.

Recalcitrant seeds have also been considered as potential subjects for cryopreservation with some hopeful results (Grout, 1986; Becwar *et al.*, 1983). Immature embryos or part of the mature embryos could be used and *in vitro* cloning could be incorporated to replicate material before or after cryopreservation. The techniques of *in vitro* fertilisation and embryo culture combined with cryopreservation may provide a novel means of conserving the germplasm of recalcitrant seed-producing species. The cryopreservation of pollen itself has applications in breeding and conservation.

Monitoring stability

The need to maintain genetic integrity in conserved germplasm is implicit. Whilst accepting that absolute stability is not the norm in nature (Scowcroft, 1985*a*; Walbot & Cullis, 1985) and may not even be desirable in evolutionary terms, *in vitro* procedures will not be acceptable if they introduce inordinate risks of genetic instability and/or selection among heterogeneous genotypes. The stability record of cryopreserved microbial, mammalian and, as indicated above, higher plant cell cultures is very good. However, organised plant cultures are only now being tested in this respect.

Morphological assessment

The *in vitro* genebank will hold very large numbers of cultures maintained under conditions where their phenotype is either modified as an essential component of the storage procedure or inaccessible to inspection during storage as in cryopreservation. Visual inspection of cultures in slow growth will check for viability loss and microbial contamination, but meaningful, monitoring of stability by eye will be difficult as vigour is intentionally being depressed.

Most species, once committed to culture, cease to resemble the parent plant, especially at the level of discrimination of varieties. Furthermore, the *in vitro* phenotype can change both randomly and progressively with time with respect to behaviour and morphology. For example, shoots of apple are found to become easier to root with the passage of time in culture (Mullins, 1987). The phenomenon of vitrification is observed in many cultures. This deformation and waterlogging of tissues is noted in response, for example, to submersion in liquid medium (John, 1986), high levels of cytokinin and disturbed (high or low) levels of atmospheric ethylene (Hussey, 1986). Vitrified cultures often show a reduced capacity to regenerate plants capable of healthy establishment *in vivo*. Although vitrification is considered not to be a genetic phenomenon, implications for short- to medium-term culture, as in slow growth, are serious. Correlation between the occurrence of vitrification and change in perioxidase isozyme profiles (Kevers & Gaspar, 1985) offer a means of detecting incipient problems but, conversely, may cast doubt upon the validity of isozyme screening of genetic stability (see below).

All of the cautions in using morphological criteria of stability in material in slow growth apply to cultures regenerated from storage by cryopreservation with the added problems of a greater degree of structural disturbance by freezing. In both storage modes, it is essential that reliable descriptors and descriptor states be determined for the *in vitro* condition. Additionally, plants should be transferred to *in vivo* growth at appropriate intervals to verify assumptions made *in vitro*.

Cytology

Some of the earliest recordings of genetic instability in culture were made on the basis of chromosome counts (D'Amato, 1975) but it is known that this approach reveals only crude genetic change. Correlations between the incidence of chromosal instability and disorganised growth are cited widely. Cytological checks of appropriately selected accessions from the *in vitro* genebank will be essential but the resolution of the information so derived should not be over-estimated particularly in the case of no apparent change.

Isozyme analysis

Analysis by electrophoresis of families of polymorphic enzymes provides a means of describing phenotypes at the level of the primary gene product (Simpson & Withers, 1986). It provides information on genetic identity that is generally less prone to perturbation by environmental conditions than is either morphology or agronomic evaluation. However, several enzymes should be examined (avoiding labile ones such as peroxidase and catalase) in any one study and isozyme analysis should be used in parallel with other methods of evaluation.

It remains to be demonstrated whether isozyme analysis can offer the resolution required to identify all types of *in vitro* related instability, although it has been used to examine somaclonal variants (Scowcroft *et al.*, 1985*b*). Within its limitations, it should be possible to establish relatively reliable *in vitro* descriptors that relate to standard storage states. Isozyme analysis is currently used to monitor cultures of cassava in slow growth storage at the Centro Internacional de Agricultura Tropical (CIAT) (W. M. Roca, pers. comm.; CIAT, 1985) and will be one of the methods used in the IBPGR/CIAT pilot *in vitro* genebank. Two-dimensional electrophoresis of proteins has yet to be evaluated as a monitor of stability in cultures.

Restriction fragment length polymorphism and DNA probes

Molecular techniques provide direct information on the genome and can be considered as definitive monitors of genetic stability (see Bernatzky & Tanksley, this volume). As yet, neither restriction fragment length polymorphism (RFLP) nor DNA probes have been applied to the analysis of stability of stored *in vitro* cultures but appropriate work is imminent in IBPGR research projects.

Other biochemical markers

Recent studies in the author's laboratory (Benson & Withers, 1986; 1987) have revealed the possibility of using gas chromatographic measurement of ethylene, ethane, other alkanes and methane to determine the nature and extent of injury in cryopreserved cultures. It is suggested here that similar approaches involving stress reactions and free radical pathology may be used to evaluate stability in stored seeds and cultures in slow growth. (Both cell membranes and nucleic acids may be targets for free radical damage).

Characterisation and evaluation

Characterisation and evaluation involve the identification and description of potentially important characters in germplasm accessions. The breeder is particularly interested in disease and stress resistance. Provision of more and improved information will be required if breeders are to make better use of genebanks (Peeters & Williams, 1984). *In vitro* techniques, particularly storage and genetic manipulation, introduce new characterisation and evaluation requirements. Realisation of breeding objectives and efficiency in the *in vitro* breeding process demand appropriate and accurate screening techniques (Wenzel *et al.*, 1987). Fortunately, as well as increasing demands, *in vitro* techniques also offer new approaches to characterisation and evaluation.

Genetic identity

Reliable characterisation by morphology alone cannot be carried out entirely *in vitro*. However, cultures may be amenable to identification by biochemical and molecular techniques. Both isozyme analysis and isoelectric focusing of fraction 1 protein have been used to identify somatic hybrids (Kung, 1983; Wetter & Dyck, 1983). Two-dimensional analysis of proteins has potential applications also; one-dimensional electrophoresis of proteins is already in use for the characterisation and evaluation of cereals, for example. Characters evaluated include bread-making and malting quality and resistance to powdery mildew (Payne *et al.*, 1979; Riggs *et al.*, 1983). An isozyme marker for resistance to bean yellow mosaic virus has been identified in *Pisum sativum* (pea; Weeden *et al.*, 1984). Stress phenomena are under investigation using two-dimensional electrophoresis which reveals specific stress-related (e.g. heat shock) proteins. This approach may help identify stress tolerant genotypes (M. Harrington, pers. comm.).

Molecular techniques such as RFLP and the use of DNA probes offer very precise means of fingerprinting the genome. Caution must, however, be observed in the interpretation of data relating to the localisation of specific newly inserted genes introduced by genetic manipulation *in vitro*. A gene will respond positively to a probe whether or not the appropriate controlling DNA sequences have also been successfully inserted. Without the latter, however, the gene will not be expressed.

An additional characterisation method that has emerged recently is the use of fourth derivative visible spectroscopy. This warrants exploration (Daley *et al.*, 1986).

Validity of* in vitro *evaluation

It is widely recognised that meaningful testing for an agronomic character can only be carried out with confidence in the field. Therefore, *in vitro* testing needs careful consideration. A number of questions must be asked. Does the *in vitro* response faithfully reflect the *in vivo* response? (This is particularly important when unorganised cultures are used.) Could *in vitro* methods give savings on cost, thus also permitting more material to be processed? Could they be more rapid or convenient? Will they be more accurate as a result of the greater environmental control that can be maintained *in vitro*?

Beyond restating that *in vitro* testing will always require verification *in vivo*, no generalisations can be made as so many of the judgements on the questions above involve examination of the particular crop to stress/pathogen relationship.

Disease resistance

Knowledge of plant/pathogen relationships would lead to the conclusion, in some cases, that disease resistance or susceptibility would not be expected to show in an *in vitro* system. The relationship between a pathogen and the host depends upon a complex interaction (see P.H. Williams, this volume). For this reason and to avoid unnecessary complexity, simple approaches that resemble the natural pathogen/host interaction warrant examination before exploring more sophisticated *in vitro* manipulations that often have a highly empirical basis. In the case of mildew of barley for example, screening can be carried out with pieces of leaf tissue taken from independently growing plants and introduced into semi- or non-sterile culture so that better environmental control can be exerted. In another example that utilises a simple *in vitro* approach, dormant apple twigs have been introduced into culture in the presence of *Phytophthora cactorum*, the causal agent of apple crown rot. This method has been used by the United States Department of Agriculture (USDA) to screen the world apple germplasm collection (Utkhede, 1986).

Screening, using organised cultures, can mimic the whole plant/pathogen relationship very successfully, for example *Gremmeniella abietina* infection of shoot cultures of *Larix* spp. and apple shoot cultures exposed to apple scab, as mentioned earlier. However, in the latter (Abdul Rahman *et al.*, 1986), the reaction between the pathogen and the host leaf was modified in culture leading to abnormal development of the appressorium and hypha which overgrew rather than penetrated susceptible genotypes. Resistant ones showed a classic hypersensitive reaction. The culture/pathogen relationship is not always as straightforward. In potato leaf discs challenged with blight (*Phytophthora infestans*), those of a tolerant genotype show overgrowth, whereas those of a sensitive genotype decline and fail to support growth of the pathogen (Wenzel *et al.*, 1987). Unorganised cultures are less suitable as whole plant reactions can be bypassed. However, callus cultures of *Pinus lambertiana* challenged with *Cronartium ribicola* show a hypersensitive reaction in resistant genotypes, as in the whole plant. This reaction is not depressed by the application of a cytokinin, unlike in some other cases (Diner *et al.*, 1984).

Reactions such as the above provide a means of identifying resistant genotypes but not necessarily of selecting positively for them. Attempts have been made to use pathogen toxin for this purpose. Recently reported examples include testing callus of soyabean (*Glycine max*) with an isolate of *Phialophora gregata* (Guan *et al.*, 1986), exposing leaf discs of *Populus* sp. exposed to conidia and extract of the stem canker pathogen *Septoria musiva* (Ostry *et al.*, 1986), and exposing callus of hop (*Humulus lupulus*) exposed to culture filtrate of the wilt pathogen *Verticillium albo-atrum* (Connell & Heale, 1986).

The exact nature of the host/pathogen relationship is important when other biochemical agents are used for screening. Methionine sulphoximine, an analogue of wildfire toxin has been used to select cells and regenerate tobacco (*Nicotiana tabacum*) plants resistant to the disease. Regenerated plants and sexual progeny showed a lack of the characteristic yellowing of leaves when exposed to the pathogen (*Pseudomonas tabaci*) but it still produced lesions on the leaf (Carlson, 1973). This indicates that selection at the cell level may only involve the direct cellular response to the pathogen toxin, leaving other deleterious effects unselected against.

Stress resistance

Screening and selection methods for stress resistance are hampered again by the nature of the plant's reaction and survival mechanism. The breeder would normally look for, say, a plant with deeper root penetration, a thicker cuticle and, possibly, altered stomatal control. These would be difficult if not impossible to determine *in vitro*, especially in unorganised cultures. Resistance mechanisms involving solute accumulation might, however, be observed in cells or callus.

Unfortunately, epigenetic change and physiological adaptation/acclimation are prominent in culture responses. Furthermore, in callus culture, inconsistencies in response *in vitro* can hamper evaluations. For example, in culture the halophytic glasswort (*Salicornia* sp.) shows a similar sensitivity to exposure to NaCl to that of the non-halophytic species cabbage, sweet clover or sorghum. However, the reactions of callus of barley, cultivated *Hordeum vulgare* and wild *H. jubatum*, do reflect those of the whole plant (Chaleff, 1983). Nonetheless, stress tolerant cells lines of a number of species have been isolated (Dix, 1986) and, in the case of tobacco, oat (*Avena sativa*) and maize, clear evidence has been given of sexual transmission (Nabors & Dykes, 1985).

In grapevine (*Vitis rupestris*) where salt tolerance was shown at the level of the unorganized culture, *in vitro* shoots died as soon as they formed roots (Lebrun *et al.*, 1985). This indicates that different mechanisms of salt tolerance are operating at the cell and organ (specifically root) levels.

Thus, selection of *in vitro* plantlets or *in vivo* regenerants will be essential, perhaps as part of a two-tier strategy. Alternatively, the selection strategy can be such that physiological adaptation is not a useful survival mechanism in the long term (Nabors & Dykes, 1985).

Experiences with slow growth storage of cultures would suggest that tolerance of low temperatures, whilst often being genotypically variable, is very labile and need not reflect clear differences at the whole *in vivo* plant level. Culture conditions, notably the growth regulators in use, may influence the response of the tissue. Abscisic acid can enhance cold tolerance just as cytokinins, auxins and GA_3 can influence the expression of disease symptoms.

Herbicide resistance

Heritable resistance to herbicides has been shown in plants regenerated from cell cultures of several species including tobacco and hybrids of *Lycopersicon* (Chaleff & Ray, 1984; Cséplö *et al.*, 1985; Dix, 1986). Herbicide resistance has been considered as one of the most promising traits to genetically manipulate *in vitro*. The level of commercial interest would suggest this to be a worthwhile area to investigate, offering advantages over conventional approaches and field selection (P.J. Dix, pers, comm.).

Selection for morphological traits

Morphology is difficult to evaluate at the culture level, particularly in the absence of clear knowledge of regulatory mechanisms. Nonetheless, two examples can be quoted wherein *in vitro* screening has been used to enhance the efficiency of field selection against undesirable traits.

The bracting defect and purple curd defect in cauliflower are both expressed variably in the field. However, by culturing small pieces of curd under appropriate conditions, the presence of the traits can be revealed clearly. In the case of the bracting defect, whereas 69 per cent of the plants could be rejected by field evaluation, this was raised to 96 per cent in culture. The respective figures for purple discoloration are even more impressive: 19 per cent and 97 per cent (Crisp & Gray, 1979). The assays are non-destructive in terms of the whole plant and are carried out relatively rapidly, taking some four weeks, leaving ample time for the parent plant to be grafted for seed production.

Primary and secondary products

It is difficult to devise a satisfactory *in vitro* screening and selection strategy for plant products. Secondary metabolism is often switched on as growth ceases, preventing selection by preferential growth in cell culture, although, in the case of a primary product such as a cereal storage protein, use of end-product inhibition can help identify over-producers of particular amino acids. A high threonine mutant of maize has been isolated by this approach (Hibberd & Green, 1982). However, problems were experienced in attempts to isolate a mutant of tobacco with increased productivity, by the application of isonicotinic acid hydrazide, which interferes with photorespiration, to callus. Stably resistant callus was successfully isolated but plants regenerated from the callus were sensitive. When callus was re-initiated from the regenerants, the resistance was again observed, revealing it to be a cellular phenomenon not expressed in the whole plant (Berlyn, 1980).

At present, direct methods seem most appropriate for the identification of specific genotypes on the basis of their products. In tobacco, *in vitro* production of nicotine by callus reflects *in vivo* performance, so selection may be carried out at the unorganised culture level. A simple squash procedure followed by manual selection would be effective (Berlin & Sasse, 1985). The use of mechanised sorting of cells or protoplasts on the basis of the product itself or using a label such as a fluorescent dye may increase the efficiency of selection (Cocking, 1986).

Utilisation through *in vitro* breeding

Breeding by conventional means is sometimes criticised as being too expensive, too uncertain and taking too long. *In vitro* techniques would be welcome if they were to remedy any of these constraints. Some techniques such as *in vitro* vernalisation simply accelerate the breeding process. Others offer alternative breeding mechanisms (Cocking, 1986; Gunn & Day, 1986).

In vitro *hybridisation*

Three means of achieving hybridisation through *in vitro* culture are available. These are most useful for the production of wide crosses. *In vitro* fertilisation can be used to achieve otherwise incompatible crosses where the barrier to fertilisation is a result of too large a pistil or of biochemical incompatibility. Successful crosses have been made within the Solanaceae and the Gramineae for example, (Gengenbach, 1984; Zenkteler, 1984). *In vitro* fertilisation may also facilitate some genetic manipulations by using the pollen tube as a vehicle for introduced genes or organelles.

Embryo rescue also utilises the sexual process but the *in vitro* stage commences after fertilisation. The immature embryo or entire ovule is transferred to culture at a stage before deleterious effects of endosperm failure threaten the embryo, which is then nursed on to maturity. A number of useful characters have been transferred by this technique (see Doré, 1986; Dunwell, 1986).

Somatic hybridisation by the fusion of protoplasts achieves wide crossing by completely bypassing the sexual process. For example, the first step in the transfer of nematode and mite resistance from the wild *Solanum sisymbrifolium* to eggplant (*S. melongena*) has been achieved thus (Gleddie *et al.*, 1986).

Pollen and anther culture

The production of homozygous diploids by pollen/anther culture followed by chromosome doubling can be more rapid than conventional methods for producing inbred lines. For example, oilseed rape (*Brassica napus*) cv. 'Mikado' entered the National List Trials in the UK only five years after the parental haploid plant was isolated (Gunn & Day, 1986).

The best-known applications of anther culture are in cereal breeding, especially the work on rice in China (Shen *et al.*, 1983). Other examples include cabbage, in which the technique is being used rapidly to fix clubroot resistance (Chiang *et al.*, 1985) and asparagus (*Asparagus officinalis*), in which super-male (*MM*) plants have been produced by the anther culture of males (*Mm*). These give all male progeny in the F_1 generation and a resultant higher yielding crop (Doré, 1987).

Other methods for producing haploids are available, such as ovary or ovule culture and chromosome elimination (the '*bulbosum*' technique; Dunwell, 1986).

Selective gene transfer

The *in vitro* methods described above mostly involve the wholesale manipulation/transfer of genomes. Newer techniques offer more exciting possibilities for identifying, cloning and selectively transferring genes (Gerlach *et al.*, 1985). Several means of transfer are available including transformation of protoplasts or tissues by plasmids or naked DNA, (sometimes enhanced in its efficiency by electroporation), and micro-injection of cells, tissues, organs and pollen.

Notable successes in the development of an *in vitro* selective gene transfer strategy include the transfer of resistance to viruses, insects and herbicides into members of the Solanaceae and of kanamycin resistance into rye (*Secale cereale*; Lörz *et al.*, 1987). Dicots are still far more amenable to manipulation than are monocots but rapid progress is being made with the latter.

Somaclonal variation

Various approaches to mutagenesis *in vitro* have been investigated, including the use of chemical mutagens and irradiation. It has been realised, over recent years, that the *in vitro* process itself may be destabilising, leading to the phenomenon of somaclonal variation (Larkin & Scowcroft, 1981). This source of variation is manifested most frequently in unorganised cultures and is widespread throughout all genera investigated (Scowcroft, 1985*a*). Many useful agronomic characters including morphology, vigour, fertility and disease resistance have been recorded in somaclonal variants. The sexual transmission of these traits is frequently confirmed (Scowcroft, 1985*a*). As yet, the mechanisms of somaclonal variation are not fully understood (Gould, 1986; Scowcroft *et al.*, 1985*a*, *b*). They appear to range from single nucleotide substitutions to changes in ploidy (Scowcroft *et al.*, 1985*b*; Peacock, this volume).

The potential application of somaclonal variation in sterile and/or clonally propagated material and to achieve gene exchange in wide crosses for which sexual recombination is not available, should not be underestimated (Novak, 1984; Scowcroft, 1985*a*). However, the equivalent potential of the phenomenon to confound efforts to conserve material *in vitro* should also be borne in mind. In both the positive and negative respects, control of variation is the key to success. Cryopreservation may emerge as a useful means of controlling somaclonal variation as it eliminates time-related change.

Conclusions

This chapter has attempted to illustrate a number of ways in which *in vitro* techniques may be used to supplement, enhance and, occasionally, replace conventional practices. In some cases, technological developments are still at the stage of examining model systems. In others, actual agricultural problems are being tackled. Initiative and impetus from users will be needed if some model procedures are to progress particularly in relation to characterisation and evaluation. However, biotechnologists need good evaluation techniques as much as do breeders; so some optimism may be justified.

The likely timescale of implementation of new technique varies. It may be considered current or imminent for some storage procedures and disease-related phenomena, for movement of germplasm, biochemical characterisation and anther or embryo culture in amenable species. With respect to *in vitro* breeding some species such as tomato that have yielded useful somaclonal variants could be in commercial production now; for others, particularly where transformation is involved, we may be looking to a 5 to 20 year horizon (Teweles, 1984). Cost may affect the extent to which some techniques are taken up. However, it is suggested here that caution should be a greater moderator. The use of new 'high-tech' procedures benefits no-one unless those procedures resolve genuine problems. A close collaboration between breeders, plant introduction officers, genetic conservationists and *in vitro* technologists will be essential to the appropriate development and implementation of these new techniques in agriculture.

Acknowledgements

The author gratefully acknowledges receipt of an SERC (UK) Advanced Fellowship. Colleagues are thanked for providing details of 'in press' publications and helpful comments on the manuscript.

References

Abdul Rahman, N., Diner, A. M., Karnosky, D. F. & Skilling, D. D. (1986). Differential responses by four coniferous tree species inoculated *in vitro* with *Gremmeniella abietina*. In *Proceedings VI International Congress of Plant Tissue and Cell Culture, Volume 1: Abstracts*, p. 401 Somers, D. A., Genenbach, B. G., Biesboer, D. D., Hackett, W. P. & Green, C. E. (eds.), International Association for Plant Tissue Culture, Minneapolis.

Banerjee, N. & De Langhe, E. A. L. (1985). A tissue culture technique for rapid clonal propagation and storage under minimal growth conditions of *Musa* (banana and plantain). *Plant Cell Reports*, **4**, 351–4.

Baulcombe, D., Flavell, R. B., Boulton, R. E. & Jellis, G. J. (1984). The sensitivity and specificity of a rapid nucleic acid hybridisation method for the detection of potato virus X and other plant RNA viruses in crude sap samples. *Plant Pathology*, **33**, 361–70.

Becwar, M. R., Stanwood, P. C. & Leonhardt, K. W. (1983). Dehydration effects on freezing characteristics and survival in liquid nitrogen of desiccation-tolerant and desiccation-sensitive seeds. *Journal of the American Society for Horticultural Science*, **108**, 613–18.

Benson, E. E. & Withers, L. A. (1986). Assessments of stability in cryopreserved cultures. In *Proceedings VI International Congress of Plant Tissue and Cell Culture, Volume 1: Abstracts*, p. 425, Somers, D. A. Gengenbach, B. C., Biesboer, D. D., Hackett W. P. & Green, C. E. (eds.), International Association for Plant Tissue Culture, Minneapolis.

Benson, E. E. & Withers, L. A. (1987). Gas chromatographic analysis of volatile hydrocarbon production by cryopreserved plant tissues culture: a non-destructive method for assessing stability. *Cryoletters*, **8**, 35–46.

Berlin, J. & Sasse, F., (1985). Selection and screening techniques for plant cell cultures. In *Advances in Biochemical Engineering, Volume 31, Plant Cell Culture*, pp. 99–132, Fietcher A. (ed.), Springer Verlag, Berlin.

Berlyn, M. B. (1980) Isolation and characterisation of isoniazid-resistant mutants of *Nicotiana tabacum*. *Theoretical and Applied Genetics*, **58**, 19–26.

Carlson, P. S. (1973). Methionine sulfoximine-resistant mutants of tobacco. *Science*, **180**, 1366–8.

Cassells, A. C. & Long, R. D. (1982). The elimination of potato viruses X,Y, S and M in meristem and explant cultures of potato in the presence of Virazole. *Potato Research*, **25**, 165–73.

Celis, J. E. & Bravo, R. (1984). *Two-dimensional Gel Electrophoresis of Proteins: Methods and Applications*, Academic Press, New York.

Chaleff, R. S. (1983). Isolation of agronomically useful mutants from plant cell cultures. *Science*, **219**, 676–82.

Chaleff, R. S. & Ray, T. B. (1984). Herbicide resistant mutants from tobacco cell cultures. *Science*, **233**, 1148–51.

Chiang, S., Frechette, S., Kuo, D., Chong, C. & Delafield, S. (1985). Embryogenesis and haploid plant production from anther cultures of cabbage. *Canadian Journal of Plant Science*, **65**, 35–8.

CIAT (1985). Tissue culture. In *Annual Report for 1984; Cassava Program*, pp. 197–217. Centro Internacional de Agricultura Tropical, Cali, Colombia.

Cocking, E. C. (1986). The tissue culture revolution. In *Plant Tissue Culture and its Agricultural Applications*, pp. 3–20, Withers L. A. & Alderson, P. G., (eds.), Butterworths, London.

Connell, S. A. & Heale, J. B. (1986). Development of an *in vitro* selection system for novel sources of resistance to *Verticillium* wilt in hops. In *Plant Tissue Culture and its Agricultural Applications*, pp. 451–9, Withers, L. A. & Alderson, P. G. (eds.), Butterworths, London.

Crisp, P. & Gray, A. R. (1979). Successful selection for curd quality in cauliflower, using tissue culture. *Horticultural Research*, **19**, 49–53.

Csépló, A., Medgyesy, A. P., Hideg. E., Demeter., Márton, L. & Maliga, P. (1985). Triazine-resistant *Nicotiana* mutants from photomixotrophic cell cultures. *Molecular and General Genetics*, **200**, 508–10.

D'Amato, F. (1975). The problem of genetic stability in plant tissue and cell cultures. In *Crop Genetic Resources for Today and Tomorrow*, pp. 333–48 Frankel O. H. & Hawkes, J. G. (eds.), Cambridge University Press, Cambridge.

Daley, L. S., Thompson, M. M., Proebsting, W. M., Postman, J. & Jeong, B.-R. (1986). Use of fourth-derivative visible spectroscopy of leaf lamina in plant germplasm characterisation. *Spectroscopy*, **1**, 27–31.

De Bry, L. (1986). Robots in plant tissue culture. *IAPTC Newsletter*, **49**, 2–22.

Diner, A. M., Mott, R. L. & Amerson, H. V. (1984). Cultured cells of white pine show genetic resistance to axenic blister rust hyphae. *Science*, **224**, 407–8.

Dix, P. J. (1986). Cell line selection. In *Plant Cell Culture Technology*, pp. 143–201, Yeoman, M. M. (ed.), Blackwell, Oxford.

Doré, C. (1987). Applications of tissue culture to vegetable crop improvement. In *Proceedings VI International Congress of Plant Tissue and Cell Culture, Volume 2*. International Association for Plant Tissue Culture, Minneapolis, in press.

Dunwell, J. M. (1986). Pollen, ovule and embryo culture as tools in plant breeding. In *Plant Tissue Culture and its Agricultural Applications*, pp. 375–404, Withers, L. A. & Alderson, P. G. (eds.), Butterworths, London.

Ellis, R. H., Hong, T. D. & Roberts, E. H. (1985). *Handbook of Seed Technology for Genebanks*. International Board for Plant Genetic Resources, Rome.

Engelman, F., Duval, Y. & Dereuddre, J. (1985). Survie et prolifération d'embryons somatiques de palmier à huile (*Elaeis guineensis* Jacq.) après congélation dans l'azote liquide. *Comptes Rendus Hebdomadaires des Séances de L'Academie des Sciences*, **301** (Série III), 3, 111–16.

Gengenbach, B. G. (1984). *In vitro* pollination, fertilisation and development in maize kernels. In *Cell Culture and Somatic Cell Genetics, Volume 1*, pp. 276–82, Vasil, I. K. (ed.), Academic Press, New York.
Gerlach, W. L., Dennis, E. S. & Peacock, W. J. (1985). Approaches to the transformation of crop plants. In *Biotechnology in International Agricultural Research*, pp. 257–67. International Rice Research Institute, Los Baños, Manila.
Gleddie, S., Keller, W. A. & Setterfield, G. (1986). Production and characterisation of somatic hybrids between *Solanum melongena* and *S. sisymbriifolium*. *Theoretical and Applied Genetics*, **71**, 613–21.
Gould, A. R. (1986). Factors controlling generation of variability *in vitro*. In *Cell Culture and Somatic Cell Genetics of Plants, Volume 3*, pp. 549–67, Vasil, I. K. (ed.), Academic Press, New York.
Gowda, A. N. S. & Narayana, R. (1986). *In vitro* studies of spike disease in *Santalum album*. In *Proceedings VI International Congress of Plant Tissue and Cell Culture, Volume 1: Abstracts*, p. 303, Somers, D. A., Gengenbach, B. G., Biesboer, D. D., Hackett W. P. & Green, C. E. (eds.), International Association for Plant Tissue Culture, Minneapolis.
Grout, B. W. W. (1986). Embryo culture and cryopreservation for the conservation of genetic resources of species with recalcitrant seed. In *Plant Tissue Culture and its Agricultural Applications*, pp. 303–9, Withers, L. A. & Alderson, P. G. (eds.), Butterworths, London.
Guan, Y. Q., Gray, L. E. & Widholm, J. M. (1986). *In vitro* screening and selection of soyabean callus resistant to the culture filtrates of *Phialophora gregata. In Proceedings VI International Congress of Plant Tissue and Cell Culture, Volume 1: Abstracts*, p. 304, Somers, D. A., Gegenbach, B. G., Biesboer, D. D., Hackett, W. P. & Green, C. E. (eds.), International Association for Plant Tissue Culture, Minneapolis.
Gunn, R. E. & Day, P. R. (1986). *In vitro* culture in plant breeding. In *Plant Tissue Culture and its Agricultural Applications*, pp. 313–36, Withers, L. A. & Alderson, P. G. (eds.), Butterworths, London.
Harper, P. C., Fordyce, W. A. & Rankin, P. A. (1986). Constraints upon the use of micropagation for the Scottish Strawberry Certification Scheme. In *Plant Tissue Culture and its Agricultural Applications*, pp. 205–10, Withers, L. A. & Alderson, P. G. (eds.), Butterworths, London.
Hibberd, K. A., & Green, C. E. (1982). Inheritance and expression of lysine and threonine resistance selected in maize tissue cultures. *Proceedings of the National Academy of Sciences* (USA), **79**, 559–63.
Hussey, G. (1986). Problems and prospects in the *in vitro* propagation of herbaceous plants. In *Plant Tissue Culture and its Agricultural Applications*, pp. 69–84, Withers, L. A. & Alderson, P. G. (eds.), Butterworths, London.
IBPGR (1986). *Design, Planning and Operation of* In Vitro *Genebanks*. International Board for Plant Genetic Resources, Rome.
John, A. (1986). Vitrification in Sitka Spruce cultures. In *Plant Tissue Culture and its Agricultural Applications*, pp. 167–74, Withers, L. A. & Alderson, P. G., (eds.), Butterworths, London.
Kartha, K. K. (1986). Production and indexing of disease-free plants. In *Plant Tissue Culture and its Agricultural Applications*, pp. 219–38, Withers, L. A. & Alderson, P. G. (eds.), Butterworths, London.
Kartha, K. K., Leung, N. L. & Mroginsky, L. A. (1982). *In vitro* growth responses and plant regeneration from cryopreserved meristems of cassava (*Manihot esculenta* Crantz). *Zeitschrift für Pflanzenphysiologie*, **107**, 133–40.

Kevers, C. & Gaspar, Th. (1985). Soluble, membrane and wall peroxidases, phenylalanine ammonia-lyase and lignin changes in relation to vitrification of carnation tissues cultured *in vitro*. *Journal of Plant Physiology*, **118**, 41–8.

Kristensen, H. R. (1984). Potato tissue culture. In *Micropropagation of Selected Rootcrops, Palms, Citrus and Ornamental Species*, pp. 25–49. FAO Plant Production and Protection Paper, 59. Rome: Food and Agriculture Organization of the United Nations.

Kung, S. D. (1983). Fraction-1 protein and chloroplast DNA as genetic markers. In *Handbook of Plant Cell Culture, Volume 1*, pp. 583–606, Evans, D. A., Sharp, W. R., Ammirato, P. V. & Yamada, Y. (eds.), Macmillan, New York.

Larkin, P. J. & Scowcroft, W. R. (1981). Somaclonal variation – a novel source of variability from cell cultures. *Theoretical and Applied Genetics*, **60**, 197–204.

Lebrun, L., Rajasekaran, K. & Mullins, M. G. (1985). Selection *in vitro* for NaCl-tolerance in *Vitis rupestris* Scheele. *Annals of Botany*, **56**, 733–9.

Lopez, M. M. & Navarro, L. (1981). A new *in vitro* inoculation method for Citrus Canker diagnosis. In *Proceedings of the International Society of Citriculture, 1981, Volume 1*, pp. 399–401.

Lopez, M. M., Arregui, J. M. & Navarro, L. (1983). Nueva técnica de inoculación con *Xanthomonas ampelina* de plantas de viña cultivadad *in vitro* para el estudio de las relaciones huésped/parásito. In *Proceedings del II Congreso Nacional de Fitopathologia*, pp. 45–50.

Lörz, H., Brown P. T., Göbel, E. & de la Pena, A. (1987). Gene transfer in cereals. In *Proceedings VI International Congress of Plant Tissue and Cell Culture, Volume 2*, International Association for Plant Tissue Culture, Minneapolis, in press.

Mullins, M. G. (1987). Propagation and genetic improvement of temperate fruits: The role of tissue culture. In *Proceedings VI International Congress of Plant Tissue and Cell Culture, Volume 2*. International Association for Plant Tissue Culture, Minneapolis, in press.

Nabors, M. & Dykes, T. A. (1985). Tissue culture of cereal cultivars with increased salt, drought and acid tolerance. In *Biotechnology in International Agricultural Research*, pp. 121–38. International Rice Research Institute, Los Baños.

Navarro, L. (1981*a*). Citrus shoot-tip grafting *in vitro* (STG) and its applications: a review. *Proceedings of the International Society for Citriculture, Volume 1*. pp. 452–6.

Navarro, L. (1981*b*). Effect of Citrus Exocortis viroid (CEV) on root and callus formation by stem tissue of Etrog Citron (*Citrus medica* L.) cultured *in vitro*. *Proceedings of the International Society for Citriculture, 1981, Volume 1*, pp. 437–9.

Navarro, L., Ballester, J. F., Juarez, J., Pina, J. A., Arregui, J. M. & Bono, R. (1981*c*). Development of a program for disease-free citrus budwood in Spain. In *Proceedings of the International Society for Citriculture, Volume 1*, pp. 70–3.

Ng, S. Y. C. (1986). *In vitro* tuberisation in *Dioscorea rotundata* Poir (white yam) – A means for germplasm exchange and propagation. In *Proceedings VI International Congress of Plant Tissue and Cell Culture, Volume 1: Abstracts*, p. 255, Somers, D. A., Gengenbach, B. G., Biesboer, D. D., Hackett, W. A. & Green, C. E. (eds.), International Association for Plant Tissue Culture, Minneapolis.

Novak, F. J. (1984). Somaclonal variation in garlic tissue culture as a new breeding system. *Eucarpia 3rd Allium Symposium 4–6 September 1984*, pp. 39–43. The Institute for Horticultural Plant Breeding (IVT), Wageningen.

Ostry, M. E., Ettingert, L., Hackett, W. P., Read, P. E. & Skilling, D. D. (1986). Development of bioassays to identify variant *Populus* cells and regenerated plants resistant to *Septoria musiva*. In *Proceedings VI International Congress of Plant Tissue and Cell Culture, Volume 1: Abstracts*, p. 303, Somers, D. A., Gengenback, B. C., Biesboer, D. D., Hacket, W. P. & Green, C. E. (eds.), International Association for Plant Tissue Culture, Minneapolis.

Payne, P. I., Corfield, K. G. & Blackman, J. A. (1979). Identification of a high-molecular-weight subunit of glutenin whose presence correlates with bread-making quality in wheats of related pedigree. *Theoretical and Applied Genetics*, **55**, 153–9.

Peeters, J. P. & Williams, J. T. (1984). Towards better use of genebanks with special reference to information. *Plant Genetic Resources Newsletter*, **60**, 22–32.

Redenbaugh, K., Paasch, B. D., Nichol, J. W., Kossler, M. E., Viss, P. R. & Walker, K. A. (1986). Somatic seeds: encapsulation of asexual plant embryos. *Biotechnology*, **4**, 797–801.

Riggs, J. T., Sanada, M., Morgan, A. G. & Smith, D. B. (1983). Use of acid gel electrophoresis in the characterisation of 'B' hordein protein in relation to malting quality and mildew resistance in barley. *Journal of the Science of Food and Agriculture*, **34**, 576–86.

Scowcroft, W. R. (1985*a*). Somaclonal variation: The myth of clonal uniformity. In *Genetic Flux in Plants*, pp. 217–45, Hohn, B. & Denis, E. S. (eds.), Springer Verlag, Vienna.

Scowcroft, W. R., Davies, P., Ryan, S. A., Brettell, R. I. S., Pallota, M. A. & Larkin, P. J. (1985*b*). The analysis of somaclonal mutants. In *Plant Genetics; UCLA Series, Volume 35*, pp. 799–815, Freeling, M. (ed.), Alan R. Liss, New York.

Shen, J.-H., Li, H.-F., Chen, Y.-Q. & Zhang, Z.-H. (1983). Improving rice by anther culture. In *Cell and Tissue Culture Techniques for Cereal Crop Improvement*, Institute of Genetics, Academia Sinica, Beijing & International Rice Research Institute, Manila, pp. 183–205. Science Press, Beijing; Gordon and Breach, New York.

Short, K. C., Warburton, J. & Roberts, A. V. (1987). *In vitro* hardening of cultured cauliflower and chrysanthemum plantlets to humidity. *Acta Horticulture*, in press.

Simpson, M. J. A. & Withers, L. A. (1986). *Characterisation of Plant Genetic Resources using Isozyme Electrophoresis: A Guide to the Literature*. International Board for Plant Genetic Resources, Rome.

Teweles, L. William & Co. (1984). Plant genetic engineering study sets timetable for improvements. *Theoretical and Applied Genetics*, **68**, 107–8.

Tovar, P., Estrada, R., Schilde-Rentschler, L. & Dodds, J. H. (1985). Induction and use of *in vitro* potato tubers. *CIP Circular*, **13**, 1–5.

Utkhede, R. S. (1986). *In vitro* screening of the world apple germplasm collection for resistance to *Phytophthora cactorum*. *Scientia Horticulturae*, **29**, 205–10.

Uyen, Nguyen Van (1985). The use of tissue culture in plant breeding in Vietnam. In *Biotechnology in International Agricultural Research*, pp. 45–8. International Rice Research Institute, Los Baños.

Van Regenmortel, M. H. V. (1982). *Serology and Immunochemistry of Plant Viruses.* Academic Press, New York.

Walbot, V. & Cullis, C. A. (1985). Rapid genomic change in higher plants. *Annual Review of Plant Physiology*, **36**, 367–96.

Walkey, D. G. A. (1987). Micropropagation of vegetables. In *Micropropagation in Horticulture: Practice and Commercial Problems*, Alderson, P. G. & Dulforce, W. M. (eds.), Institute of Horticulture/University of Nottingham, Nottingham, in press.

Wanas, W. H., Callow, J. C. & Withers, L. A. (1986). Growth limitation for the conservation of pear genotypes. In *Plant Tissue Culture and its Agricultural Applications*, pp. 285–90, Withers, L. A. & Alderson, P. G. (eds.), Butterworths, London.

Wang, P. J. & Hu, C. Y. (1982). *In vitro* mass tuberisation and virus-free seed-potato production in Taiwan. *American Potato Journal*, **59**, 33–7.

Weeden, N. F., Provvidenti, R. & Marx, G. A. (1984). An isozyme marker for resistance to bean yellow mosaic virus in *Pisum sativum*. *Journal of Heredity*, **75**, 411–12.

Wenzel, G., Foroughi-Wehr, B., Deimling, S. & Schumann, R. (1987). Breeding for disease-resistant crop plants by cell culture techniques. In *Proceedings VI International Congress of Plant Tissue and Cell Culture, Volume 2.* International Association for Plant Tissue Culture, Minneapolis, in press.

Wetter, L. & Dyck, J. (1983). Isoenzyme analysis of cultured cells and somatic hybrids. In *Handbook of Plant Cell Culture, Volume 1*, pp. 608–28, Evans, D. A., Sharp, W. R., Ammirato, P. V. & Yamada, Y. (eds.), Macmillian, New York.

Wheelans, S. K. & Withers, L. A. (1984). The IBPGR international database on *in vitro* conservation. *Plant Genetic Resources Newsletter*, **60**, 33–8.

Williams, R. R. & Taji, A. M. (1986). Tissue culture applications in plant conservation. In *Proceedings VI International Congress of Plant Tissue and Cell Culture, Volume 1: Abstracts*, p. 428, Somers, D. A., Gengenbach, B. G., Biesboer, D. D., Hackett, W. P. & Green, C. E. (eds.), International Association for Plant Tissue Culture, Minneapolis.

Withers, L. A. (1979). Freeze-preservation of somatic embryos and clonal plantlets of *Daucus carota*. *Plant Physiology*, **63**, 460–7.

Withers, L. A. (1985*a*). *Minimum requirements for receiving and maintaining tissue culture propagating material.* FAO Plant Production and Protection Paper, 60, Food and Agriculture Organization of the United Nations, Rome.

Withers, L. A. (1985*b*). Cryopreservation of cultured cells and meristems. In *Cell Culture and Somatic Cell Genetics of Plants, Volume 2*, pp. 254–316, Vasil, I. K. (ed.), Academic Press, New York.

Withers, L. A. (1986*a*). Cryopreservation and genebanks. In *Plant Cell Culture Technology*, pp. 96–140, Yeoman, M. M. (ed.), Blackwell, Oxford.

Withers, L. A. (1987). *In vitro* methods for germplasm collecting in the field. *Plant Genetic Resources Newsletter*, **69**, 2–6.

Withers, L. A. & King, P. J. (1980). A simple freezing unit and cryopreservation method for plant cell suspensions. *Cryoletters*, **1**, 213–320.

Withers, L. A. & Williams, J. T. (1982). (eds.) *Crop Genetic Resources – The Conservation of Difficult Material.* Paris: IUBS/IGF/IBPGR, IUBS Series B42.

Withers, L. A. & Williams, J. T. (1985*c*). Research on the long-term storage and exchange of *in vitro* plant germplasm. In *Biotechnology in International Agricultural Research*, pp. 11–24. International Rice Research Institute, Los Baños.

Withers, L. A. & Williams, J. T. (1986*b*) *IBPGR Research Highlights*: In Vitro *Conservation*. Rome: International Board for Plant Genetic Resources.

Yepes, L. M. & Aldwinckle, H. S. (1986). Screening apple cultivars for resistance to *Venturia inaequalis in vitro*. In *Proceedings VI International Congress of Plant Tissue and Cell Culture, Volume 1: Abstracts*, p. 304, Somers, D. A., Gengenback B. G., Biesboer, D. D., Hackett, W. P. & Green, C. E. (eds.), International Association for Plant Tissue Culture, Minneapolis.

Yidana, J. A., Withers, L. A. & Ivins, J. D. (1987). Developments of a simple method for collecting and propagating cocoa germplasm *in vitro*. *Acta Horticulturae*, in press.

Zenkteler, M. (1984). *In vitro* pollination and fertilisation. In *Cell Culture and Somatic Cell Genetics, Volume 1*, pp. 269–75, Vasil, I. K. (ed.), Academic Press, New York.

20
Screening for resistance to diseases

P.H. WILLIAMS

Introduction

Experience by plant breeders over the past century has demonstrated that plant germplasm collections are rich sources of resistances to pests and pathogens. The lineages of many important resistant cultivars can be traced to germplasm collections (Peterson, 1975). Yet most collections largely remain unexplored for their potential as reservoirs of resistance (Harlan, 1977). In the future, effective utilisation of genetic resources in resistance breeding programmes will depend increasingly upon an understanding of the concepts and technology employed in identifying, and deploying useful levels of resistance. Central to the utilisation of resistance in breeding programmes is a thorough understanding of what is known as the *disease screen.*

Disease screen

Screening is the process of growing segregating populations of plants or propagules under established environmental conditions such that individuals exhibiting desired phenotypes may be identified and selected for further propagation. An essential characteristic of the disease screen is that the environmental parameters be such as to permit reliable and repeated selection of the desired host phenotypes. The screen consists of three major interacting components, the plant, the pathogen and the environment, each of which may be described, quantified, and used in numerical estimations of value in plant breeding.

In a typical breeding programme, the disease screen yields phenotypes which provide a basis for understanding the genetics of resistance. Breeding strategies based on this knowledge will lead to the development and evaluation of resistant lines, and ultimately to the deployment of resistant cultivars. Each of the many stages in the resistance breeding process in dependent on the reliable reproduction of the desired *interaction phenotypes* provided repeatedly via the disease screen.

Interaction phenotype (IP), is a general term that encompasses all phenotypes resulting from the interaction of the plant, a symbiont and the environment. The concept of IP enables more precise descriptions of interactions which may be quantified in terms of positive or negative

Fig. 20.1. Disease resistance breeding scheme and the disease screen. Dotted lines and arrow feed back from various stages in the resistance breeding programme to where the disease screen must be executed

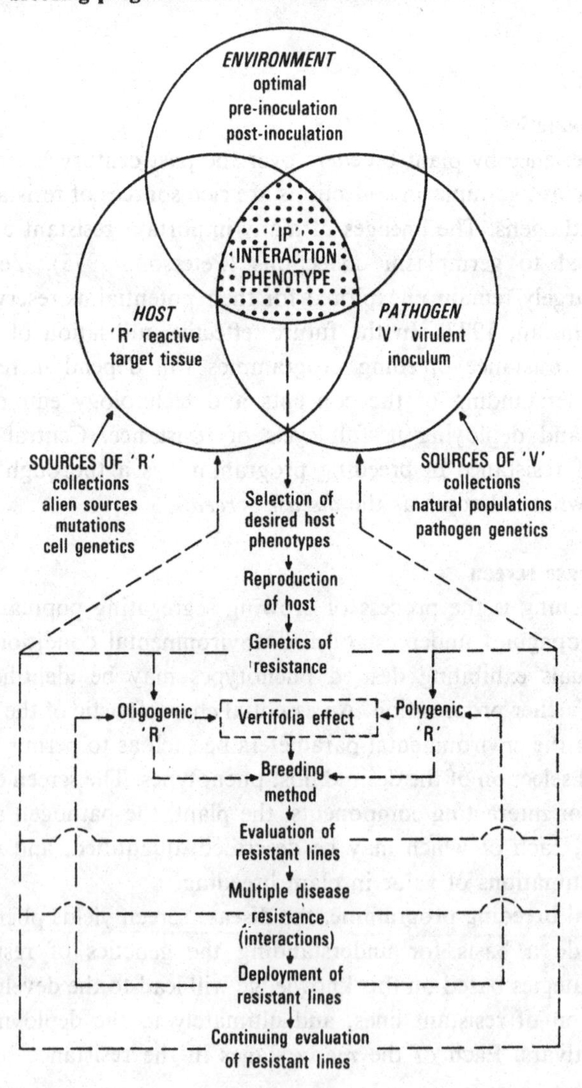

effects on the host or the symbiont. Under the concept of IP, diverse manifestations of the interaction, ranging from the beneficial effects of mutualistic symbionts to the expression of disease symptoms or signs of a pathogen, may be accommodated (Table 20.1). Interaction phenotypes are designed by descriptors that are meaningful to the context in which a particular plant or crop is being considered. Farmers generally view their crop from the perspective of damage, yield reduction and economic loss; epidemiologists may be more interested in inoculum production as it affects epidemic development, whereas physiologists may examine the host response leading to reduced colonisation or disease (Table 20.2).

Host and pathogen populations interact over time in a range of environments and exhibit a wide range of IPs. Variation in the expression of the IP will result from variation in the environment or in the individual components of host and pathogen populations. An appreciation of the potential expression of the IP may be gained by examining models representing host and pathogen interacting over a range of environments (Figure 20.2). In model A, the components may be viewed as interacting within the limits that are normally found in natural or undisturbed ecosystems in which stabilising selection is a predominant force guiding the genetic interrelations between host and pathogen populations. Model B depicts an agricultural system, or perhaps an experimental disease

Table 20.1. *Manifestations of the interaction phenotype*

Preinfection
- infection frequency reducing

Postinfection
- signs of the symbiont (pathogen)
  - propagule type
  - total number of propagules
  - rate of propagule production
  - distribution of propagules
  - viability and longevity of propagules
- symptoms of disease (expression of symbiosis)
  - hypersensitive response
  - necrosis, chlorosis
  - wilt
  - tissue maceration, collapse
  - stunting, overgrowth, malformation
  - delay in growth, flowering, fruiting
  - loss of flowering, fruiting
  - enhanced growth, enhanced respiration
  - green islands, metabolite accumulation
- mixed signs and symptoms
  - infection type, reaction type, lesion type, aegricorpus

Table 20.2. *Terms of reference for considering the interaction phenotype*

Agronomic	– yield loss or increase – damage – delay in maturation
Epidemiologic	– infection efficiency, frequency (infections/spore) – infection type (host and pathogen) – incubation period, latent period – infectious period (period of sporulation) – spore yield production (spores/leaf area) – cumulative sporulation (spore/colony/time) – sporulation capacity (spore/colony/life time) – spore type
Economic	– loss, gain
Physiological	–
Biochemical	–
Cytological	–
Genetic	–
Mathematical	–
Morphological	–
Developmental	–

Fig. 20.2. Models depicting the interaction of host, pathogen and environment in producing the interaction phenotype (IP). A. Fullest expression of IP in the natural ecosytem. B. Limited expression of IP in agroecosystem of experimental disease screens. C. Highly limited expression of the IP in experimental or *in vitro* screens. D. and E. Examples of failure to produce an IP due to inappropriate control over one or more of the interacting components.

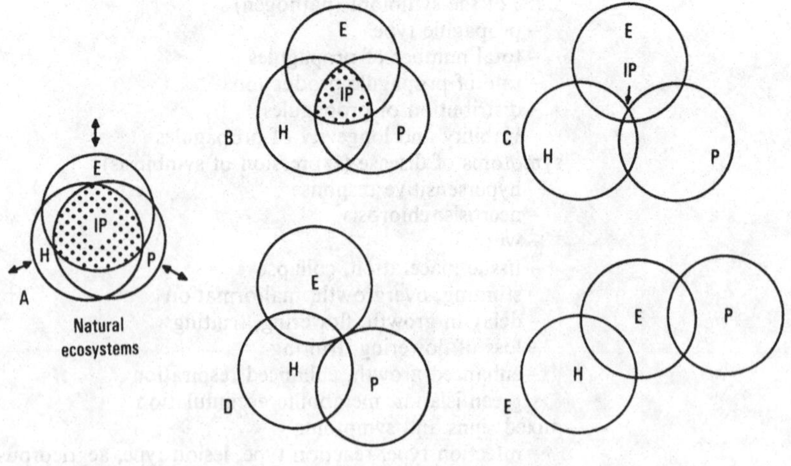

screen, in which only a portion of the host and pathogen populations are interacting in more narrowly defined environments. Thus, only a fraction of the full potential of the IP is represented. Model C depicts highly controlled experimental systems in which only restricted components of the host, pathogen, and environment are involved in the production of a limited portion of the IP. Such a model could represent many forms of seedling, tissue and *in vitro* cell selection systems in which many aspects of the environment are carefully regulated for the purposes of attaining efficiency and reproducibility. In determining the specific parameters of host, pathogen, and environment to be used in the experimental screen, it is essential that the resulting IP be representative of, or highly correlated with, resistance expressed in the crop.

Execution of Model D would result in no IP. In the hands of an inexperienced researcher, and in the absence of susceptible controls, such an experiment could lead to the erroneous conclusion that the host was immune. This model reveals the absence of an environment suitable to support infection by the pathogen. Model E again represents the case where no IP is produced but where the environment is satisfactory for both host and pathogen; in this case the pathogen has been inappropriately positioned on the host to permit infection. Such models may be of value in interpreting results derived from experimental disease screens.

Standardisation of the interaction phenotype

In the disease screen it is essential that criteria for evaluating the IP adequately accommodate the range of phenotypic expression expected. The descriptors should be clear, relevant to the context in which the IP is being viewed, and easily recorded. In general, when a wide range of host genotypes is examined against representative examples of a pathogen species, wide variation in components of the IP are observed. Historically, breeders and pathologists have developed descriptors and scales to categorise the observed variation into classes that have served their particular needs. Commonly, breeding objectives have required the categorisation of phenotypes into simple resistant or susceptible classes. Where the interactive effects of genes results in the incomplete expression of resistance, the addition of one or more intermediate class facilitates genetic analysis. With many of the main food crops, particularly cereals, where resistance breeding is underway worldwide, researchers have developed more detailed descriptors and complex scaling systems involving quantification of various components of the IP (such as pustule size and number, amount of host chlorosis and necrosis, and distribution of lesions on the host). However, for most diseases on most crops, the development and use of descriptors has lacked any standardisation. Once the limits and ranges of effective operation for each of the essential parameters in the disease screen are known, then a protocol can be developed which will have a high degree of precision and repeatability. The resultant of a carefully constructed disease screening protocol will be IPs that reflect the genetic variation inherent in the host population in relation to the genotypes of the pathogen. Codification of the IP, therefore, becomes the means by which comparisons among selections can be made. In the past, the lack of standardisation of IP descriptors has resulted in substantial losses of potentially useful resistant germplasm from breeding programmes and loss of information on plant genetic resources.

For each particular characteristic of the IP, a range of expression is possible. Experience has shown that the genetic control underlying the IP does not differ from that conditioning other plant phenotypic characteristics. In considering principals upon which IP descriptors may be standardised, it is important to provide a scaling system which will accommodate the full range of expression. The 10 point scale of 0–9 is convenient in that it recognises the need to describe and quantify a range of expression possible, Table 20.3. The description and measurement of continuous variation in large populations can be tedious and is often influenced by subjective forces such as mental exhaustion in the person evaluating the screened plants. The 10 point scale is usually easier to use than more complex scaling systems that tax the concentration and endurance of the evaluator. When variation falls into discrete classes, the

Table 20.3. *Descriptors and scaling for the non-metric*[a] *quantification of the interaction phenotype involving resistance and susceptibility*

0 = immune, absence of any interaction
1 = very high resistance
2 = high resistance
3 = resistant
4 = moderate resistance (weak susceptibility)
5 = intermediate resistance/susceptibility
6 = moderate susceptibility (weak resistance)
7 = susceptible
8 = high susceptibility
9 = very high susceptibility

[a] Non-metric means that the trait or traits are not quantified in specified units of measure.

10 point scale is adaptable and can be used by selecting numbers chosen to reflect the best approximation of the phenotype expressed by each class, e.g. classes described as *resistant*, *intermediate* and *susceptible* can be scaled 3, 5, 7 respectively. As greater variation in the IP becomes resolvable, either through increased familiarisation with the material or though the introduction of more variable genotypes, additional descriptions and scales, e.g. 2, 4, 6, 8 can be added.

An important function of the scaling system is the precise documentation of the phenotypic make-up of a collection or seed stock. The most informative way of designating the genetic potential of a collection is to record the number of individuals in each IP class. These numerical data can be used in the calculation of IP indices which, depending on the method of calculation, could reflect the IP potential of a collection. Such indices do not give precise information on the frequency or range of IP classes in a collection and as such are of limited value.

In constructing IP descriptors and rating scales it is important to determine precisely what components of the IP will be described. IPs are multidimensional in that they may consist of more than one quantifiable manifestation of the host (symptoms) and/or the pathogen (signs). Descriptors therefore must be chosen carefully so as to identify as clearly as possible characteristics being used for comparison and classification. The perspectives of the evaluation can be important in the choice of descriptors, e.g. a breeder selecting for components of disease epidemic rate reduction in wheat powdery mildew may use criteria relating to the amount and rate of spore production (Table 20.2), whereas a breeder selecting for a reduction in yield loss may measure the amount of damage to the plant.

For IP descriptors to be comparable it is important that the conditions of the disease screen be carefully documented. Precise specifications of host growth, target tissue, inoculum preparation, inoculation procedures, environmental regimes and time of the evaluation after inoculation are all important when producing comparable descriptions.

Although the use of the 0–9 scale is an efficient way of introducing standardisation in disease screening, a major drawback of the system is that it tends to partition into discrete classes phenotypic variation that may be continuous. Furthermore, the scaling of IP is not likely to reflect adequately the quantitative resultants of the IP. Where accurate quantitative estimates of the IP are required, more precise means of quantification can be devised. Direct spore counts can be obtained from experimental systems and provide important data on the epidemiologic potential of various forms of resistance. Direct quantification of the IP should be considered more desirable than descriptive estimators, providing that the measurements can be made easily. With improvement in the design of electronic scanners linked to computerised analysers, greater precision and efficiency in the quantification of the IP is becoming possible.

The absence of standardisation of descriptors stems largely from a lack of understanding of the interplay of physical, chemical, and biological processes underlying the development and expression of the IP in the disease screen.

The host

Effective breeding for resistance in most crops requires a knowledge of the genetics, ecology, developmental and reproductive biology, physiology and cultivation of the crop. The rapidly developing technologies in plant cell and molecular biology also have increasing potential for exploitation in resistance breeding (see Withers and also Peacock, this volume).

Sources of resistance

Collections of adapted and unadapted plant germplasm are rich sources of resistance to many diseases. The intrinsic value of these collections is that they provide reserves of, as yet, undetected resistances. In the vast majority of cases, when resistance has been sought in collections, it has been found (Harlan, 1977). Conventional wisdom has led breeders to seek resistance firstly among well-adapted genotypes within the primary gene pool (Leppik, 1970), then among more distantly related primitive and wild forms in the primary and secondary gene pools, and finally among more remotely related species.

Efforts to generate resistance using chemical or physical mutagenesis have been extensive (IAEA, 1977). Nonetheless, although the idea of induction of resistant mutants in well-adapted cultivars has merit, mutation breeding for resistance has met with relatively limited success (Simons, 1979).

The potential for recovering resistant plants regenerated from populations of cells grown in tissue or suspension culture has stimulated research in many countries (Daub, 1986). Technologies designed to recover resistant variants from mutagenised and non-mutagenised cell populations and from cell fusion products will require the development of selective screens that will be applied at various stages in the regeneration process from the protoplast to the mature plant. An essential criterion for evaluation of the resistant regenerants lies in the demonstration of transmission and expression of the resistant phenotype in subsequent generations of sexually produced progeny.

Target tissues

When new sources of resistances are being sought from germplasm collections, no presumptions should be made as to how resistance will be manifest during the development of the crop. For this reason it is important that the initial evaluation of collections be made in the field under epidemic conditions with as broad a representation of pathogenic variation and as favourable environmental conditions for epidemic development as possible.

Following the identification of effective resistance in the field, screening methods should be devised that would evaluate the potential for that resistance to be expressed under the more controlled environments in the laboratory. In addition to enabling greater precision in the production of the IP through the control of the environment and the delivery of inoculum, laboratory methods may provide the opportunity of screening for multiple disease resistance. By partitioning the plant into various areas to be targeted for the placement of inoculum, each plant in a population may be evaluated for resistance to many pathogens, or to different strains of a pathogen. When working with forms of resistance which are expressed both in seedlings and mature plants, closely grown seedlings

Fig. 20.3. Target tissue locations on a *Brassica* seedling for 10 pathogens. Ab = *Alternaria*, Ac = *Albugo*, Ar = *Aphanomyces*, Ec = *Erwinia*, Foc = *Fusarium*, Pb = *Plasmodiophora*, Pl = *Phoma*, Pp = *Peronospora*, TuMB = Turnip Mosaic Virus, Xc = *Xanthomonas*

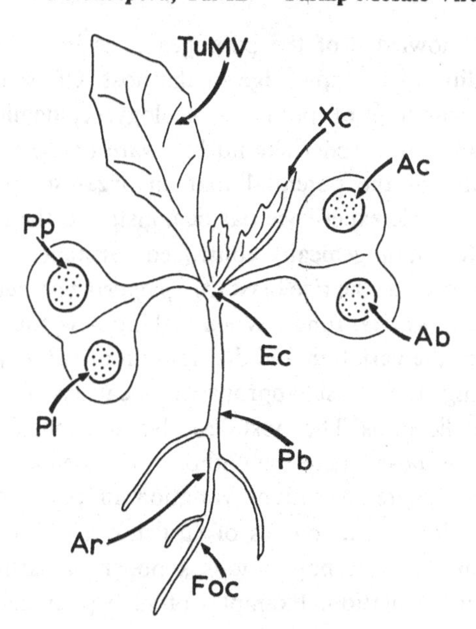

targeted with different pathogens provide a highly efficient means of identifying multiple disease resistant phenotypes (Fig. 20.3).

Whether laboratory methods will be effective in the disease screen, depends largely on whether the particular resistance expressed in the field will also be expressed at other stages in the plant's development. Resistance is the diverse manifestation of a wide array of different physiological and morphological mechanisms many of which are under complex genetic control. Expression of resistance is sometimes highly tissue-specific and restricted to certain organs or to expression at certain developmental or physiological stages of the plant growth. In spite of this, resistance to many pathogens is expressed generally in the tissues of the plant, from the young seedling through to the mature plant. In these instances, the development of multitargeted seedling screens can be an effective approach in resistance breeding.

It is too early yet to know whether *in vitro* screens using cell and protoplast technologies will be generally useful in resistance breeding. For certain types of pathogens such as viruses and some bacteria and fungi whose primary pathogenic determinants are host-genotype specific toxins, it may be possible to devise effective *in vitro* selection systems. For the majority of pathogenic symbionts which have developed complex relationships in particular niches on their host, it is unlikely that *in vitro* selection schemes will be able to identify usable resistance.

The pathogen

Comprehensive knowledge of the pathogen is as important to effective resistance breeding as is knowledge of the host. Of particular importance is an understanding of pathogen ecology, epidemiology, reproduction and genetics. The breeder should be aware of the existing pathogenic variability and of the potential that an organism has for generating new variants. Ample experience has demonstrated the futility of selecting exclusively for monogenically controlled resistance against highly variable organisms such as the rice blast (*Magnaporthe oryzae*) and late blight (*Phytophthora infestans*) fungi. When seeking new sources of resistance, a knowledge of the variation and distribution of pathotypes is important in determining the most appropriate locations for field screening of germplasm collections. The most desirable locations for field screening of collections are those where the pathogen is endemic and/or seasonally epidemic, and where maximum variation in pathogenicity exists in the local population. Such sources of virulence are commonly found where diversity in the host populations supports a pathotypic diversity in the pathogen population. Examples of such pathogenically diverse populations would be the population of *P. infestans* found in the Toluca Valley of Mexico where the A and B mating types of the fungus permit recombination of virulence, and the highly diverse populations of *Puccinia* spp. found on the wild Gramineae and *Rhammus* spp. in the mountainous terrain of the Middle East (Dinoor & Eshed, 1984).

In the absence of naturally occurring pathogenic diversity, plant pathologists have created disease nurseries to be used in field screening of crops. Cereal breeders commonly use rust nurseries surrounded by hedges of the alternate host species to provide inoculum of recombinant pathotypes. Pathologists have developed specialised root-rot plots in which mixtures of pathotypes are maintained through the cropping of suitable combinations of resistant and susceptible hosts.

The balance and mix of pathotypes found in field nurseries can be monitored by including in the plots, differential cultivars containing genotypes that are capable of identifying the various components of pathogenicity in the population. This is known as *virulence analysis* and is an important part of the screening process. In field screens with airborne pathogens the proportion of virulence genes in the populations of a pathogen may change significantly during the growing season (Wolfe & Schwarzbach, 1978).

Disease nurseries developed for the selection of resistance to root rot complexes require particularly careful surveillance by pathologists knowledgeable in soil microbiology, chemistry and physics (Lumsden *et al.*, 1976). Such plots are likely to contain several pathotypes of different root invading species each of which may exert an influence on the IP at different times in the plant's development. In such screens it may be important to determine the relative contributions of various species to the IP.

The determination of where plant germplasm collections may be most effectively evaluated may be influenced by the quarantine laws of a country. The establishment of field nurseries carries measured risks which should be carefully assessed. If field screening involves the endemic population, risks are likely to be low. If new pathotypes are being introduced to a location where the potential for new pathogenic recombinants exists then careful risk assessment should be made, and if necessary, screening under containment required.

Collections

Isolates of pathogens for the disease screen may be collected from the field or obtained from collections (APS, 1981). National microbial collections serve as type culture collections and as long-term repositories for a limited number of major genetic variants of plant pathogens. Specialised and working collections usually contain a wider range of variants than national collections, and depending on their particular objectives, working collections may represent most of the pathotypes found in the natural populations. The pathogenicity of some organisms may be reduced or lost upon storage. In such cases, inoculum may have to be collected from the field.

Inoculum

Inoculum consists of pathogen propagules, in many forms, capable of establishing infections. Inoculum should be quantifiable both in terms of its viability and its infectivity. The preparation of inoculum prior to quantitative delivery to the target tissue is known as *formulation* and is an important step in the screening procedure. Conditions of formulation vary widely with various pathogens. Many viruses and bacteria are stabilised in buffers. Zoospores often require pure water and some ascospores need an exogenous supply of carbon to become infective. Rusts and powdery mildew spores require no aqueous carrier but are sometimes diluted in talcum powder or volatile oils. Inoculum of organisms which do not produce infective propagules may be formulated as colonised units of organic substrate such as seeds, pieces of straw or wood chips. Inoculum should be formulated such that carefully quantified numbers of infective propagules may be delivered to the target tissue.

An important aspect of formulation is in the establishment of a dosage-response curve in which the relationship of inoculum dosage to IP response is determined. Potentially useful resistance may be lost when inoculum is applied at such high doses that the plant's resistance is overcome. By carefully establishing a dosage–response curve on various host tissues, in various environments, the optimal dosage for identification of desired IP may be determined.

Environment

An understanding of the complex role of the environment in contributing to the IP is equally as important as the knowledge of host and pathogen. Over the course of time, the pathogen and host have co-evolved within a range of physical, chemical and biological parameters comprising many interacting subcomponents of the environment (Table 20.4). Though there may be one set of environmental conditions which represents the optimum for the expression of the IP, normally a range of conditions exists under which a particular IP can be expressed. The successful execution of a disease screen is dependent upon an understanding of the range and limits of the various limiting environmental parameters under which the disease screen can be carried out.

In the field, all components of the environment are operational. Depending on the pathogen and the host, different subcomponents of the environment have varying roles in influencing the IP. Temperature, radiation, water, oxygen and minerals are critical factors in the growth of the plant. For pathogens adapted to the rhizosphere, soil temperature, moisture, acidity, and the mix of other organisms living in the rhizosphere may be important determinants of whether infection is successfully established. In field screens there is usually little opportunity for environmental regulation, rather those major limiting parameters of growth and IP expression such as temperature, soil moisture, relative humidity and radiation can be monitored. Such records may be useful in the interpretation of results of the screen.

The primary reason for bringing the disease screen into the laboratory is to achieve greater control over the environment (Walker, 1965). A wide range of equipment which will both monitor and regulate various components of the environment can be constructed. The range of influence of various physical and chemical parameters may be explored in relation to the expression of the IP. Likewise, components of the biological environment can be eliminated or manipulated so as to examine their interactions with the pathogen and host.

By systematically examining the various components of the environment, those essential to the expression of the IP can be determined and incorporated into efficient experimental protocols that will ensure a high degree of repeatability in the disease screen (Williams, 1981).

Rather than being viewed as fixed in time, the disease screen can be considered as a dynamic progression of the host, the pathogen and the interaction phenotype. In the context of time the environment must also be viewed as a progression of varying parameters each optimised to meet the needs of a particular stage in the screen. Over the time course of the disease screen, consideration must be given to as many as eight potentially

Table 20.4. *Environmental parameters in the disease screen*

Physical	*Chemical*	*Biological*
Temperature	Water	Symbionts
Radiation	Oxygen	– parasites
Gravity	Inorganics	– antagonists
Air velocity	Organics	– predators
Soil structure	Buffer, pH, CO_2	– mutuals
	Exchange capacity	

distinct environments (Figure 20.4). Prior to inoculation both the host and the pathogen often require separate environments for their propagation (the host and pathogen propagational environments, 1 and 3 in Figure 20.4). Prior to inoculation the pathogen propagules may be formulated in ways that ensure optimised distribution and infectivity on the host (the pathogen preparational environment, 4 in Figure 20.4). For some screens, preconditioning of the plants with stresses such as high or low light or moisture, may be a prerequisite to inoculation (the host preparational environment, 2 in Figure 20.4). Following inoculation special environmental conditions are usually required to ensure that infection of the host will take place (the infection-inducing environment, 5). The incubation period may be carried out in an environment (the incubational environment, 6) which favours the colonisation of the host by the pathogen. Pathogens such as the downy mildews require high relative humidity for the production of sporophores and for the full development of the IP (the IP-inducing environment, 7). After the process of selection from the disease screen, plants often have sustained infection from more than one pathogen and frequently require treatment with

Fig. 20.4. Progression of different environments required in the disease screen.

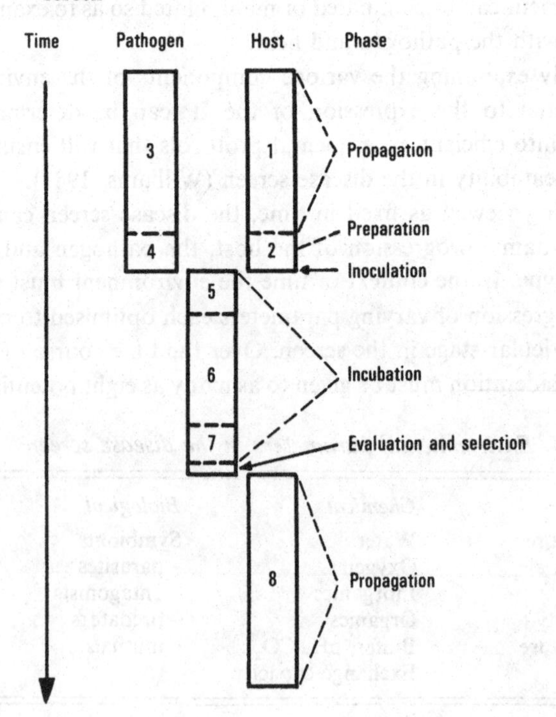

systemic fungicides and other specialised conditions to ensure their recovery (the propagational environment, 8).

The use of controls in the disease screen

An essential component of the disease screening procedure is the inclusion of plants of known phenotypic response as controls. Controls provide the necessary assurance that the many variables of environment, pathogen and host are operational within the range that will provide satisfactory IP responses. Of the various controls that can be included in the disease screen, the susceptible phenotype is the most important. This susceptible control should be as homogeneous as possible, genetically, and provide a highly susceptible IP against which segregants may be measured. Controls should be distributed throughout the test plot in sufficient numbers to ensure adequate coverage of any potential lack of homogeneity in the environment or in the distribution of inoculum. Adequate assurance within the test is provided if controls represent 10 per cent of the total number of plants. In laboratory screens, controls should accompany each group of plants that constitute a growing unit. Additional kinds of controls will provide additional useful information to the screen. Controls representing other known IPs can be useful in assignment of IP ratings to the test plants. Controls containing specific genes for differential resistance can serve as indicators of changes in virulence within the pathogen population.

When questions are raised as to the degree of resistance expressed in a particular genotype, accession, or breeding line, it is convenient to describe these resistances in comparison to other known stocks or cultivars. It is useful therefore, to include cultivars of known IP in the disease screen. This strategy can help resolve the common problem of describing to farmers how levels of resistance in new cultivars compare with existing cultivars.

Multiple disease resistance screen

In the past, most breeding programmes have focused on the incorporation of resistance to the major limiting disease confronting the crop in a region at a particular time. Today, however, breeding objectives require that multiple disease resistances (MDR) be incorporated into crop varieties. MDR screening has become increasingly complex as new sources of resistance to existing pathogens and resistance to new diseases are sought and combined.

The principle relating to the precise control of the host, pathogen and environment are highly relevant when developing methodologies for the targeting of two or more pathogens on a single plant. Of particular importance in MDR screening is the precise targeting of carefully quantified inoculum doses such that the desired IP can be produced. In selecting combinations of pathogens to be inoculated simultaneously on a host, particular care must be given to the optimising of the infectional and incubation environments.

Of importance in the MDR screen is evaluating the potential for interactions that could occur between the host and the various pathogens applied (Ouchi, 1983). The potential interactive effects that would lead to spurious IPs should be carefully pretested. Examples of the possibility of interactions of two pathogens (A and B) on four different host genotypes leading to induced resistance or susceptibility are depicted in Fig. 20.5. When several pathogens are being targeted on a single leaf or other tissue, the potential interactive effects of inoculum placement of each pathogen in relation to the others may be important (Wyszogrodzka *et al.*, 1987).

When environmental requirements for infection and incubation of two pathogens preclude simultaneous inoculation of two pathogens, then sequential inoculation may be possible. Following establishment of one pathogen, a second may be inoculated and incubated under a different environment. As with simultaneous inoculations the potential for interactions must be examined.

Conclusions

By understanding the principles underlying screening for disease resistance, more effective use can be made of plant genetic resources in the development of improved crop varieties. In the disease screen, the environmental parameters affecting the host–pathogen interaction must lie within the limits for expression of an interaction-phenotype that is representative of the underlying host genotype. Improving the control

Fig. 20.5. Hypothetical interactions leading to spurious interaction phenotypes (either induced resistance or susceptibility) on resistant or susceptible host genotypes

		Host genotype			
		1	2	3	
Pathogen	A	R(S)	R(S)	S(R)	S(R)
	B	R(S)	S(R)	R(S)	S(R)

R = resistant; (R) = induced resistance
S = susceptible; (S) = induced susceptibility

over the environment can lead to greater reproducibility and increased efficiency in the selection of resistant segregants. During the course of the disease screen, different environmental parameters may assume the limiting role in determining the outcome of the interaction phenotype. In screening plant germplasm of unknown resistance, care must be taken to ensure that exposure to disease potential occurs at the appropriate stages during plant development. Likewise it is important to ensure that the pathogen isolates used represent the virulence found in natural populations of the pathogen. Virulence analysis is an appropriate way of determining the adequacy of the isolates being used in a disease screen. An understanding of pathogen inoculum dosage in relation to the host-interaction-phenotype is important in evaluating germplasm for disease resistance. Excessive pathogen pressure in a disease screen may mask useful levels of host resistance. The converse may also be true and lead to spurious conclusions regarding a host's potential for resistance.

The inclusion in the screen of control plants of known susceptible phenotype provides assurance that the many variables of environment, pathogen, and host are operating within the range that will provide a satisfactory interaction phenotype.

The use of widely accepted, easily used, quantifiable descriptors for the interaction phenotype, (e.g. the 0–9 scaling system) will enable breeders to exchange their results more efficiently and accurately. Scaling systems should be constructed to accommodate the potential for continuous variation within the widest possible range of interaction phenotypes.

Multiple pathogen screening is a more effective approach to evaluate plant germplasm resources. By understanding the various principles associated in the execution of single pathogen screens, multiple pathogen screens can be developed that may involve either the simultaneous or sequential inoculation of many pathogens (or several pathotypes of one or more pathogens) on a single population of plants. Careful orchestration of the various parameters of the environment, pathogen and host are required. In multiple pathogen screens, consideration must also be given to the potential for interactions that result in phenotypes which are not representative of the underlying host genotype.

References

APS (1981). *Microbial Collections of Major Importance to Agriculture.* American Phytopathological Society, St. Paul.

Daub, M. E. (1986). Tissue culture and the selection of resistance to pathogens. *Annual Review of Phytopathology*, **24**, 159–86.

Dinoor, A. & Eshed, N. (1984). The role and importance of pathogens in natural plant communities. *Annual Review of Phytopathology*, **22**, 443–66.
Harlan, J. R. (1977). Sources of genetic-defense. *Annals New York Academy of Sciences*, **287**, 345–56.
IAEA (1977). *Induced Mutations Against Plant Diseases*. International Atomic Energy Agency, Vienna.
Leppik, E. E. (1970). Gene centers of plants as sources of disease resistance. *Annual Review of Phytopathology*, **8**, 323–44.
Lumsden, R. D., Ayers, W. A., Adams, P. B., Dow, R. L., Lewis, J. A., Papavi-zas, G. C., & Kantzes, J. G. (1976). Ecology and epidemiology of Pythium species in field soil. *Phytopathology*, **66**, 1203–9.
Ouchi, S. (1983). Induction of resistance or susceptibility. *Annual Review of Phytopathology*, **21**, 289–315.
Peterson, C. E. (1975). Plant introductions in the improvement of vegetable cultivars. *HortScience*, **10**, 575–9.
Simons, M. D. (1979). Modification of host-parasite interactions through artificial mutagenesis. *Annual Review of Phytopathology*, **17**, 75–96.
Sprague, G. F. (1980). Germplasm resources of plants: their preservation and use. *Annual Review of Phytopathology*, **18**, 147–65.
Walker, J. C. (1965). Use of environmental factors in screening for disease resistance. *Annual Review of Phytopathology*, **3**, 197–208.
Williams, P. H. (1981). *Screening Crucifers for Multiple Disease Resistance*. Dept of Plant Pathology, University of Wisconsin, Madison.
Williams, P. H. (1983). Conservation of plant and symbiont germplasm. In *Challenging Problems in Plant Health*, pp. 131–44, Kommedahl, T. and P.H. Williams, (eds.), *American Phytopathology Society*, St. Paul.
Wolfe, M. S., & Schwarzbach, E. (1978). Patterns of race changes in powdery mildews. *Annual Review of Phytopathology*, **16**, 159–80.
Wyszogrodzka, A., Williams, P. H. & Peterson, C. E. (1987). Multiple pathogen inoculation of cucumber (*Cucumis sativus* L.) seedlings. *Plant Disease*, **87**, in press.

21

Restriction fragments as molecular markers for germplasm evaluation and utilisation

R. BERNATZKY AND S. D. TANKSLEY

Introduction

Molecular markers are beginning to be used in many aspects of plant genetics and breeding (Burr *et al.*, 1983; Tanksley, 1983). They have been applied to basic studies of taxonomy, variability of populations and mating systems. Breeding schemes have been developed to exploit molecular markers as gene tags (through tight linkage) thus facilitating gene transfer from wild germplasm. Molecular markers, both isozymes and DNA sequences, are useful because they are naturally occurring, have very few negative effects on phenotype, are co-dominant and are not subject to environmental influence.

Isozyme analysis, although relatively inexpensive and easy to handle, is not as useful as DNA markers due to the lower level of polymorphism and limited number of loci. DNA markers, based on restriction fragment length polymorphisms (RFLPs), are unlimited in number and show higher levels of variation. This class of genetic marker, therefore, offers a powerful tool for establishing the relationships of plants and for augmenting classical breeding methods to transfer genes from wild germplasm.

Identification of restriction fragment length polymorphisms

The method used to detect polymorphism at the DNA level is known as Southern blotting and hybridization and is briefly described as follows. Total genomic DNA is isolated from an individual, digested with a given restriction enzyme and then size fractionated on an agarose gel. The DNA is then transferred from the gel to a permanent filter in a single stranded state. This filter can then be hybridised in solution to a specific radioactively labelled DNA fragment (the probe) which is also made single stranded. After washing off the excess probe, the filter is placed against X-ray film. The location of the homologous sequence can then be observed as 'bands' on the film. This is illustrated in Fig. 21.1. An RFLP results when there is variation for the position of restriction sites between individuals.

Fig. 21.2 illustrates the method for two individuals of the genus *Lycopersicon* (tomato), *L. esculentum* (cv. VF36) and *L. pennellii* (LA 716). In Fig. 21.2A, the DNA from these plants was digested separately with five restriction enzymes and probed with a cloned DNA sequence. This clone, derived from a random messenger RNA (mRNA) and termed complementary DNA or cDNA, represents a single locus probe. As can be seen in this figure, these two individuals can be distinguished with three of the five enzymes. Fig. 21.2B shows the same individuals digested with *Eco* RI and probed with a cloned sequence that codes for the actin protein – a multigene family probe.

Typically, restriction enzymes are used which recognise specific 4–6 base pair DNA sequences. Variation in the length of a restriction fragment can result from either the loss of a restriction site or the creation

Fig. 21.1. A schematic representation of the detection of single-copy sequences from total genomic DNA. The top portion illustrates two chromosomes, A and B, which differ in the position of a restriction site (arrow). When these two DNAs are digested with the restriction enzyme and separated electrophoretically, the result is the pattern at lower left. In reality, there are very many fragments produced by digestion of total genomic DNA, making the visualisation of any one fragment impossible. To overcome this, the electrophoresed DNA is transferred in a single-stranded state to a permanent filter (a Southern blot). A cloned DNA fragment homologous to this region (the probe) is radioactively labelled and hybridised in solution to the filter. This is then placed against X-ray film and the pattern at lower right is produced. This variation is known as a restriction fragment length polymorphism

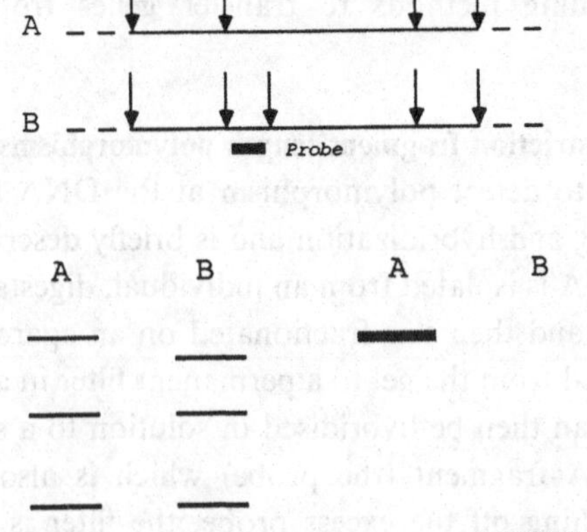

of a new one. This can occur through base substitution at the site or by insertions or deletions whose end points fall within the site. Length variations can also occur from insertions or deletions which fall between sites or from genome arrangements which overlap sites. Through Southern analysis it is not possible to determine which of these genetic events have given rise to a particular polymorphism. However, a similar length change observed with a number of restriction enzymes might suggest an insertion or deletion. In maize, detailed maps of restriction sites are now available for several loci. At the *Sh1* locus (Burr *et al.*, 1983) and *Adh1* locus (Johns

Fig. 21.2. (A) Tomato DNA digested with various restriction enzymes and probed with a random cDNA clone. 1. *Dra* I digest; left lane *L. esculentum*, right lane *L. pennellii*. 2. *Eco* RI. 3. *Eco* RV. 4. *Hind* III. 5. *Xba* I. (B) Tomato DNA digested with *Eco* RI and probed with a clone for actin (a gift from R. Meagher). Left lane is *L. esculentum* and right lane is *L. pennellii*. Molecular weights were derived from lambda DNA digested with *Hind* III

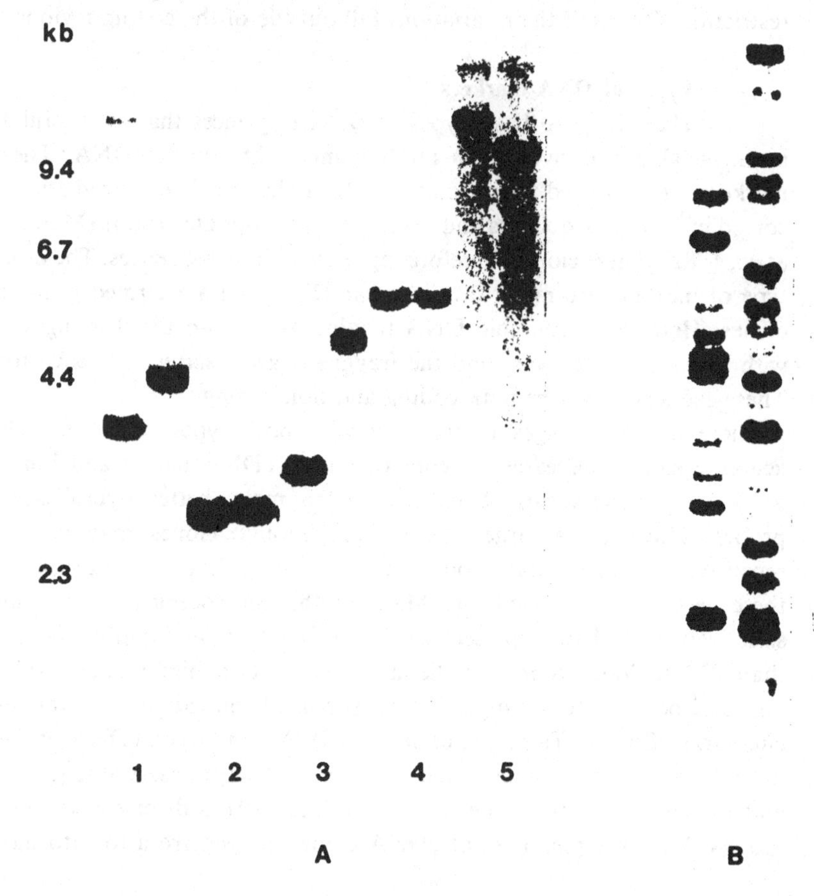

et al., 1983) the variation between inbred lines appears to be due largely to base substitutions creating or destroying sites. In contrast, insertions and deletions were primarily responsible for length variants at the *Wx* locus (Wessler & Vargona, 1985).

Restriction site polymorphisms tend to fall in non-coding rather than coding regions of DNA. For example, Johns *et al.* (1983) compared restriction maps for 7 alleles at the *Adh1* locus and found that the maps were identical within and adjacent to the coding region outside of this region there was a tremendous amount of variability for restriction sites. In a study of three loci within a single tomato genome that code for the small submit of ribulose bisphosphate carboxylase (Rbcs), Pichersky *et al.* (1986) found higher levels of variation outside than inside the region that codes for the final protein product. Our work based on single locus random cDNA clones in tomato (Bernatzky & Tanksley, 1986*a*, *b*) also supports this notion. These probes represent coding sequences, having DNA which generally hybridises to only one fragment, suggesting that the restriction sites (and their variation) fall outside of the coding regions.

Types of DNA markers

There are two basic types of DNA sequences that are useful as markers. One is derived from mRNA and is known as cDNA. These markers are produced from isolated mRNA that has been enzymatically copied into DNA sequences and cloned into appropriate vectors (Maniatis *et al.*, 1982). These clones therefore represent coding sequences. The other type of markers are made from nuclear DNA and are termed genomic clones. Here, total genomic DNA (coding and non-coding) is digested with a restriction enzyme and the fragments are inserted into a vector. These sequences will be both coding and non-coding.

There are advantages to working with both types of clones. The genomic clones are easier to construct than cDNA clones and longer genomic sequences can be selected which make better hybridisation probes. However, a large fraction of genomic clones may contain repetitive sequences that produce complex hybridisation patterns and these need to be screened out. Many of the non-coding fragments are selectively neutral and represent sequences that are more rapidly diverging than cDNA clones. Some may lie in regions that are highly polymorphic and can be used to distinguish closely related individuals (i.e. tomato cultivars) (Keyes & Tanksley, unpublished). Alternatively, cDNA probes have the distinct advantage of representing relatively conserved sequences and this enables them to be used as markers across diverse taxonomic groups. For example, a set of cDNA clones derived from tomato have been used to compare linkage maps in both pepper (*Capsicum*) and tomato (Tanksley & Bernatzky, in preparation).

RFLPs as a measure of genetic diversity

DNA markers that are based on RFLPs uncover more differences between individual plants than do protein markers. Proteins are most often detected electrophoretically and allelic variation is dependent on replacement of charged amino acids. However, variation in the DNA coding sequence that results in the replacement of neutral amino acids will not alter the electrophoretic mobility of proteins. It is also possible that some charged amino acid replacements (i.e. a positive charge for a positive charge) will not alter protein mobility appreciably. Nucleotide substitutions at the third base of amino acid codons will not always result in amino changes (silent substitutions). Also, variation in non-coding nucleotide sequences such as introns or flanking regions will not alter the final protein product but may result in restriction site changes.

The variation in DNA coding sequence can be as much as 10 fold higher than for the corresponding protein sequence. Kreitman (1983) studied 11 *Adh* alleles from wild populations of *Drosophila melanogaster*. Among the 43 changes observed in nucleotide sequence only one resulted in an amino acid replacement. Comparing the DNA sequences of the three Rbcs genes in tomato Pichersky *et al.* (1986) found DNA sequence divergence of 9–11 per cent (34–40 substitutions/369 bases) while the amino acid sequences differed by only 0.8–3.2 per cent (1–4 substitutions/123 amino acids).

Another comparison of the level of variation of proteins versus DNA sequence can be made with isozymes and random cDNA probes. In tomato, a survey of isozyme differences between *L. esculentum* (cv. VF36) and *L. pennellii* (LA716) revealed that 23 of 40 (58 per cent) of the loci were different (Tanksley, unpublished). However, using random cDNA probes, RFLP analysis uncovered differences at 74 of 75 (8 %) loci of these same plants after their DNA was digested with between 6 and 8 restriction enzymes (Bernatzky & Tanksley, 1986a and unpublished data). Four of 30 (13 per cent) isozymes were different between *L. esculentum* (cv. VF36) and *L. cheesmanii* (LA 1401), a closely related species. In contrast, 6 of the 7 (86 per cent) random cDNA clones detected differences between these taxa when between 5 and 11 restriction enzymes were used. A similar study in pepper found only 13 of 32 (40 per cent) of the isozymes to be different between *Capsicum annuum* (cv. CA 133) and *C. chinense* (cv. CA4). Using a subset of the tomato cDNA clones and 8 to 11 restriction enzymes, 90 per cent of the 71 probes were capable of distinguishing these accessions (Tanksley & Bernatzky, in preparation). To summarise the RFLP data in another way, for any randomly chosen cDNA probe, a given restriction enzyme will uncover differences between *L. esculentum* and *L. pennellii* 80 per cent of the time, 22 per cent of time between *L. cheesmanii* and *L. esculentum* and will have a 34 per cent chance of finding differences between *C. annuum* and *C. chinense*. Use of additional restriction enzymes would probably reveal differences at most loci for such interspecific surveys.

Clearly then, the potential for detecting variability between individual plants is higher for DNA restriction fragments than for their protein products. What then, are the limits of variation detected by RFLP analysis? Between species diversity is high, but what about intraspecific variation? And how do RFLPs compare among different genera of crop plants?

Intraspecific differences for restriction fragments is low in cultivated tomato (although isozyme polymorphism is practically zero). Helentjaris *et al.* (1985) studied 22 random cDNA clones with two restriction enzymes on a number of tomato cultivars and wild species. They found that all of the probes could distinguish some of the species but only three could detect differences between two morphologically distinct *L. esculentum* cultivated varieties (a cherry and a processing type). Additional restriction enzymes failed to reveal any more differences. In our laboratory, 20 varieties and landraces of *L. esculentum* were digested with one restriction enzyme and probed with DNA clones that code for actin, Rbcs and the major chlorophyll a/b binding protein (Keyes & Tanksley, unpublished). These probes monitored variation at 14 independent loci for the restriction enzyme used. Only five of the accessions could be unambiguously distinguished.

RFLPs in maize present quite a different story. Early work on the *Adh1* locus (Johns *et al.*, 1983) and *Sh1* locus (Burr *et al.*, 1983) suggested that intraspecific variation for maize was very high. Variation for restriction sites outside of the transcript unit of *Adh1* was substantial among seven alleles chosen from different maize lines and approached the level of variation observed between different *Drosophila* species for *Adh* (Langley *et al.*, 1982). However, the *Adh1* alleles were not randomly selected, but were known to be variable at the protein level. At the *Sh1* locus, Burr *et al.*, (1983) could distinguish four alleles among 6 common inbreds using two restriction enzymes. Using a combination of probes for the *Sh1* and *Wx* loci and two restriction enzymes, Evola *et al.* (1986) could differentiate five random inbred lines of maize. The question arose whether these particular loci were unusually polymorphic or if they were representative of the genome as a whole. Comparisons that were then made using random cDNA probes also showed polymorphism among inbreds to be high (Burr *et al.*, 1983; Helentjaris *et al.*, 1985; Evola *et al.*, 1986). For example, 9 out of 16 random cDNAs that hybridise to unique loci could detect polymorphism among inbreds using a single restriction enzyme (Burr *et al.*, 1983).

To quantify the variation further, Evola *et al.* (1986) used nucleotide diversity and heterozygosity indices of Engels (1981). These values are based on nucleotide substitutions (derived from colinear restriction maps) and ranged from 0.048 at the *Sh1* locus to 0.082 for the *Adh1* locus of the maize inbreds. These estimates are large when compared to values of 0.003 at the serum albumin locus in humans (Murray *et al.*, 1984) or 0.006 for the *Adh* locus of *Drosophila melanogaster* (Langley *et al.*, 1982). In tomato a similar, although crude, estimate can be made using a set of 20 random cDNAs and scoring fragments that represent single loci. We find values of 0.067 for *L. pennellii* and, using a smaller set of clones (7), we estimate a value of 0.015 for *L. esculentum* by *L. cheesmanii* for random cDNA sequences. These estimates are very rough, however, since the diversity statistic requires values of base substitutions such as could be derived from the restriction maps of the *Adh1 and Sh1* loci. Our estimates come from polymorphism per restriction enzyme for a given locus and do not account for the extact nature of the variation. But making the assumption that the differences arise from single base pair substitutions, it is quite evident that divergence for restriction sites among maize inbreds is as great as that among different species of *Lycopersicon*. Although the genetic data for pepper are not as definitive as they are for tomato, the expected level of restriction site divergence for *C. annuum* by *C. chinense* ranges between the two tomato comparisons.

It is not obvious why the degree of divergence in maize is so high in contrast to tomato and pepper (or animals for that matter). Johns *et al.* (1983) make a number of suggestions, one of which is that the observed coding regions may be in the midst of highly mobile repetitive DNA. Whether this high level of variation is unique to maize is unknown. Data from other crops, although limited, suggests that the high level of polymorphism in maize is exceptional. Working with lettuce, Landry & Michelmore (1986, submitted) have found that RFLPs for this crop is intermediate to maize and tomato. They are able to distinguish lettuce types fairly readily (i.e. butter versus crisphead, same species) using a number of restriction enzymes and either random cDNA or genomic clones as probes. RFLPs for a diverse collection of small white beans (*Phaseolus vulgaris*) from Colombia appears limited, but preliminary results in *Brassica napus* suggest that restriction site variation may be quite useful (Helentjaris, pers. comm.). In barley, 4 out of 8 lines could be differentiated using a cloned sequence for α-amylase, where as many as 9 fragments hybridised when digested with a single restriction enzyme (Bernatzky & Jorgensen, unpublished). Additional analysis of these barley lines with a clone for Rbcs did not differentiate them further although an additional five fragments were scored.

Considering these varied observations, each crop and its relatives need to be independently assessed to determine at what taxonomic level RFLPs are useful.

Molecular markers for germplasm evaluation and utilisation

Where other indicators are not available, molecular markers can provide an objective estimate of basic taxonomic relationships of germplasm. The variation detected with these markers is not subject to environmental effects and is therefore a reliable index of plant genotype. Although chloroplast DNA has been most widely used in biosystematic studies (Palmer 1987, for review), similar efforts with nuclear RFLPs will probably yield a higher level of resolution and find applications from species classification to variety identification.

We do not always know at present those characters that will be important in future breeding programmes (i.e. resistance to new diseases or tolerance to stresses in new geographical areas). Diversity at the molecular level may be a reasonable estimate of phenotypic diversity when specific selection criteria are not available (Burr *et al.*, 1983). Preliminary collections can be analysed to determine geographical centres of diversity and can indicate those areas that would benefit from additional collection while reducing the maintenance of large numbers of genetically similar samples. High levels of diversity can be maintained throughout the genome if the chromosomal locations of the molecular markers are known. If particular chromosome segments are of interest, markers which map to these regions can be used to identify diverse sources in collections.

One of the major obstacles for the utilisation of wild material is in the introgression of desirable wild alleles into a cultivated background while at the same time eliminating unwanted material from the wild donor. If this procedure is based on phenotypic selection of the character and relies on repeated backcrossing to remove the wild background, it can be quite time consuming and costly. For this reason there is reluctance among breeders to go into wild or exotic germplasm.

Molecular markers, however, can provide an efficient means for the selection and introgression of characters (Burr *et al.*, 1983; Tanksley, 1983). Once linkage is established between the gene(s) responsible for a trait and one or more molecular markers, then selection of these associated markers and the elimination of unlinked wild alleles can substantially reduce the number of generations and the size of populations required to isolate a trait in a desirable cultivated background. This is especially true of recessive characters which require progeny testing to identify heterozygotes. Since molecular markers are codominant, recessive traits can be co-selected without the need for phenotypic expression. Many agronomic traits are complex and a saturated linkage map can be used to identify the discrete genetic components underlying these quantitative characters. The appropriate linked markers can then be similarly selected to exploit the components of multigenic traits found in wild materials. A well-populated linkage map of molecular markers can also be used to uncover cytogenetic anomalies inherent in wide crosses and can help to define efficient methods for handling such crosses.

The use of DNA markers in germplasm studies is limited primarily by cost and by the relatively high level of technical expertise required to perform these procedures. However, as the technology advances, the cost and complexity of these methods will continue to be reduced. Isozyme analysis, though, is inexpensive and the techniques are simple. Although isozymes are limited in the level of polymorphism and in the number of detectable loci, they can provide a good primary database to estimate diversity and to being to develop molecular linkage maps. Therefore the judicious use of both types of analyses should be emphasised in the further understanding of the variation to be found in germplasm collections.

References

Bernatzky, R. & Tanksley, S. D. (1986*a*). Toward a saturated linkage map in tomato based on isozymes and random cDNA sequences. *Genetics*, **112**, 887–98.

Bernatzky, R. & Tanksley, S. D. (1986*b*). Majority of random cDNA clones correspond to single loci in the tomato genome. *Molecular and General Genetics*, **203**, 8–14.

Burr, B., Evola, S. V. & Burr, F. A. (1983). The application of restriction fragment length polymorphism to plant breeding. In *Genetic Engineering: Principles and Methods*, Vol 5, pp. 45–59, Setlow, J. K. & Hollaender, A. (eds.), Plenum Publishing Corporation, New York.

Engels, W. R. (1981). Estimating genetic divergence and genetic variability with restriction endonucleases. *Proceedings of the National Academy of Sciences*, **78**, 6329–33.

Evola, S. V., Burr, F. A. & Burr, B. (1986). The suitability of restriction fragment length polymorphisms as genetic markers in maize. *Theoretical and Applied Genetics*, **71**, 765–71.

Helentjaris, T., King, G., Slocum, M., Siedenstrang, C. & Wegman, S. (1985). Restriction fragment polymorphisms as probes for plant diversity and their

development as tools for applied plant breeding. *Plant and Molecular Biology*, **5**, 109–18.
Johns, M. A., Stromer, J. N. & Freeling, M. (1983). Exceptionally high levels of restriction site polymorphism in DNA near the maize *Adh1* gene. *Genetics*, **105**, 733–43.
Kreitman, M. (1983). Nucleotide polymorphism at the alcohol dehydrogenase locus of *Drosophila melanogaster*. *Nature*, **804**, 412–17.
Langley, C. H., Mongomery, E. A. & Quattlebaum, W. F. (1982). Restriction map variation in the *Adh* region of *Drosophila* second chromosome. *Proceedings of the National Academy of Sciences*, **29**, 5631–5.
Maniatis, T., Fritsch, E. F. & Sambrook, J. (1982). *Molecular Cloning*, Cold Spring Harbor Laboratory, New York.
Murray, J. C., Mills, K. A., Demopulos, C. M., Hornung, S. & Motulsky, A. G. (1984). Linkage disequilibrium and evolutionary relationships of DNA variants (restriction fragment length polymorphisms) at the serum albumin locus. *Proceedings of the National Academy of Sciences*, **81**, 3486–90.
Palmer, J. D. (1987). Chloroplast DNA evolution and biosystematic uses of chloroplast DNA variation. *American Naturalist*, **103**, S6–S29.
Pichersky, E., Bernatzky, R., Tanksley, S. D. & Cashmore, A. R. (1986). Evidence for selection as a mechanism in the concerted evolution of *Lycopersicon esculentum* (tomato) genes encoding the small subunit of ribulose-1,5-bisphosphate carboxylase/oxygenase. *Proceedings of the National Academy of Sciences*, **83**, 3880–4.
Tanksley, S. D. (1983). Molecular markers in plant breeding. *Plant Molecular Biology Reports* **1**, 3–8.
Wessler, S. R. & Vargona, M. J. (1985). Molecular basis of mutations at the waxy locus of maize: Correlation with the fine structure map. *Proceedings of the National Academy of Sciences*, **82**, 4177–81.

22
Molecular biology and genetic resources

W.J. PEACOCK

Introduction

Molecular biology, and especially recombinant DNA methodology, has impacted significantly on many aspects of plant biology in recent years. Initially plant molecular biology was concerned with gaining knowledge as to what active plant genes were at a molecular level, but now this method of experimental analysis has extended into plant physiology, plant pathology and other areas. In plant physiology, for example, classical modes of analysis of photosynthesis, action of plant growth regulators and flowering initiation are now being supplemented and integrated with molecular level analyses. In plant pathology the intracellular basis of host/pathogen interactions is being examined and new approaches to disease resistance have opened up in an exciting way. Will molecular biology have a comparable impact on plant genetic resource work? Will it influence the way in which we store germplasm, or modify the principles used in deciding what to collect, and is it likely to improve our abilities to utilise stored plant germplasm?

Plant genetic resources collections for the most part are stores of germplasm of agriculturally important species, usually in the form of seed samples. Seed is stored under conditions which maximise long-term viability. The stored samples attempt to provide a representative collection of genetic variation of the species for future needs of plant improvement programmes. Initially, emphasis was placed on the collection of landraces, and locally adapted, primitive cultivars of the major food crops. Collections were focused in those geographical areas where landraces were threatened by the various activities of man. Prime attention was paid to the centres of diversity for the major crop species.

Emphasis in collecting is now changing to include collecting of wild species related to the important crop species. This has happened because of the effectiveness of the initial period of collecting of landraces, but also because it has been realised that wild species frequently have genes which are needed in plant improvement programmes. In cereal improvement programmes, for example in wheat, there have been a number of important examples where a gene from a wild grass species has provided a needed disease resistance gene. In wheat, as well as in other crops, breeders and geneticists have devised powerful procedures for dealing with the difficulties of wide crosses. In general, when wide-crossing is practised there is usually a period of pre-breeding; genetic material from a wild species is most valuable to a breeder when it has been transferred into a genetic background which approximates to that in current use in crop cultivars. There are usually problems when large segments of a genome of a wild species remain in a hybrid. Back-crossing procedures are efficient in diluting-out the content of the wild genome but ultimately the limit is set by the frequency of recombination needed to isolate required genes out of transferred chromosomal segments.

In many plant improvement programmes the need is for a specific gene, or allele of a gene, and not for a large number of genes. This particularly applies to those crop, pasture or horticultural species where there has been a history of plant improvement. Breeders involved with such species, when asked for their major needs or objectives, will generally indicate resistance to specific pests or diseases, and tolerance to particular environmental or nutritional stresses. Often single genes provide a solution to these sorts of objectives. In species less developed in their programme of adaptation to our agricultural environments larger segments of genomes, presumably with adaptive gene complexes, are likely to be of significance.

There is no doubt that many plant characteristics have a multigene inheritance pattern but I am convinced that as our abilities to track the effects of particular loci increase, cases of polygenic inheritance will generally resolve to a small number of genes with major effects coupled with lower level interactions of those genes with other parts of the genome. Already in agriculture there are good examples of single genes which are important components of increases in a complex character like yield; the vernalisation genes and dwarfing genes of wheat have significant effects on plant development, and, in turn, can have significant effects on yield.

The importance of one to a few genes in improvement programmes for any given species is discussed because, in the near future, recombinant DNA methodology is likely to contribute to plant breeding primarily in the identification, monitoring and transfer of one, or at most of a few, genes at a time. This underlines the consideration of the possible impact of molecular biology, on the collection, storage and future use of germplasm.

Gene transfer

It is now possible to add a segment of DNA to the genome of a plant so that it is incorporated into chromosomal DNA and is subsequently inherited according to normal Mendelian rules. The most widely used method of gene transfer utilises a naturally occurring vector system. The soil dwelling organism *Agrobacterium tumefaciens* causes crown gall disease in many dictoyledonous plants. It does this by inserting a segment of DNA from one of its plasmids into plant chromosomal DNA. This segment, T-DNA, contains genes which interfere with auxin and cytokinin synthesis, and this results in tumorous growth of the infected tissue. The T-DNA segment has now been modified to remove the tumour-inducing genes together with other genes concerned in the production of amino acid analogues normally needed for growth of the bacteria. In its most sophisticated form T-DNA has been stripped so that only the key end-sequences remain. A gene for antibiotic resistance is usually inserted between these termini so that a selection regime can be used to identify transformed cells. Kanamycin is lethal to most plant cells but not if they have amino-glycoside phosphotransferase activity from an introduced gene. This bacterial resistance gene modifies the antibiotic so that the cells survive.

Any other gene inserted into the T-DNA segment alongside the resistance gene is co-transferred during the *Agrobacterium*/plant interaction. All of this DNA manipulation is carried out in *Escherichia coli*, which is handled easily in the laboratory. The engineered T-DNA segment, on a small plasmid capable of multiplying in *E. coli*, is then transferred into *Agrobacterium* where another small plasmid carrying the genes necessary for the transfer operation exist. The infection process has also been streamlined, *Agrobacterium* cells carrying the binary vector system can be co-incubated with plant tissue culture cells or with surface-sterilised leaf discs, or other small segments of plants. Infection occurs with high frequency. Previously, *Agrobacterium* was applied to wounds on the stem or other plant parts. The advantages of the leaf disc method are that it is simple, and in many plants, cells from around the cut edge of the disc can be induced to regenerate into plants. Each regenerated plant traces back to a single cell of the leaf disc, so that if the leaf disc is grown under selection pressure the regenerated plants have a high probability of originating from transformed cells. Much of this development took place with *Agrobacterium tumefaciens* infecting *Nicotiana tabacum*. The time-scale with this plant species is such that small, rooted, transformed plants are available from 6–8 weeks after the transformation event and large numbers may be produced readily.

There are other methods of gene transfer. Direct microinjection into the nucleus of protoplasts has been used successfully in *Medicago sativa*. In this case transformation frequency is so high (up to 50 per cent) that it is possible to introduce genes without the need to use a selection regime to select transformants. Another method used with protoplasts is that of electroporation. In this technique, which works well with a wide variety of species of both monocotyledonous and dicotyledonous plants, an electric current is used to make temporary pores in the protoplast membrane and DNA from the surrounding solution can enter the cell. High frequencies of transformation can be achieved with this method. There are similar procedures using various chemical and osmotic shocks rather than the electric current to interfere temporarily with membrane integrity.

All of these methods are useful for gene transfer only in those species from which plants can be regenerated from calli derived from single protoplasts. There is an increasing number of species in which tissue culture methodology has overcome the regeneration barrier and recently the first cereal plant regeneration from protoplasts has been reported with the success achieved with rice. Cereals, which have been refractory to transformation methodologies up to now, have also succumbed to yet another method of gene transfer. Rye plants have been transformed by injection of DNA into the immature inflorescence. The method of entry of DNA into the premeiotic cells in the anthers is not fully understood but it is thought to be via the conductive tissue.

Although transformation frequencies are not high in all of the systems operating, there is an impressive increase in the number of species for which gene transfer is now a reality. Among the important agricultural species success has been reported with broadleaved species such as tobacco, cotton, and sunflower and now leguminous species such as soyabean and *Medicago* have been successfully transformed. There had been, up to recently, a difficulty with the major monocotyledonous cereals but now with the first transformation reported for rye and with successful regeneration from protoplasts in rice, it seems likely that working methods will be available for all of the cereals in the near future.

Gene transfer and regeneration of plants from transformed cells is likely to be achieved for any species in which a concerted research effort is applied. This means that any desired gene that is available can be introduced into a plant genome and can be introduced without any other unwanted genetic material, as is usually the case with any hybridisation procedures, whether between close cultivars or in wide crosses. Gene segments can be specifically tailored and introduced without alteration. As yet there is no method in plants for targeting the chromosomal entry point for a gene, but chromosomal location seems not to be of overriding significance with regard to the normal functioning of the gene. There is quantitative variation in the level of gene activity between independent transformants, and where this has been looked at critically some of the effect can be attributed to chromosomal location. Some transformed plants show little activity of an introduced gene and may represent instances where the gene has been incorporated into the middle of a heterochromatic region and is not expressed. But only a few transformants need to be screened to have one or more with the desired level of gene activity. In the longer-term it is conceivable that methods will be worked out for targeting a gene to a particular place in the genome. Gene replacement, rather than gene addition, may then be a practical target.

Genes and gene regulations

The fact that I have been discussing gene transfer as a reality, implies that our knowledge of plant genes and their regulation is sufficient to enable experiments to be designed in which functional genetic units are transferred. This is the case. We know the basic parameters of organisation of plant genes. They have a three-part anatomy; a head region, which contains most of the control apparatus; a coding region in which is located the series of codons, in triplet code, specifying the amino acid chain of the polypeptide product of the gene; and a tail region, which contains other controls, primarily those to indicate the termination of transcription and translation – two essential steps in gene expression.

The coding language is virtually universal; and gene sequences taken from bacteria or animals are capable of being transcribed and translated correctly in a plant. There is, however, one major difference between most bacterial genes and most plant – and for that matter, animal – genes. The eukaryote genes are, in general, interrupted. The meaningful coding sequences, the exons, are interspersed with a variable number of introns or sequences which are not meaningful for determination of part of the amino acid chain of the gene product. The introns are transcribed in the initial primary transcript but are then spliced out in the production of the functional mRNA. The exons join so that the coding regions become contiguous. Intronless genes, say from bacteria, do work in plants; and so too do cDNA genes. The latter are gene copies made from mRNA in which all of the introns have been spliced out. However, in some genes it is known that one or more introns are essential for high and normal activity of that gene. For example, intron one in the alcohol dehydrogenase 1 gene of maize contains sequences which are important for high level expression.

Most control sequences are in the head region of a gene and it is on this segment of plant genes that most research has centred in the past couple of years. In the region in front of the gene are DNA sequences which are critical for regulated expression of the gene. Plants are dependent, both for their development and for daily operation, on differential gene activity. Very few genes are constitutive, being expressed in all cells of a plant. Most genes are expressed in only a subset of cells; these may be cells in a particular tissue, or cells at a particular stage of development. Some genes are activated by a particular environmental stimulus. Plants have to cope with changing environmental conditions and some genes are induced by these changes. Thus genes may have both developmental and environmental programming of their activity. The important point is that most of these controls are in the head region of the gene, immediately upstream of the coding region, and if the head region is included, along with the coding region, in a gene transfer, then the gene has a strong chance of being expressed in a recipient plant in an appropriately regulated fashion.

The fact that genes have normal regulation in gene transfer experiments, even where the transfers are trans-specific or even trans-generic, shows us another important feature of plant genes. Many of the control signals are conserved and are similar in plants widely separated in a phylogenetic sense. We, in our laboratory, have specific experimental experience with the *Adh1* gene of maize; this is expressed in tobacco, although poorly; but the conservation of the control circuitry is sufficient that with the addition of another element, an enhancer, which increases the frequency of gene transcription, the expression of the maize gene becomes functionally normal in tobacco. This conservation of genetic circuitry is of great importance to plant improvement programmes. Breeders can plan to isolate a gene from one species and insert it into another, expecting it to be expressed in the correct tissue at the correct time; or that it will be induced by the appropriate environmental stimulus.

The short sequence regions that are critical for the control of gene expression, are likely to function, in most cases, through a mechanism dependent on an association of a specific region with a specific regulatory protein. This protein is transcribed from another gene. Our data on the anaerobic induction of the maize *Adh1* gene indicate that there are transacting factors that must interact with the anaerobic regulatory element, and other activating proteins controlling tissue-specific expression of the gene. Gene transfer experiments have demonstrated that developmental controls are conserved. For example a seed storage protein from *Phaseolus*, the French bean, is expressed in transgenic tobacco in the developing seed, as it should be. The tobacco gene control system recognises the control signals written in *Phaseolus* DNA! Another example is that the leghaemoglobin gene from soyabean, when transferred into another legume genus, *Lotus*, is expressed only in nodule tissue.

Probably there are many control sequence/protein interactions determining the overall regulation of any gene's activity. This could mean that for any given essential gene product, like an enzyme or basic structural protein, there may be several other gene products needed to control its production. This could mean astronomical numbers of genes in plants, or alternatively, the distinct possibility that many of these control proteins operate on more than one gene. For example, genes that are co-ordinated in their expression, either at a specific developmental time in a specific tissue, or in response to a specific environmental condition may all have common control elements. Already there are data pointing to common control sequences both for developmental and environmental gene activity programmes. With the growing knowledge of these control circuits, specific gene constructs have been designed for gene transfer experiments which result in precise and limited expression of an added gene in a recipient plant. Gene transfer is taking its place as another tool for the plant breeder.

Genes for agronomic traits

The two previous sections outlined the status of knowledge of what a gene is, how its expression is controlled and how it can be transferred into a recipient plant. But what of our capacity to isolate and modify those genes that are likely to be important in plant improvement programmes, genes which specify important agronomic characteristics? The first genes which were isolated were genes which could be identified readily through their product protein or mRNA. These were genes which coded for prevalent products or which were expressed at specific times or conditions of induction such that plus-minus comparisons of proteins or mRNAs could be made to help identification. This latter method was used in the isolation of the gene for alcohol dehydrogenase in maize, and prevalent mRNAs and proteins were the basis of the isolation of genes coding for seed storage proteins and RuBP carboxylase in legumes and cereals. Techniques are improving rapidly and new ones are being introduced all the time. For example it is now possible, from a small amount of labelled protein in a gel band, to produce enough amino acid sequence data to permit the design of an oligonucleotide that can be synthesised and used as a probe for detection of a gene among a cDNA or genomic library. This was the technique used in isolating the plant haemoglobin gene from the non-legume, *Parasponia*.

Among the various technologies being devised to help with gene isolation is the use of transposable elements. These are DNA sequences capable of movement in their host genome. They are best known from the work of Barbara McClintock in maize. If a transposable element relocates into a gene, the chances are good that it will inactivate that gene. If gene inactivation produces a significantly different phenotype from the active state then the gene in question can be identified with a probe for the sequence of the transposable element. A problem with this technology is that naturally occurring transposable elements occur as families of elements and the identification of the particular transposable element sequence which has jumped into the needed gene can present a problem. There are ways around this, and the technology has already been used successfully in maize and in *Antirrhinum* to isolate specific genes.

Maize transposable elements have been shown to be mobile in *Nicotiana* and there are now a number of laboratories trying to use homologous or heterologous transposable elements to locate disease resistance and other genes. If resistance genes can be isolated then analysis of their control and coding regions should lead to a better knowledge of the mechanisms of disease resistance and perhaps to the design of robust resistance systems in the important crop species.

A key point to the identification of genes is that there must be knowledge of the gene product or of a phenotype specifically associated with the gene. This is often not the case for genes that are perceived to be important agronomically. Obviously it is not the case for any character that is considered to be polygenic, but as discussed above, our methods of analysis are becoming more powerful. As we improve in our ability to track and isolate genes with major effect, we are likely to increase the opportunities to isolate and transfer genes of significance for a wide range of agronomic characteristics.

If the technologies of gene transfer become highly efficient then it may be possible to transfer all genes of a donor genome into a recipient plant species. We would then rely upon a suitable selection technology to pick the needed event of gene transfer from the populations of such transfers. This form of shotgun-gene transfer has currently worked only for the recovery of a gene for which a really efficient selection system exists (antibiotic resistance). The present technologies of selection and transformation are not efficient enough to make shotgun cloning a reality for most genes, but this situation will change.

Gene Libraries

As was mentioned above, the total genetic information of a plant, its total genome, can be isolated readily and used in gene transfer experiments. The long DNA molecules of each chromosome can be cut up into suitably sized segments in such a way that all genes can be guaranteed to be in the population of segments, each gene in a complete form. This is usually done by cutting the DNA by a partial digestion with a restriction enzyme. The resultant segments of DNA can be stored easily and indefinitely as purified lyophilised DNA. Storage in a small, preferably evacuated, tube can be at room temperature. The genome of all the plant species of the world could easily be stored in a single small room if this, in fact, was a sensible thing to do. The gene segments are generally used in gene transfer experiments after each has been incorporated into a small DNA molecule capable of multiplying in a bacterial cell. The multiplying agent may be a plasmid molecule or a bacteriophage molecule. The total library of a plant's genes, contained in plasmids or bacteriophage molecules, can be readily stored in one small test tube. Storage conditions can ensure the gene library will not deteriorate over extremely long periods of time.

Does it make sense to store all germplasm in this form? The answer is clearly no for most plant species – DNA storage only allows for the recovery of single genes, not of genomes or genome segments. The recovery of a specific gene depends upon our ability to identify the activity of that gene and select for it in the recipient plant species. The advantage of storing the total DNA of a plant in cells of a seed is that all genes from the genome will function, and function in a regulated fashion, so that the needed phenotypes can be identified readily.

Despite the limitations of the detection and isolation of genes from stored DNA it seems sensible to me that we at least consider the need to make DNA storages for those plant species where there is no other convenient long-term storage of germplasm. There are species which can only be conserved as vegetative tissues, or in field gene banks (e.g. plantations), and other species which can only be stored by frequent seed regeneration procedures. Enough DNA for storage can be readily isolated from a few grams of plant tissue. Duplication and amplification of stored genes can be carried out any time by making a gene library as I have indicated above. There is no limit to the level of amplification, although there are some problems in that some gene segments could be preferentially amplified relative to others, but these are points of minor technical significance.

Even though there is no practical difficulty in the isolation of DNA we still have the problem of sampling the available genetic variation. This applies for DNA sampling in the same way as it does for sampling of germplasm by any other means of storage, such as seed storage. The principles worked out according to the rules of population genetics can, and should be applied, to collections of plant material for DNA storage.

What plant material should be used for the isolation of DNA? For the most part it is, again, inconsequential. The most convenient tissues are usually tissues where there is a high frequency of cell division and hence a high density of nuclei. From what we know of plant development, DNA isolated from the nuclei of a root or of a leaf is identical to the DNA obtained from the gametic cells or from meristematic regions.

The major limitation of gene library storage, that it provides for the use of only those genes which in some way can subsequently be selected for in a recipient plant (or by other means) will become less of a problem as our knowledge of plant genes and genomes increases with further research. I propose that our level of knowledge of plant genes and their controls, and our level of expertise in gene isolation and transfer, is already such that DNA storage is justified in all of those species where there is not a convenient long-term seed storage potential.

Gene pools

One of the ways in which DNA libraries are likely to be used is that they will be able to be searched for genes with probes made from existing gene sequences. I made the point earlier than the control circuitry of genes is conserved widely through the plant kingdom. If a control signal with a particular characteristic is needed, for example a control which is more responsive to low or high temperatures, then it is highly probable that a control that is sufficiently homologous to the needed form of the control will be obtainable from an existing plant species and will be able to be used as a probe to isolate the needed one from a DNA library. The same argument applies to the coding sequences of genes. Plant species differ from each other in their phenotypes and characteristics, not because they have large numbers of genes that are unique from species to species, but because the particular form, alleles, of each gene, and their controls, differ from one species to another.

One of the ways in which DNA libraries are likely to be probed with high efficiency is with nucleic acid probes prepared from a specific subset of the recipient species genes. For example, a probe mix could be prepared from the subset of genes active under conditions of pathogen attack. These probes will selectively search stored gene libraries for related gene sequences and could detect a better resistance gene. A gene library will be able to be examined in many 'compartmentalised' ways.

Probing a gene library with specific DNA or RNA probes can successfully pull out homologous, but different, coded forms of a gene, but there are limitations to the level of sequence differentiation permitting the process to be effective. The higher the degree of sequence conservation the easier it is to use probe technology for gene isolation. If the sequences differ too much for the probe detection method, it should be possible to approach stepwise the gene of the target species through successive probings between phylogenetically related species, thus bridging the initial phylogenetic gap.

At present there is not the technology available for plants to transcribe DNA sequences *in vitro*. Presumably this technology will become available and will provide for even greater use of stored DNA libraries. Already it is possible to take RNA transcripts and translate them *in vitro*, or to express genes in bacterial systems. There are circumstances where a small amount of translated gene product will prove an effective detection system for a needed gene sequence. I mention this as an example of the way we can expect recombinant DNA and related technologies to increase the value of genetic material stored as DNA or as gene libraries. The rapidity with which these technologies are developing suggests that now is the time to seriously contemplate the initiation of appropriate DNA storages.

So far, I have suggested that DNA, or gene, libraries are sensible for plant species for which there is no other current means of germplasm storage. Are there other situations which we should consider? As I mentioned earlier, there is an increasing awareness that related wild species may be important sources of genetic variation for an agricultural plant species. One of the principal reasons is that the wild species have evolved in the environments and under the stresses that may now, or in the future, apply to the agricultural species. Presumably the wild species have the necessary genetic information to cope. Obvious examples are tolerances to particular soil and nutritional conditions and tolerances to other environmental stresses such as low night temperatures or high day temperatures. Genetic resistances to pathogens and pests generally exist in polymorphic condition in populations of plants which have evolved along with their pests and pathogens. Knowledge of seed physiology of wild plant species is often small and it may be that many of these species will prove difficult with regard to seed regeneration and storage. Germplasm of wild species seems to me to be another situation where DNA storage should be considered. Particularly with the more developed agricultural species, many genetic traits needed from related wild species are encoded by single genes. These will be increasingly available from stored gene libraries.

Just as molecular biological techniques add to, but do not supplant, other methodologies of plant breeding, so too are they additive in plant genetic resources conservation. DNA libraries will, whenever possible, be adjuncts to other germplasm collections.

Collecting

In planning germplasm collecting we need to consider the likely limitations to the performance of agricultural species. As any given agricultural species is moved into increasingly marginal or different environments there will be new limitations to its performance. The prediction of the environments and the likely stresses to which an agricultural species will be subjected should provide for an intelligent ecogeographic collection procedure for germplasm. Although currently this is being thought about in terms of collection of wild species related to the crop species, gene transfer technology argues that we may need to think beyond the immediate relatives of a crop species. It may be that an unrelated plant species, highly adapted to a particular ecogeographic region, will prove a valuable source of genes or gene regulatory controls needed in a crop species. It may be that the only viable approach to preserving such an extended horizon of genetic variation is through DNA storage. Of course there is no need to set up germplasm storages, DNA or seed, if there is no threat to the continued existence of genetic variation in the natural environment. There are other reasons for thinking about an extension of the gene pool to species unrelated to a crop species. We may wish to introduce a new strategy to the genetic system of the crop species, for example, we may wish to manage it as a hybrid crop. We could transfer elements of incompatibility systems or of male sterility systems from other plant species, and this is not as fanciful as it at first sounds. We currently are gaining molecular-level information on genes and gene products that are active in such breeding system controls, and the rationale for transfer of at least some components of the systems into unrelated species, already exists. Other examples can be given where new attributes could be added to our relatively few important food, fruit and fibre species. Resistance to a pest may be endowed on a plant species by using a gene obtained from an entirely different form of life. This is already being attempted where bacterial genes have been transferred to provide plants with the capability of manufacturing a particular insecticidal protein. The difficulties in considering extended gene pools, and how to cope with their conservation, are immense. Where do we draw the lines?

Germplasm storage safeguards

There are numerous problems with germplasm storage, even for those plant species in which there is long-term seed longevity. One factor is that regeneration procedures need to be carried out in ways that minimise the loss of the genetic variation which existed in the original sample. Another factor is accidental loss of a germplasm storage facility. This has been an important consideration in setting up duplicate storages. Both of these factors perhaps give grounds for another form of storage. When a collection of germplasm is made, should that accession be stored in the form of DNA as well as seeds? DNA is so easily extracted and conserved that I believe this should at least be discussed carefully. One of the factors that might need consideration with regard to this form of safeguard for genetic resources conservation is that even now, requests for germplasm are often requests for gene sequences available in those storages. An increasing number of plant molecular biology laboratories are turning to germplasm collections, and for a variety of reasons. Frequently all they need are DNA samples rather than seeds, which are germinated for use in DNA preparation. Perhaps the efficiency of operation of gene banks could be increased if their accessions are stored in a DNA library as well as in a seed library.

Conclusions

My assessment to the questions posed in the introduction of this paper is that molecular biology is, and should be, having an impact on genetic resources conservation. Already the technologies of gene isolation and transfer, and the increasing knowledge of plant genes and their controls, most importantly their relation to plant processes and development, are such that storage of genes in DNA segments, as well as of storage of genomes is a sensible objective. This certainly applies to plant species for which there are no suitable long-term storage procedures for storage of their genomes; the species that rely on vegetative reproduction and the species that have recalcitrant seeds. It is also my opinion that, as there is increasing realisation of the value of genes contained within the secondary and tertiary gene pools of our agriculturally important species,

DNA storage will have an increasing role. It seems inevitable, to me, that as our understanding of processes of plant development and plant function increases at the molecular level, we will have increasing opportunities of using individual gene segments and individual units of gene control in striving to meet the challenges of plant improvement programmes.

Index

Numbers in *bold face* refer to the page numbers of a chapter dealing with the subject.

Printed in the United States
By Bookmasters